References to pp -
13/64

ACTIONS OF RADIATIONS
ON
LIVING CELLS

ACTIONS OF RADIATIONS
ON
LIVING CELLS

BY THE LATE
D. E. LEA

SECOND EDITION

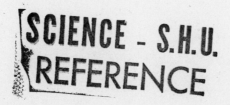

CAMBRIDGE
AT THE UNIVERSITY PRESS
1956

PUBLISHED BY
THE SYNDICS OF THE CAMBRIDGE UNIVERSITY PRESS

London Office: Bentley House, N.W.1
American Branch: New York

Agents for Canada, India, and Pakistan: Macmillan

First Edition 1946
Second Edition 1955
Reprinted 1956

First printed in Great Britain at the University Press, Cambridge
Reprinted by photo-lithography by Jarrold & Sons Ltd., Norwich

CONTENTS

Prefaces   *page* ix

List of Illustrations   xiii

Chapter I.   Physical Properties and Dosimetry of Different Radiations   1

Ionization, excitation, and 'point-heat', p. 1; Ultra-violet light, p. 3; X-rays, p. 6; γ-rays, p. 13; β-rays and cathode-rays, p. 15; α-rays, p. 16; Protons and neutrons, p. 19; Dosimetry, p. 21; Primary ionization, ranges, and energy dissipation, p. 23; Secondary ionization, clusters, and δ-rays, p. 26; Number and total range of ionizing particles per unit volume of tissue, p. 30

Chapter II.   Chemical Effects of Ionizing Radiations, and Possible Mechanisms of Biological Action   33

The ionic yield, p. 33; Gas reactions, p. 34; Liquids and solids, p. 37; Solutions, p. 39; Decomposition of water, p. 40; Indirect action in aqueous solution, p. 42; Ionic yields in aqueous solution, p. 45; Chemical mechanism of the indirect action in aqueous solution, p. 47; Spatial distribution and recombination of the active radicals, p. 48; Kinetics of indirect actions in aqueous solution, p. 52; Competition between two solutes; protective action, p. 55; Reduced yield in dilute solutions, and with densely ionizing radiations, p. 57; Direct and indirect actions of radiation, p. 60

POSSIBLE MODES OF BIOLOGICAL ACTION OF RADIATION, p. 64

Cell poisons, p. 64; Activated water reactions, p. 64; Direct action on large molecules, p. 65; Localization of ionization, p. 66; The target theory, p. 66; Spread of the effect of an ionization, p. 67

Chapter III.   The Target Theory   69

Recognition of the single-ionization type of action, p. 71; Shape of survival curve, p. 72; Time-intensity factor, p. 78; Dependence on type of radiation, p. 79; Relation between target size and inactivation dose, p. 80; Secondary ionization, p. 88; The multi-target theory, p. 90; α-rays and γ-rays, p. 91; The assumptions of the single-ionization target theory, p. 92; The validity of the simple model, p. 98

Chapter IV.   The Inactivation of Viruses by Radiation   100

The viruses, p. 100; The sizes of the viruses, p. 102; The estimation of virus activity, p. 104; Technique of virus irradiation experiments, p. 106; Direct and indirect actions of radiation on viruses, p. 108

EVIDENCE THAT VIRUS INACTIVATION IS DUE TO A SINGLE IONIZATION, p. 111

Exponential survival curves, p. 112; Independence of inactivation dose on intensity, p. 114; Dependence of inactivation dose on ion-density of radiation, p. 116; Relation between virus size and inactivation dose, p. 116; The status of the viruses, p. 123; Inactivation of viruses by ultra-violet light, p. 124

Chapter V. Genetical Effects of Radiation — page 126

The mechanism of heredity, p. 126; Mutations, p. 131; The nature of genes and mutations, p. 133; Chromosome structural changes, and the position effect, p. 136; The production of visible mutations by radiation, p. 140; Recessive lethal mutations in *Drosophila*, p. 144; Relation between lethals and chromosome structural change, p. 154; Dominant lethals in *Drosophila*, p. 161; Deductions concerning the size of the gene, p. 172; Cosmic rays and mutations, p. 180; Genetical effects of ultra-violet light, p. 181

Chapter VI. The Production of Chromosome Structural Changes by Radiation — 189

Experimental materials, p. 189; Structural changes and physiological changes in chromosomes, p. 192

TYPES OF STRUCTURAL CHANGE, p. 197

Chromosome breaks, p. 199; Chromatid breaks, p. 201; Isochromatid breaks, p. 202; Chromosome intrachanges, p. 205; Chromatid intrachanges, p. 207; Interchanges, p. 207; Relative frequency of interchanges and intrachanges, p. 211; Location of breaks in chromosomes, p. 214; Frequency relations, p. 217; Modifying factors, p. 220; Dependence of the yield of structural changes on radiation intensity, p. 225; Dependence of yield on dose, p. 229; Relative efficiencies of different wave-lengths and types of ionizing radiation, p. 237; Coefficients of aberration production in *Tradescantia*, p. 240; Structural changes induced by ultra-violet light, p. 242

Chapter VII. The Mechanism of Induction of Chromosome Structural Changes — 245

Reasons for believing that a break is caused by a single ionizing particle, p. 246; Distance apart at the moment of breakage of breaks which exchange, p. 249; Proportion of breaks which restitute in *Tradescantia*, p. 253; Dependence of the yield of aberrations in *Tradescantia* upon the duration of exposure, p. 262; The relative efficiencies of different radiations, p. 269; Isochromatid breaks, p. 277; Recapitulation, p. 279

Chapter VIII. Delayed Division — 282

Introduction, p. 282; Delay of first cleavage in sea-urchin eggs, p. 284; Stage of division which is subject to delay, p. 293; Rapidly dividing tissues, p. 295; Increase of cell size, p. 300; The relative efficiencies of different radiations in causing delay of division, p. 303

Chapter IX. Lethal Effects — 307

Death precipitated by division, p. 307; The cause of death at division, p. 311

LETHAL MUTATIONS, p. 313

The inactivation of viruses, p. 313; The killing of bacteria, p. 316; The killing of bacteria interpreted as lethal mutation, p. 324

LETHAL CHROMOSOME STRUCTURAL CHANGES, p. 328

Drosophila melanogaster, p. 329; *Tradescantia*, p. 332; The bean root, p. 335; Lethal physiological changes in chromosomes, p. 337; Hereditary partial-sterility, p. 338; Recapitulation and application to rapidly dividing animal tissues, p. 341

Appendix I. Supplementary Calculations *page* 345

Appendix II. Textual Revisions and Additions 364

MUTATIONS, *p.* 364
The influence of temperature, *p.* 364; Mutation yields in different cells, *p.* 364
DOMINANT LETHALS, *p.* 365
CHROMOSOME STRUCTURAL DAMAGE, *p.* 366
Location of breaks, *p.* 366; Frequency relations, *p.* 366; Modifying factors, *p.* 367; Sensitivity at different stages, *p.* 369; Dose relations, *p.* 370; Relative efficiencies of different types of radiation, *p.* 371; Evaluation of breakage frequency, *p.* 372; Alternative derivation of G, *p.* 373
BACTERIA, *p.* 373
Effect of intensity, *p.* 373; Metabolic disturbances, *p.* 374

Bibliography 375

Additions to Bibliography 396

Author Index 399

Subject Index 404

Note. A marginal asterisk (*) refers to Appendix II.

PREFACE TO THE FIRST EDITION

My intention in writing this book has been to give an account of certain of the simplest and most fundamental actions of X-rays and other ionizing radiations on living cells. I have not attempted to survey the whole field of the biological effects of radiations. Instead, I have thought that a useful purpose would be served by giving a rather detailed discussion of the mechanism of those actions of radiation which are sufficiently well understood for such a treatment to be profitable at the present time.

After introductory chapters describing the relevant physical properties and chemical effects of ionizing radiations, the bulk of the book is occupied by the discussion of the effects of radiations on viruses, and on the genes and chromosomes of higher cells. In the concluding chapter the killing of cells by radiation is discussed in so far as it can be understood in the light of the preceding chapters.

One of the difficulties in writing the book has been to decide what background to assume the reader to possess. In an endeavour to make the book to some extent understandable by all classes of reader, I have written the physical chapter in an elementary fashion, and where it seemed practicable, have prefaced my accounts of the various biological actions of radiation by a brief description of the properties and method of handling of the organisms concerned. While in general these introductory sections are brief, I have thought it advisable to provide a rather detailed introduction to the chapter on the genetical effects of radiation. The rather technical vocabulary of the geneticist is not well understood by non-specialists, and in consequence many workers studying the biological effects of radiation do not adequately appreciate the very considerable contributions which have been made to this subject by geneticists.

The study of the action of radiations on viruses, genes, and chromosomes has reached a stage where the experiments are mainly quantitative, and the interpretations therefore necessarily to some extent involve elementary mathematics. When developing such mathematical interpretations I have kept the algebraic detail in the background as far as possible, and have provided graphs and tables which enable the experimentalist to interpret his results along the lines suggested without any technical mathematical ability being required.

I have been at pains in the physical chapter to provide an adequate amount of numerical data concerning the amount and spatial distribution of ionization in tissue exposed to the various radiations, such numerical data being constantly required in quantitative interpretations of the various biological actions of radiation. The tables in Chapter I and the Appendix have been specially computed for this book. While the physical principles involved in the dissipation of energy by ionizing radiations in their passage through matter are understood, numerical data of the sort tabulated are not so accurately known as one would wish. The tables are likely, therefore, eventually to need revision in the light of more exact information. It is hoped that in the meantime they will be found of value by workers in this field of research.

I should not have been in a position to write this book had I not, during the past ten years, been enjoying the collaboration of a number of friends and colleagues in studies of various biological effects of radiation. I cannot adequately express my indebtedness to these collaborators, namely Dr D.G. Catcheside, the late Dr R.B. Haines, Dr M.H. Salaman, Dr K.M. Smith, and my colleagues at the Strangeways Laboratory. I am especially indebted to Dr L.H. Gray and Dr F.G. Spear, in frequent discussions with whom, my ideas on the subject have taken shape.

For providing me with material for the plates, or for permitting the reproduction of published photographs, I am indebted to my collaborators already mentioned, and to Dr J.G. Carlson, Prof. P.I. Dee, Dr I. Lasnitzki, Dr R. Markham, Dr A. Marshak, Dr C.F. Robinow, Dr J.E. Smadel, Prof. C.T.R. Wilson, the British Institute of Radiology, the Radiological Society of North America, the Rockefeller Institute for Medical Research, and the Royal Society. Plate II E is a photograph taken at the National Physical Laboratory, and is reproduced by courtesy of the Director, Crown copyright reserved. I should like to thank Mrs D.E. Lea and Mr V.C. Norfield for preparing the figures and the plates respectively. Finally, I acknowledge gratefully the support of the British Empire Cancer Campaign and the Prophit Trust.

D. E. L.

STRANGEWAYS LABORATORY
Cambridge, July 1944

PREFACE TO THE SECOND EDITION

This book has been out of print for some considerable time. In attempting to meet the demand for a second edition, it was considered desirable to make this challenging presentation of certain aspects of radiobiology available again in its original form, which bears on every page the mark of Lea's clear mind and incisive logic. The textual revisions, and some of the additions which Lea himself had written into his own copy of the book, have, however, been incorporated. These additions appear as Appendix II, and the attention of the reader is directed to them by a marginal asterisk in the text.

As a physicist Lea was acutely aware of the concentrated dissipation of energy along the tracks of individual particles, which is a unique feature of the ionizing radiations considered as cytotoxic agents, and was especially interested in exploring the implications of this fact for radiobiology. In his writings, however, at least as far back as 1941, he makes the distinction, which is explicitly discussed in the opening paragraph of Chapter III, between those forms of biological damage to which these concepts can and cannot usefully be applied. In his treatment of those forms of biological damage which he believed to be initiated by individual ionizing acts, he had the foresight to recognize and to allow for the spread of the energy absorbed within a molecule over distances of a few millimicrons, and for the effects of energy absorbed in the aqueous phase surrounding a biological structure.

Had he lived, the present edition would undoubtedly have been enriched in many ways, and particularly by a development of his ideas regarding the physical processes which underlie the production of chemical change in aqueous solution.

As different forms of radiobiological damage yield one by one to quantitative experimental investigation, it is pertinent to enquire what role, if any, the quantum nature of the primary physical processes plays in determining the reactions of the living cell to radiation. This book records how such an enquiry was carried through by a brilliant mind in relation to virus inactivation, gene mutation, and the production of various forms of chromosome structural damage. As such, I believe it is of permanent value to the student of radiobiology, however far our knowledge of the phenomena treated by Lea may transcend that which was available to him.

I am indebted to Dr J. W. Boag and Dr J. G. Neary for correcting a small error which ran through some of the data presented in Table 10, and for the amendment of Lea's treatment of the overlapping factor F as it appears in Appendix I.

L. H. GRAY

MOUNT VERNON HOSPITAL
NORTHWOOD
MIDDLESEX
May 1954

ILLUSTRATIONS

Plate   page
I. Distribution of ionization produced by different radiations   *facing* 10
II. Viruses   „   102
III. Chromosome structural changes   „   192
IV. Bacteria; viruses; abnormal division figures   „   314

Figure
1. Linear and exponential dose relations   38
2. Independence of absolute yield upon concentration of solute   43
3. Number of collisions per second made by an active radical   53
4. Diminution of ionic yield at low concentrations   58
5. Methods of plotting experimental data   73
6. Survival curves   76
7. Associated volume method of calculation   84
8. Relation between 37 % dose and target diameter
9. Relation between 37 % dose and molecular weight
10. Relation of 37 % dose to 37 % dose with γ-rays
11. Relation of 37 % dose to 37 % dose with X-rays   *between 88 & 89*
12. Target with indefinite boundary   97
13. Inactivation of plant viruses by radiations   107
14. Dependence of inactivation dose on protein concentration   109
15. Survival curves of irradiated plant viruses   112
16. Survival curves of irradiated animal viruses   113
17. Survival curves of irradiated bacteriophages   114
18. Relation between virus diameter and inactivation dose of X- and γ-rays   117
19. Relation between virus diameter and inactivation dose of α-rays   119
20. Visible mutations induced by X-rays as a function of dose   143
21. Sex-linked recessive lethal mutations induced by X-rays   145
22. Yield of sex-linked recessive lethals at different intensities   147
23. Induction of sex-linked recessive lethals by different radiations   148
24. Analysis of sex-linked recessive lethals   157
25. Survival of eggs fertilized by irradiated sperm   161
26. Proportion of viable sperm having chromosome aberrations   168

ILLUSTRATIONS

Figure		page
27.	Dominant lethals as a function of dose	168
28.	Depression of sex-ratio	171
29.	Efficiencies of different wave-lengths of ultra-violet light	188
30.	Structural changes in unsplit chromosomes	194
31.	Structural changes in split chromosomes	195
32.	Influence of temperature on structural changes	221
33.	Dependence upon duration of exposure of yield of aberrations	227
34.	Breaks per cell as a function of dose	230
35.	Yield of two-break aberrations as a function of dose	231
36.	Normal division figures as a function of dose	236
37.	Yield of ultra-violet-induced deletions as a function of incident energy	243
38.	Diminution of yield of exchanges with increased duration of exposure	266
39.	Chromosome exchanges as a function of dose	268
40.	Relations between ion-density and probability of breakage	271
41.	Passage of an electron through a chromatid	274
42.	Primary chromatid breaks as a function of wave-length	275
43.	Cleavage delay in *Arbacia*	286
44.	Cleavage delay as a function of time of irradiation	287
45.	Cleavage delay as a function of dose	288
46.	Decay of cumulative dose	290
47.	Mitotic activity at various times after irradiation	296
48.	Dependence upon intensity of dose required for a given delay	297
49.	Division delay as a function of dose	299
50.	Inhibition of division in *Bact. coli*	301
51.	Inhibition of division in bean root-tips	302
52.	Reduction of mitotic count as a function of dose	304
53.	Germination of pollen irradiated at different stages	308
54.	Dividing and degenerating cells after irradiation	309
55.	Survival curves of irradiated bacteria	318
56.	Survival curves (theoretical)	320
57.	Killing of *Bact. coli* by soft X-rays	323
58.	Relation between chromosome structural changes and lethal effect	334
59.	Hereditary partial-sterility	339
60.	Illustrating target calculation	353
61.	Relative efficiencies of ionizing particles of different ion-densities	354

Chapter I

PHYSICAL PROPERTIES AND DOSIMETRY OF DIFFERENT RADIATIONS

Ionization, excitation, and 'point-heat'

The radiations with which we are concerned are the α-, β- and γ-radiations of radioactive substances, X-rays, protons and neutrons. These may be grouped together as ionizing radiations. Occasionally we deal with ultra-violet light which is a non-ionizing radiation. Since the term ionization will be used very frequently, we begin by explaining its meaning. An atom consists of a positively charged nucleus and a surrounding constellation of negative electrons, the whole being electrically neutral. The principal means of energy dissipation by an ionizing radiation in its passage through matter is the ejection of electrons from atoms through which it passes. An atom so *ionized* is left positively charged, and is referred to as an *ion*. It is possible that some actions of radiation of biological significance are due to this separation of electrical charge, but in most cases it is more plausible to attribute it to chemical change resulting from the ionization. For, *when an atom is ionized the molecule of which it is a part almost certainly undergoes chemical change.* Knowing that the chemical bonds which hold a molecule together are constituted by electrons shared between the two atoms joined by the bond, it is to be expected that the removal of such a bonding electron from a molecule will lead to its dissociation or other chemical change. The removal of electrons other than bonding electrons may also be expected to lead to chemical change, since the energy involved in ionization, 10 electron-volts[1] or upwards depending on the atom ionized and the level in it from which the electron is ejected, exceeds the energy required to remove an atom from the molecule.

A second method by which radiations dissipate energy in tissue is by *excitation*. This means the raising of an electron in an atom

[1] The electron-volt (eV.) = $1 \cdot 602 \times 10^{-12}$ erg, is a unit of energy of suitable magnitude for dealing with energy changes in single atoms or molecules. 1 eMV. = 10^6 eV.; 1 ekV. = 10^3 eV. 1 eV. per molecule = $23 \cdot 05$ kilogram-calories per gram-molecule. Thus the statement that the energy of dissociation of the C—H bond is 94 kilocalories per gram-molecule means that the energy required to dissociate a single bond is about 4 eV.

or molecule to a state of higher energy, and is a less drastic process than the complete ejection of an electron. Ultra-violet light, as well as ionizing radiations, is capable of causing excitation. In simple reactions (e.g. inorganic gas reactions) ultra-violet-induced excitations are not much less effective than X-ray-induced ionizations in causing chemical change. There is some evidence, however, that in the decomposition of large organic molecules, an excitation is a good deal less efficient than an ionization.[1]

In biological actions of radiation in which quantitative data are available both for ionizing radiations and for ultra-violet light, particularly the inactivation of viruses and the killing of bacteria, an excitation by ultra-violet light is much less effective than an ionization. Thus it appears probable that when we are dealing with an ionizing radiation, excitation may usually be neglected as a cause of biological effect by comparison with ionization.

The electron which is ejected from an atom in the process of ionization eventually becomes attached to another atom and makes it a negative ion. As far as the physical measurement of ionization is concerned, the positive and negative ions are equally significant, and one usually speaks of the production of an ion-pair. But since the energy involved in the attachment of an electron to an atom to form a negative ion is usually even less than the energy of excitation, it is probably safe to neglect negative-ion formation as a factor of biological importance. Thus when we speak of an ionization we refer to the production of a positive ion by the ejection of an electron. This ejected electron may have sufficient energy to ionize on its own account before it is brought to thermal energy and finally attached, and this secondary ionization is important and will be discussed later. But again, it is the positive ions and not the negative ions which are of biological importance.

Practically all the energy dissipated by radiation in tissue ultimately becomes degraded to heat energy. Thus a dose of 10^5 roentgens is sufficient to raise the temperature about $0.25°$ C. This small temperature rise for a large dose of radiation means that temperature change is quite inadequate to explain the biological effects of ionizing radiations in the way in which tempera-

[1] Cp. Jordan, P. (1938a).

ture rise accounts for most of the biological effects of, for example, short-wave wireless waves. On the other hand, the highly localized nature of the energy dissipation means that the energy which eventually causes a rise of temperature of 0·25° C. in the tissue as a whole is initially confined to a small proportion of the atoms, and might be considered to produce a large rise in temperature of these atoms. This is the basis of the 'point-heat' theory.[1] The concept of an ionization as a hot spot is less satisfactory than the concept of an atom ionized leading to chemical change in its molecule. However, in the course of the degradation of ionization and excitation energy to thermal energy, the possibility must be borne in mind of molecules near to a high concentration of ionization suffering chemical change even although not themselves ionized.[2]

Ultra-violet light

In quantitative experiments it is necessary to use monochromatic ultra-violet light, since the biological effectiveness per unit of energy varies very much with wave-length and in particular is very low above 3000 A. The physical measurement will usually give the energy incident upon the irradiated preparation in ergs per square centimetre. If we are irradiating small objects such as bacteria or viruses, and can obtain a suspension in a non-absorbing medium free from absorbing impurities, a satisfactory procedure is to use a stirred suspension sufficiently deep or sufficiently concentrated to absorb completely the incident ultra-violet light. The total incident energy may then be divided either by the total number of suspended organisms or by the total volume of protoplasm in suspension to obtain the absorbed energy either in ergs per organism or in ergs per cubic micron.

In other cases we may arrange to use a thickness of irradiated material sufficiently small to absorb only a small proportion of the incident radiation. If we know the absorption coefficient of the irradiated material, or better, the absorption coefficient of that part of it absorption in which we believe to account for the biological effect, we may calculate the energy absorption in this part in ergs per cubic centimetre. If I is the incident intensity in ergs/cm.^{2}, ρ the density of the absorbing substance in g./cm.^{3}, and μ the absorption coefficient in cm.$^{-1}$ defined by the relation

[1] Dessauer, F. (1923).   [2] Jordan, P. (1938c).

$\mu = \dfrac{1}{x} \log_e \dfrac{I_0}{I} = \dfrac{2 \cdot 3}{x} \log_{10} \dfrac{I_0}{I}$, where I is the intensity transmitted by a layer x cm. thick, I_0 being the initial intensity, then μI is the energy absorption in ergs/cm.³ and $(\mu/\rho)\,I$ the energy absorption in ergs/g. If (as happens, for example, in the irradiation of *Drosophila* sperm in the male) it is not possible to avoid excessive absorption, either by too great thickness of the irradiated material or of intervening tissue, then quantitative experiments are not possible unless measurement can be made of the intensity of radiation reaching the specimen under the conditions of the experiment.

To give some idea of the magnitude of the absorption coefficients, and the variation between different materials, we list in Table 1 the absorption coefficients[1] of a number of substances for 2536A., this wave-length being chosen since it is in the biologically most effective region, and also is readily obtainable nearly monochromatic. In the references cited the absorption coefficients for other wave-lengths may be found.

Ultra-violet light for biological experiments is usually obtained from a mercury lamp. If the efficiency of different wave-lengths is being compared, then a quartz mercury arc in conjunction with a large-aperture quartz monochromator is the usual equipment.[2] An intensity of about $2 \cdot 5 \times 10^4$ ergs/cm.²/sec. is possible with a lamp dissipating 1 kW. If a monochromator is not available, or if one is not interested in the relative efficiency of different wave-lengths, the best type of lamp to use is the low-pressure discharge lamp[3] containing neon and emitting 85–95 %

[1] If the absorption of a substance has been measured in solution, then the result will often be expressed as the extinction coefficient α in the formula $\log_{10}(I_0/I) = \alpha c x$, c being the concentration of solute in mg./cm.³ Or it may be expressed as the molar extinction coefficient ϵ in the formula $\log_{10}(I_0/I) = \epsilon c x$, where c is now the concentration in gram-molecules per litre of a substance of molecular weight M. In order to reduce the absorption coefficients to a common basis suitable for calculating energy absorption from knowledge of the ultra-violet intensity, all have been reduced to μ or μ/ρ, the absorption coefficient of the pure solute, by assuming that the absorption coefficient of a solution is proportional to the concentration of the absorbing solute. Thus $\mu/\rho = 2300\alpha = 2300\epsilon/M$.

[2] Cp. for example, Gates, F.L. (1929a); Benford, F. (1936); Uber, F.M. & Jacobsohn, S. (1938); Uber, F.M. (1940); Cannon, C.V. & Rice, O.K. (1942).

[3] Melville, H.W. (1936); Steacie, E.W.R. & Phillips, N.W.F. (1938); Heidt, L.J. (1939); Peel, G.N. (1939).

of its radiant energy in the line 2536A. The monochromatism may be further improved, if desired, by the use of gaseous filters of chlorine and bromine,[1] or liquid filters,[2] or by using the radiation from the lamp to excite resonance radiation.[3] A convenient lamp[4] takes the form of a 30 cm. length of quartz tubing about 8 mm. diameter wound in a close spiral of about 2·5 cm. diameter.

TABLE 1. Absorption coefficients for wave-length 2536A.

Material	μ cm.$^{-1}$	μ/ρ g.$^{-1}$ cm.^{2}	ϵ	% transmission	Reference
Plant virus protein	—	10,000	—	—	1
Tyrosine	—	3,800	300	—	2
Tryptophan	—	36,000	3,200	—	2
Bacterial protoplasm	3600	—	—	—	3
Bacterial nucleoprotein	—	32,000	—	—	4
Bacterial ribose nucleic acid	—	87,000	—	—	4
Trypsin	–	1,900	30,000	—	5
Ribonuclease	—	900	6,000	—	6
Maize-pollen contents	1900	—	—	—	7
Maize-pollen wall	—	—	—	30	7
Vitelline membrane of hen's egg	—	—	—	8	8
Abdominal wall of *Drosophila*	—	—	—	67	9

1 Bawden, F.C. & Pirie, N.W. (1938).
2 Holiday, E.P. (1936).   3 Gates, F.L. (1930).
4 Lavin, G.I., Thompson, R.H.S. & Dubos, R.J. (1938).
5 Uber, F.M. & McLaren, A.D. (1941).
6 Uber, F.M. & Ells, V.R. (1941).   7 Uber, F.M. (1939).
8 Uber, F.M., Hayashi, T. & Ells, V.R. (1941).
9 Durand, E., Hollaender, A. & Houlahan, M.B. (1941).

At 10 W. this gives an intensity of about 10^4 ergs/cm.^{2}/sec. at 10 cm. distance, or about 5×10^5 ergs/cm.^{2}/sec. on a specimen inside the spiral.

It is important to understand the difference in spatial distribution between the energy dissipation by ultra-violet light and by an ionizing radiation such as X-rays. With ultra-violet light, the absorption coefficient depends on the molecular structure, and is, for example, different for nucleic acid and for protein. The dose in ergs/cm.^{3} absorbed energy may be very different, for

1 Peskoff, N. (1919); Oldenburg, O. (1924); Villars, D.S. (1926); Heidt, L.J. (1939); Svedberg, T. & Pedersen, K.O. (1940); Mitchell, J.S. (1942).
2 A convenient filter is a quartz cell 1 cm. thick containing an aqueous solution of nickel sulphate (20%) and cobalt sulphate (6%). Houston, R.A. (1911); Bäckström, H.L.J. (1940); Bowen, E.J. (1942); Lavin, G.I. (1943).
3 Thomas, L.B. (1941).
4 Procurable from the Thermal Syndicate Ltd., London.

example, in different parts of an irradiated chromosome depending on the degree to which they are loaded with nucleic acid, and at different stages of the division cycle. No such differences exist with X-rays, apart from an increased energy absorption in and near to bone or other components containing atoms of elevated atomic number, since the absorption of X-rays by atoms is not affected by their chemical combinations. A further difference is that with ultra-violet light the excited atoms are distributed spatially at random in a homogeneous tissue irradiated by a uniform intensity, with no tendency for excited atoms to occur in groups, produced simultaneously, or concentrated in a linear path. This is because each excited atom is produced by the complete absorption of a single quantum of ultra-violet light and the quanta are emitted independently. With ionizing radiations the ionizations are localized along the paths of ionizing particles, and thus a number of ionizations may be concentrated in a *cluster* or a *column* of ionization.

The energy of a single quantum of ultra-violet light is connected with the wave-length (λ) in Angstroms by the relation: Energy in electron volts $= 12,400/\lambda$. Thus the wave-length 2536 A. has quantum energy 4·89 eV.

X-rays

X-radiation like ultra-violet light is an electromagnetic radiation emitted in quanta, but the difference in wave-length (0·05–10 A. for X-rays against 2000–3000 A. for ultra-violet light) results in there being little similarity in practice. The absorption coefficient of X-rays depends not on the chemical combination of the absorbing atoms but only on their atomic number. On account of the greater penetrating power of X-rays it is not usually convenient to measure the total energy *incident* on a surface, but to measure the energy *absorbed* in a given volume. In practice the energies involved are too small for a thermal method of measuring energies to be used, and use is made of the fact that when the absorption takes place in air, the latter becomes conducting and the saturation current through a given volume of air is a measure of the rate of energy absorption in that volume of air. The *roentgen*—defined as 'the quantity of X- or γ-radiation such that the associated corpuscular emission per 0·001293 g. of air produces, in air, ions carrying 1 electrostatic unit of quantity of electricity of either sign '—is the unit of dose employed.[1] It corresponds to the liberation of $2 \cdot 082 \times 10^9$ ion-pairs per cm.³ of air at 0° and 760 mm. pressure, involving an energy dissipation of $0 \cdot 1083$ erg/cm.³ of air (taking $32 \cdot 5$ eV. per ion-pair as the mean energy dissipation in air). Since the roentgen is already a unit of energy absorbed, there is no question of multiplying dose in roentgens by absorption coefficient. What is loosely spoken of as 'intensity' in roentgens per minute is strictly dose-rate, being a rate of energy absorption and not a rate of energy incidence.

For the convenience of the physical measurement the roentgen is defined in terms of energy absorption per unit volume of air. In the interpretation of experiments one is interested in the energy absorption per unit volume of tissue, which for the same incident intensity of radiation will be about 1000 times greater on account of the greater density of tissue. The actual factor varies with different wave-lengths, since the ratio of the absorption coefficients of tissue and air varies somewhat with wave-length. It also depends on the elementary analysis of the tissue though not on the chemical nature of the compounds in which the elements are combined. The way in which one calculates the amount of energy absorbed per gram of tissue per roentgen of radiation is explained in the Appendix, and in Table 2 the results of a calculation of this sort are given for water, for dry virus protein, and for an undried soft tissue. The percentage composition by weight assumed for the virus protein is H 7, C 49, N 16, O 25, P 1, S 0·5 and ash 1·5, the ash being treated as having an average atomic number of 16 for the purpose of the calculation. The wet tissue is taken as having a percentage composition by weight H 10, C 12, N 4, O 73, Na 0·1, Mg 0·04, P 0·2, S 0·2, Cl 0·1, K 0·35 and Ca 0·01. With most experimental materials it will be found satisfactory to use the figures given in Table 2, either for water or for virus protein or for wet tissue, according to which approximates best to the composition of the tissue actually being irradiated. Thus when irradiating a dried virus preparation the figures for virus protein will be used. When irradiating micro-organisms in aqueous suspension the figures for water are appropriate, while the figures for wet tissue may be used when one is dealing with undried tissue in bulk.

[1] $0 \cdot 001293$ g. of air occupies 1 cm.³ at 0° and 760 mm. pressure.

TABLE 2. Energy absorption in tissues of various composition for 1 r. of various radiations
(for 1 v-unit in the case of neutrons)

Radiation	Ergs/g.			eV./10^{-12} g.			Energy units			Ionizations/10^{-12} g.		
	Water	Virus protein	Wet tissue	Water	Virus protein	Wet tissue	Water	Virus protein	Wet tissue	Water	Virus protein	Wet tissue
γ-rays: 0·0152 A.	93·12	89·84	92·13	58·12	56·08	57·51	1·000	0·965	0·989	1·789	1·727	1·771
X-rays: 0·0506 A.	92·93	89·54	92·03	58·01	55·89	57·44	0·998	0·962	0·988	1·786	1·721	1·769
0·0809 A.	92·23	89·23	91·68	57·63	55·70	57·23	0·992	0·958	0·985	1·775	1·715	1·762
0·1348 A.	89·75	87·54	90·21	56·02	54·65	56·31	0·964	0·940	0·969	1·725	1·683	1·734
0·1618 A.	88·07	86·19	89·19	54·97	53·80	55·67	0·946	0·926	0·958	1·693	1·657	1·714
0·2022 A.	85·81	84·03	87·78	53·56	52·45	54·79	0·922	0·902	0·943	1·649	1·615	1·687
0·3033 A.	83·21	80·32	85·97	51·94	50·14	53·66	0·894	0·863	0·923	1·599	1·544	1·652
0·4853 A.	83·24	77·96	85·64	51·96	48·66	53·46	0·894	0·837	0·920	1·600	1·498	1·646
1·54 A.	86·13	75·47	86·49	53·76	47·11	53·99	0·925	0·810	0·929	1·656	1·451	1·662
4·15 A.	102·31	81·53	96·31	63·86	50·89	60·12	1·099	0·876	1·034	1·966	1·567	1·851
8·32 A.	96·94	63·06	89·01	60·51	39·36	55·56	1·041	0·677	0·956	1·863	1·212	1·711
Neutrons: D+D	107·98	75·17	94·48	67·40	46·92	58·98	1·160	0·807	1·015	1·926	1·341	1·685
Li+D	107·22	76·67	98·32	66·93	47·86	61·37	1·151	0·823	1·056	1·912	1·367	1·753
β-rays: 1 eMV.		95·36			59·52			1·024			1·833	
Protons: 0·5 eMV.		109·10			68·10			1·172			1·946	
1 eMV.		107·22			66·93			1·151			1·914	
2 eMV.		106·03			66·18			1·139			1·892	
3 eMV.		105·49			65·85			1·133			1·883	
4 eMV.		105·19			65·66			1·130			1·877	
5 eMV.		104·97			65·52			1·127			1·873	
6 eMV.		104·79			65·41			1·125			1·870	
8 eMV.		104·56			65·26			1·123			1·866	
10 eMV.		104·39			65·16			1·121			1·863	
α-rays: 1 eMV.		112·32			70·11			1·206			2·005	
2 eMV.		109·07			68·08			1·171			1·947	
3 eMV.		107·87			67·33			1·158			1·925	
4 eMV.		107·21			66·92			1·151			1·913	
5 eMV.		106·77			66·65			1·147			1·906	
6 eMV.		106·45			66·44			1·143			1·900	
8 eMV.		106·02			66·18			1·139			1·892	
10 eMV.		105·70			65·98			1·135			1·887	
α-rays of Rn+Ra A+Ra C′ completely absorbed, i.e. radon in solution		109·26			68·20			1·173			1·949	

If the material being studied can be prepared in films of a few microns thickness, as with viruses, it is convenient to use long wave-length X-rays (1–10 A.), since these may be obtained in high intensity from inexpensive apparatus, and have also the considerable advantage from the point of view of subsequent calculations that they are obtainable nearly monochromatic. The wave-lengths 1·5, 4·1 and 8·3 A. listed in Table 2, are obtainable in this way. They are reduced to half their intensity by passage through layers of water of thicknesses respectively 695, 34·6 and 4·9 μ.

If more penetrating radiations are required an X-ray tube of the type manufactured for X-ray therapy will usually be used. Such X-ray tubes emit a range of wave-lengths. The shortest wave-length emitted is related to the peak kilovoltage on the tube by the relation: wave-length in Angstroms = 12·4/kilovoltage, the quantum energy in electron-kilovolts of this shortest wave-length being equal to the kilovoltage on the tube. The wave-length emitted at greatest intensity is usually about twice the wave-length of the short-wave limit, and the longest wave-length emitted depends on the thickness of filter employed since long waves have a higher absorption coefficient than short. Thus an X-ray tube operated at 160 kV. with filters of 0·7 mm. Cu and 1·2 mm. Al emits a continuous range of wave-lengths from 0·078 to about 0·4 A., the 'effective' wave-length, conventionally defined as that wave-length which would, if monochromatic, be reduced to half value by the same thickness of copper absorber as actually reduces the mixed radiation to half value, being 0·15 A. The corresponding quantum energies for the shortest, longest and effective wave-lengths are 160 ekV., about 30 and 83 ekV.[1]

The dissipation of energy in tissue irradiated by X-rays of fairly long wave-length (e.g. 1·5 A.) takes place as follows. Some of the quanta which enter the tissue are absorbed by atoms. The energy of a quantum, 8 ekV. for an X-ray of 1·5 A., is completely absorbed and results in the ejection of an electron. The innermost electrons of the atom are most effective in this type

---

[1] Fuller information on the effective wave-lengths of X-rays emitted from tubes operating at various kilovoltages and filtrations may be found in the article by Taylor, L.S. in Duggar, B.M. (1936), *Biological Effects of Radiation*.

of absorption (*photoelectric* absorption), and an appreciable amount of energy is required to detach the electrons, depending on the atomic number of the absorbing atom. It is 285 eV. for a carbon atom, 528 eV. for an oxygen atom, more for heavier elements, and may be taken as about 0·5 ekV. on the average for tissue. Thus the photoelectron on leaving the atom has energy $8 - 0 \cdot 5 = 7 \cdot 5$ ekV. (or in general, $12 \cdot 4/\lambda - 0 \cdot 5$ ekV.). This ejected electron has a range of $1 \cdot 5 \mu$ in tissue and enough energy to produce about 230 ionizations. Thus the ionization in tissue irradiated by X-rays of 1·5 A. is not distributed at random but is localized along tracks as in Plate I F.[1] This localization is very important in the theory of the biological action of radiations, and will enter into a good deal of the subsequent calculations. The distribution of ionizations along the electron tracks is described later.

The energy of a few hundred electron-volts needed to detach an electron from the atom which absorbs the quantum of radiation eventually appears as ionization. The mechanism may either be the emission of a second electron from the same atom (Auger effect), or, less frequently, the emission of a long wavelength X-ray quantum, which after travelling a distance of the order of 1μ in the tissue is itself absorbed by another atom with the production of a photoelectron having practically the whole energy of the long wave-length quantum. Thus the absorption of the original quantum of 8 ekV. in the tissue gives rise usually to two electrons, one of about 7·5 ekV. and the other of about 0·5 ekV. This division of the energy of the quantum between two electrons is rarely important, and we shall often, in Table 3 for example, calculate as if the energy of the original quantum were dissipated by a single photoelectron of the full energy 8 ekV.

With shorter wave-lengths, e.g. 0·1 A., absorption still takes place to some extent by the photoelectric effect, but a second

---

[1] The photographs in Plate I are Wilson-chamber photographs (from Wilson, C.T.R. 1923; Dee, P.I. 1932; Curie, P. 1935) in which the ions in a gas have been made visible by the condensation of water drops upon them. The scale has been adjusted so that the photographs represent approximately correctly the distribution of ions in irradiated tissue. Thus Plate I F may be taken to represent the distribution of ionization in a section of tissue $15 \times 12 \times 1 \mu$ thick, irradiated by a dose of 10 r. of X-rays of wave-length 1·5 A. No significance is to be attached to the *width* of the tracks.

A. α ray with δ-rays

E. Recoil eletrons and photoelectrons produced by X-rays of wavelength 0·1–1 A.

B. Proton.

C. Increasing ion-density towards end of electron track.

D. Fast electron.

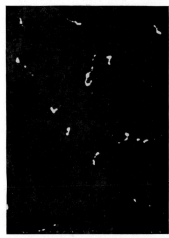

F. Photoelectrons produced by X-rays of wavelength 1·5 A.

10 μ tissue

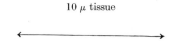

PLATE I. Distribution of ionization produced by different radiations

mechanism also contributes, this being the production of Compton recoil electrons. In this process a quantum of radiation is scattered by an electron with reduced energy (i.e. increased wave-length), the balance of energy being transferred to the electron which is ejected from its atom. Depending on the angle at which the quantum is scattered, the energy of the Compton recoil electron may vary from zero up to a maximum which is a fraction of the quantum energy depending on the wave-length (0·327 for 0·1A., 0·108 for 0·4A.). Thus when X-rays of 0·1A. are irradiating tissue, in addition to a certain number of photo-electrons of energy 124 ekV., there will be Compton electrons of all energies between 0 and 40 ekV. Remembering further that X-rays of 0·1A. cannot be obtained monochromatic (except by means not usually practicable for biological experiments), it will be seen that calculations based on experiments using these wave-lengths are more complicated than experiments with longer wave-lengths. In Table 3 are listed the relative numbers of Compton and photoelectrons liberated in tissue (here taken as H_2O) by different wave-lengths of X-rays, together with the energy of the photoelectrons and the maximum and mean energies of the Compton electrons. Plate IE shows recoil electrons (the shorter tracks) and photoelectrons (the longer tracks).

The atom from which the photoelectron or Compton electron is ejected is of course ionized; there is no reason, however, to attribute any special virtues to an ionization caused by absorption of an X-ray quantum, and there are very many more ionizations caused by the impact of recoil electrons or photoelectrons. Thus if an effect, e.g. breakage of a chromosome, occurs at a particular point in a cell there is no reason to attribute this to absorption of an X-ray quantum at that point, though it may very well be due to the passage of a photoelectron or recoil electron through the point.

In using Table 2 the composition of tissue relevant is that of the tissue in which the electrons arise. Thus when dealing with micro-organisms suspended in water, the linear dimensions of the organisms being much smaller than the range of the electrons, and their total volume being only a small fraction of the total volume of the water, most of the electrons which traverse the organisms will have arisen in the water and not in the organisms,

TABLE 3. Recoil and photoelectrons liberated in water by X- and γ-rays

$\frac{h\nu}{mc^2}$	Radiation wave-length λ in A.	Quantum energy $h\nu$ in ekV.	Proportion of total number of electrons which are		Proportion of total electron energy which appears in		Photo-electron energy in ekV.	Recoil electron energy in ekV.		Mean electron energy, photo- and recoil electrons in ekV.
			Recoil electrons	Photo-electrons	Recoil electrons	Photo-electrons		Maximum	Mean	
0·01	2·4265	5·108	0·006	0·994	0·000	1·000	5·11	0·10	0·05	5·08
0·02	1·2132	10·217	0·045	0·955	0·001	0·999	10·22	0·39	0·19	9·76
0·03	0·8088	15·325	0·138	0·862	0·004	0·996	15·32	0·87	0·43	13·27
0·04	0·6066	20·433	0·275	0·725	0·014	0·986	20·43	1·51	0·75	15·02
0·05	0·4853	25·542	0·424	0·576	0·032	0·968	25·54	2·32	1·15	15·20
0·06	0·4044	30·650	0·558	0·442	0·063	0·937	30·65	3·28	1·62	14·47
0·07	0·3466	35·758	0·665	0·335	0·106	0·894	35·76	4·39	2·15	13·41
0·08	0·3033	40·866	0·746	0·254	0·167	0·833	40·87	5·64	2·79	12·46
0·09	0·2696	45·975	0·805	0·195	0·237	0·763	45·98	7·01	3·46	11·73
0·10	0·2426	51·083	0·849	0·151	0·316	0·684	51·08	8·51	4·20	11·28
0·11	0·2206	56·191	0·881	0·119	0·397	0·603	56·19	10·13	4·99	11·08
0·12	0·2022	61·300	0·905	0·095	0·476	0·524	61·30	11·86	5·83	11·10
0·15	0·1618	76·625	0·948	0·052	0·672	0·328	76·62	17·68	8·68	12·25
0·18	0·1348	91·949	0·968	0·032	0·797	0·203	91·95	24·34	11·89	14·45
0·21	0·1155	107·27	0·979	0·021	0·872	0·128	107·27	31·73	15·56	17·48
0·24	0·1011	122·60	0·986	0·014	0·916	0·084	122·60	39·76	19·50	20·99
0·27	0·0899	137·92	0·990	0·010	0·942	0·058	137·92	48·36	23·72	24·92
0·30	0·0809	153·25	0·992	0·008	0·959	0·041	153·25	57·47	28·22	29·20
0·33	0·0735	168·57	0·994	0·006	0·970	0·030	168·57	67·02	32·93	33·75
0·36	0·0674	183·90	0·995	0·005	0·977	0·023	183·90	76·98	37·89	38·59
0·39	0·0622	199·22	0·996	0·004	0·983	0·017	199·22	87·30	43·04	43·64
0·42	0·0578	214·55	0·997	0·003	0·986	0·014	214·55	97·95	48·33	48·85
0·45	0·0539	229·87	0·997	0·003	0·989	0·011	229·87	108·89	53·83	54·29
0·48	0·0506	245·20	0·998	0·002	0·991	0·009	245·20	120·10	59·49	59·90
0·6	0·04044	306·50	0·999	0·001	0·996	0·004	306·50	167·18	83·55	83·82
0·8	0·03033	408·66	0·999	0·001	0·998	0·002	408·66	251·48	127·80	127·96
1·0	0·02426	510·83	1·000	0·000	0·999	0·001	510·83	340·55	175·95	176·05
1·2	0·02022	613·00	1·000	0·000	0·999	0·001	613·00	432·71	227·21	227·30
1·4	0·01733	715·16	1·000	0·000	1·000	0·000	715·16	526·96	280·67	280·72
1·6	0·01517	817·33	1·000	0·000	1·000	0·000	817·33	622·73	336·52	336·58
1·8	0·01348	919·49	1·000	0·000	1·000	0·000	919·49	719·60	393·92	393·93
2·0	0·01213	1021·66	1·000	0·000	1·000	0·000	1021·66	817·33	452·56	452·56

and hence one should use the columns of Table 2 appropriate to water. On the other hand, if having obtained the energy dissipation in ergs/g. we wish to convert to ergs/cm.^{3} by multiplying by the density, the density which is relevant is that of the tissue in which the ionization is actually produced. Thus if we have a suspension in water of micro-organisms of density 1·2 g./cm.^{3}, we convert a dose measured in roentgens to ergs/cm.^{3} in the organisms by multiplying by 1·2 the figure given in the second column of Table 2 for the radiation concerned.

It will be noticed in Table 3 that a change in the wave-length of X-rays from 0·15 to 0·4A. does not make a great deal of difference in the mean energy with which the electrons are projected in the tissue. This is due to the replacement of electron recoil by photoelectron production as the principal means of absorption as we change from 0·15 to 0·4A. In consequence, experiments designed to see whether there is a variation of biological effect with wave-length should not be confined to wavelengths in this range, which unfortunately is the range most readily available with standard X-ray equipment of the therapy type. If the nature of the experimental material permits, a much more valuable test is a comparison of soft X-rays (1A. or greater) with medium X-rays (i.e. X-rays in the range 0·1–0·4A.), and γ-rays.

γ-rays

The γ-rays used in biological work are those of radium and its products, or sometimes radon and its products, sealed in a closed container with sufficient thickness of wall to absorb α- and β-rays (at least 0·5 mm. platinum or equivalent). The radiations, which are natural X-rays of short wave-length, are emitted in a number of discrete wave-lengths ranging, with this filtration, in quantum energy from about 0·2 to about 2·2 eMV.

Absorption in tissue is practically entirely by Compton recoil, and electrons of all energies from zero up to about 2·0 eMV. are thus produced.

By the use of a lead filter between the radium and the irradiated material it is possible to reduce the intensity of the longer, more easily absorbed, wave-lengths, and so reduce the average wave-length and increase the proportion of the ionization in the irradiated tissue which is due to electrons of high energy. In

Table 4[1] is given the energy distribution of the recoil electrons produced in tissue by radium γ-rays with different degrees of filtration. It is seen, for example, that with a filter of 0·5 mm. of platinum only, 22% of the ionization is produced by recoil electrons of less than 0·25 eMV., but that when the filtration is increased by 2·5 cm. Pb, only 5% of the ionization is so produced. The efficiency of a lead filter in removing the longer wave-lengths and so enabling the biological effectiveness of different

TABLE 4. Electrons liberated in tissue by radium γ-rays

The figures give the percentages of the total electron energy which are dissipated by electrons liberated with the stated initial energy.

Range of electron energies (eMV.)	0–0·05	0·05–0·1	0·1–0·25	0·25–0·5	0·5–1	1–2	Total
Filter 0·5 mm. Pt	1·5	3·8	17·1	24·9	26·3	26·4	100·0
0·5 mm. Pt + 1 cm. Pb	0·4	1·2	7·1	20·3	33·0	38·0	100·0
0·5 mm. Pt + 2·5 cm. Pb	0·2	0·6	4·0	13·1	34·1	48·0	100·0

γ-ray wave-lengths to be compared is reduced if scattered or secondary radiations from the filter contribute much to the radiation falling on the irradiated object, since such scattered radiation is of longer wave-length than the primary. In experiments of this sort the diameter of the lead filter should be as small as possible by comparison with the distance between the source and the irradiated material.

γ-rays, like X-rays, are measured in terms of roentgens, and the factors for converting doses in roentgens into energy dissipation in tissue are given in Table 2. As an alternative to measuring the dose-rate, it may often be calculated from the known content of the radium sources making use of the fact that at 1 cm. from

[1] Table 4 is calculated as follows. By integration of the Klein-Nishina formula one can calculate the energy distribution of recoil electrons produced by γ-rays of any specified wave-length. The intensities of the principal lines in the γ-ray spectra of Ra, Ra B and Ra C have been given by Ellis, C.D. & Aston, G.H. (1930) and Stahel, E. & Johner, W. (1934). By use of the absorption coefficients of these wave-lengths in platinum and lead, which have been computed by Sizoo, G.J. & Willemsen, H. (1938) and Kaye, G.W.C. & Binks, W. (1940), one can deduce the intensities of the different lines after any filtration. Carrying through the energy distribution calculation for each wave-length in turn, and combining in the proportions appropriate, one finally arrives at the figures given. They do not allow for any scattered radiation from the filter reaching the irradiated material.

a point source of 1 mg. of radium filtered by 0·5 mm. of platinum the dose rate is 8·3 r./hr.[1] The allowance to be made for absorption in intervening tissue-like material is from 2·5 to 5% per cm., depending on the extent to which scattering reduces the effective absorption coefficient.[2] For accurate estimation of dose an ionization measurement is to be preferred.

The electrons which dissipate energy in a tissue irradiated by γ-rays may, on account of their high energy, have a range in tissue of several millimetres. When irradiating small objects it is desirable that at least 3 mm. of tissue-like material should surround the relevant part of the objects to serve as a source of electrons.

β-rays and cathode-rays

X-rays and γ-rays produce their effects in tissue by the projection at high speeds of electrons already in the tissue. The same effects would be produced if electrons of similar speed could be produced in other ways. In the case of experimental materials which are usable in thin layers, electrons may be introduced from an external source. Either β-rays, which are fast electrons emitted from radioactive materials, or cathode-rays, which are artificially accelerated electrons,[3] may be used. The most convenient source of β-rays is a thin-walled vessel containing radon gas, the β-rays being emitted by Ra B and Ra C which deposit on the walls. A wall thickness of 0·05 mm. of glass or aluminium suffices to remove the α-rays, and the γ-rays which are also emitted produce negligible ionization by comparison with the β-rays. The distribution of energy of the β-rays emitted by such a source is given in Table 5, the figures in which are approximate only. The second line of the table gives the relative numbers of β-rays of different energies entering the tissue. If one is interested in the proportion of the total energy dissipation which is produced by β-rays of various energies, the result is very different according to the

[1] Mayneord, W.V. & Roberts, J.E. (1937); Gray, L.H. (1937); Kaye, G.W.C. & Binks, W. (1937).

[2] Cp. Mayneord, W.V. & Honeyburne, J. (1938).

[3] Apparatus for the production of cathode-rays suitable for biological experiments has been described by Coolidge, W.D. (1926); Cauchois, Y. (1932); Haskins, C.P. (1938); Cooper, F.S., Buchwald, C.E., Haskins, C.P. & Evans, R.D. (1939); Morningstar, O., Evans, R.D. & Haskins, C.P. (1941).

thickness of the layer of tissue used. If the layer is only a few microns thick, then the total energy dissipation in the thin layer is divided among β-rays of different energy in the manner shown in line 3, in which the effect of the slower β-rays is enhanced by

TABLE 5. Energy distribution of β-rays from Ra B+C

β-ray energy (eMV.)	0–0·1	0·1–0·25	0·25–0·5	0·5–1	1–3	Total
Relative number of β-rays of stated energy as percentage of total number	8	29	30	19	14	100
Energy dissipation in thin layer of tissue by β-rays of stated energy, as percentage of total energy dissipation in the layer	21	40	22	10	7	100
Energy dissipation in thick layer of tissue by β-rays of stated energy, as percentage of total energy dissipated in the layer	1	10	22	27	40	100

the greater number of ionizations they produce per micron path. If, however, a thicker layer is used, e.g. a layer of fluid several mm. in thickness and kept stirred, the β-rays of higher energy will become of more importance as shown in the fourth line.

While the definition of the roentgen does not contemplate its use for radiations other than X-rays and γ-rays, an obvious extension would be to describe as 1 r. of β-rays that dose which in 0·001293 g. of air liberates ions carrying 1 e.s.u. of charge of either sign. Table 2 gives the energy dissipation in tissue corresponding to 1 r. of β-rays defined in this way.

α-rays

α-rays are the nuclei of helium atoms emitted spontaneously by radioactive materials and also obtainable artificially with the cyclotron. Their range in tissue is very short and they are therefore generally only usable with materials obtainable in thin films. α-rays produce many more ionizations per micron path in tissue than do electrons, and the comparison of the relative efficiencies, per ionization, of electrons and α-rays is very valuable for testing theories of the biological action of radiations. Hence in experimental work it is worth while going to considerable trouble to overcome the experimental difficulties attendant upon the use of this feebly penetrating radiation.

For irradiation of thin films of material by α-rays emitted from an external source, polonium serves as a convenient source. It is prepared from old radon seeds,[1] and emits α-rays of energy 5·298 eMV. with practically no other radiation. α-rays of this energy have a range[2] of about 39μ in tissue of unit density.

Alternatively, 'active deposit' sources may be prepared by exposing a metal disk to radon.[3] A source made in this way decays in a few hours and is only suitable therefore for short exposures. After the lapse of half an hour, required for the complete decay of Ra A which emits less energetic α-rays, the α-rays emitted are those of Ra C′ which have an energy of 7·680 eMV. and a range[2] in tissue of unit density of about 71μ. This source emits β-rays and γ-rays as well as α-rays. The ionization per unit volume due to the β-rays is, however, only about 1 % of that due to the α-rays and that due to the γ-rays is quite negligible.

TABLE 6. Ra C′ α-particles

Total absorber between source and surface of tissue	α-ray energy at 3μ depth in tissue in eMV.	Relative ion density at 3μ depth
1 cm. air	6·77	1·00
1 cm. air + $8·8\mu$ Al	5·52	1·13
1 cm. air + $17·6\mu$ Al	4·03	1·42
1 cm. air + $26·4\mu$ Al	1·98	2·09

Experiments on the dependence of the efficiency of α-rays on their energy are most conveniently made by using an active deposit source in conjunction with screens of aluminium foil.[4] By passage through the foil the energy of the α-ray is reduced and the ion density (i.e. number of ionizations per micron path) increased. Table 6 gives data for the ion density and α-particle

[1] For methods of preparation see for example, Rutherford, E., Chadwick, J. & Ellis, C.D. (1930); Whitaker, M.D., Bjorksteadt, W. & Mitchell, A.C.G. (1934); Kanne, W.R. (1937).

[2] These ranges are calculated from the accurately known mean ranges in air and the relative stopping powers of air and water given in Table 12, p. 25. The measured ranges in liquid water are lower and may be wrong, viz. Po 32μ (Michl, W. 1914) and Ra C′ 60μ (Philipp, K. 1923). According to Philipp the range of Ra C′ α-rays in water vapour corresponds to 67μ at density 1 g./cm.³

[3] For apparatus for doing this see, for example, Rutherford, E., Chadwick, J. & Ellis, C.D. (1930); Naidu, R. (1934).

[4] Cp. Zirkle, R.E. (1935).

energy after passage through various thicknesses of aluminium foil.

By an obvious extension of the definition of the roentgen, we may define 1 r. of α-rays as that dose which liberates in 0·001293 g. of air ions carrying 1 e.s.u. of charge of either sign. Table 2 gives the energy dissipation in tissue corresponding to this dose.

When the experimental material is not usable in thin films, two methods exist for the liberation of α-rays in the body of the tissue, enabling α-irradiation to be made without the necessity of employing thin films. The first method is to use radon gas dissolved in the tissue by immersing the tissue in a solution of radon in water enclosed in an air-tight vessel.[1] The α-rays are emitted by the radon itself and by its products Ra A and Ra C' in equal numbers. In addition, to every three α-particles emitted two β-rays are emitted, and if the tissue is thick enough for their complete absorption, as will usually be the case, the total energy dissipation due to the β-rays is about 5% of that due to the α-particles. γ-rays will usually add negligibly to the energy dissipation. Table 7 gives the data of the energies.

TABLE 7. α- and β-particles from dissolved radon

				Energy dissipation in tissue containing 1 mc. of radon per g.				
Particle	Emitter	Energy of particle plus recoil nucleus in eMV.		r./ sec.	Ergs/g./ sec.	eV./10^{-12} g./sec.	Energy units/ sec.	Ionizations 10^{-12} g./sec
α-ray	Rn	5·586 ⎫						
α-ray	Ra A	6·110 ⎬	19·522	10·59	1157	722·3	12·43	20·64
α-ray	Ra C'	7·826 ⎭						
		Max.	Mean	Mean				
β-ray	Ra B	0·65 ⎫	0·23 ⎫					
β-ray	Ra C	3·15 ⎬	0·76 ⎭	0·99	0·62	58·7	36·6	0·630	1·13

The physical measurement will usually consist of the determination of the radioactive content of the tissue in millicuries per gram. Reference to Table 7 enables this to be converted to energy dissipation in the tissue.

A second method[2] of obtaining α-particles liberated in the bulk of the tissue is to impregnate the tissue with a salt of

[1] See Gray, L.H. & Read, J. (1942d) for a full account of this method.
[2] Cp. Kruger, P.G. (1940); Zahl, P.A. & Cooper, F.S. (1941).

lithium or boron and irradiate with slow neutrons. α-rays are emitted as a product of the disintegration of the lithium or boron by the neutrons. Methods of dosimetry have not yet been worked out.

Protons and neutrons

Protons are hydrogen nuclei moving at high speed and are not emitted by radioactive materials but are obtainable as a beam from a cyclotron. They have not as yet been used in biological work, but should be usable with thin layers of tissue. They give a number of ionizations per micron path intermediate between electrons and α-rays, and should be valuable in the same type of experiment for which α-rays are used. Defining a roentgen of protons as for α-rays the energy dissipation in tissue corresponding to 1 r. of protons is listed in Table 2.

In the same way that electrons already existing in the tissue are projected at high speed when it is irradiated by X- or γ-rays, hydrogen nuclei already in the tissue are projected at high speed as protons when the tissue is irradiated by fast neutrons. The neutrons also cause the projection of carbon, oxygen and other nuclei, and dissipate some energy in other ways, but most of the energy they dissipate in tissue is by the process of proton projection.

A roentgen of neutrons, if defined as the dose liberating in 0·001293 g. of air ions carrying 1 e.s.u. of charge, would not be a very suitable unit for neutron dosimetry, since the mechanism of ionization in air is by projection of nitrogen and oxygen, and in tissue mainly by projection of hydrogen nuclei. The factor for conversion of 1 r. defined in this way to energy dissipation in tissue would be several times greater than the factors for other radiations, and would vary with neutron energy since the ratio of the collision cross-sections of hydrogen and of the other gases varies with neutron energy. The disintegration of nitrogen by the neutrons would also complicate the determination of the factor. These difficulties can be largely avoided by defining the unit of neutrons in terms of a material containing a proportion of hydrogen comparable to that in tissue. To realize the unit experimentally it is necessary simply to measure the ionization in the air in a small chamber the walls of which have the appropriate content of hydrogen, the dimensions of the chamber being small enough, or the gas pressure low enough, to ensure that the ionization is almost entirely due to nuclei projected from the walls, and not to nuclei originating in the air.

Actuated by these considerations, Gray[1] has proposed the following unit for neutron dosimetry, provisionally designated by the symbol $1v$. 'The dose of neutrons at a point shall be $1v$ when the associated corpuscular radiations produced in water liberate in air 1 e.s.u. of charge of either sign per 0·001293 g. of air.' The energy dissipation in various tissues corresponding to $1v$ of neutrons has been set out in Table 2.

The n-unit used in America is an arbitrary unit based on the reading given by a commercial X-ray dosemeter (the Victoreen) when exposed to neutrons. It is believed that 1 n-unit equals 2·5 v-units, but the conversion factor varies slightly with different instruments.[2]

Neutrons used in biological work up to the present have been obtained from three types of generator:

(a) A cyclotron employing the Be + D reaction with deuterons of 8 eMV. or more.

(b) An ion tube employing the Li + D reaction with deuterons of about 1 eMV.

(c) An ion tube employing the D + D reaction with deuterons of about 0·3 eMV.

Neutrons from any of these reactions liberate protons in tissue with a variety of energies. The energy distributions given in Table 8 are very approximate but may be used as a basis of calculation.[3]

A few experiments have been reported in which biological materials have been irradiated by slow neutrons. Slow neutrons

[1] Gray, L.H. (private communication); Gray, L.H. (1944 a).
[2] Aebersold, P.C. & Lawrence, J.H. (1942).
[3] They are calculated as follows. Since the scattering of neutrons by protons is isotropic in centre of gravity coordinates, it follows that of the recoil protons produced by neutrons of energy E_0, the proportion which are projected with energy between E_1 and E_2 is $(E_2 - E_1)/E_0$, while the proportion of the total recoil proton energy carried by protons of this range of energies is $(E_2^2 - E_1^2)/E_0^2$. The data of Bonner, T.W. (1941) and of Bonner, T.W. & Brubaker, W.M. (1935, 1936) are used in calculating the various neutron energies E_0 emitted by the different radiations. The $D + D$ neutrons are supposed observed in the direction of the deuteron beam, the Li + D and the Be + D neutrons at right angles to the deuteron beam.

have too little energy to ionize by setting protons into motion by simple collision, which is the principal process by which fast neutrons dissipate energy in tissue. They are however captured by the atomic nuclei of various elements. A nucleus which has captured a neutron emits an α-particle, a proton, an electron, or a γ-ray depending on the element concerned. From the theoretical point of view there is not much interest in studying the

TABLE 8. Distribution of initial proton energies in neutron experiments.

Reaction	Deuteron energy in eMV.	Proton energy in eMV.									Total
		0–1	1–2	2–3	3–4	4–6	6–8	8–10	10–12	12–14	
D+D	0·3	31	31	31	7	—	—	—	—	—	100
		10	28	47	15	—	—	—	—	—	100
Li+D	0·9	51	19	12	4	6	4	2	1·5	0·5	100
		11	15	14	11	17	12	9	7	4	100
Be+D	8	12	12	12	12	24	19	9	—	—	100
		2	5	7	10	29	31	16	—	—	100

Of the two lines of figures given for each reaction, the upper line gives the percentages of the total *number* of protons which have initially the indicated energy, and the lower line gives the percentages of the total proton *energy* dissipated by protons of the indicated initial energy.

biological effect of slow neutrons, since the effects obtained will presumably be no different from those obtained when tissue is irradiated by α-rays, protons, electrons or γ-rays as the case may be. There may, however, be some practical importance attaching to the use of slow neutrons, in that the tissue dose can be made greater in one region than another by impregnating it with lithium or boron (see p. 18).

Dosimetry

The roentgen is a unit of dose internationally accepted for γ-rays and X-rays, and capable of obvious extension to cover most of the other ionizing radiations, in the manner which has been indicated separately for each radiation. It is a unit chosen primarily for convenience in physical measurement, and while 1 r. of any radiation represents the same amount of ionization in air it does not always represent for different radiations the same ionization or energy dissipation in the tissue. It is necessary therefore, in comparing the efficiencies of different radiations, to be able to convert roentgens into ionization in tissue or into energy dissipation in tissue. There is no difficulty in principle in converting roentgens into energy dissipation in tissue, and if the elementary analysis of the tissue is known the conversion can probably be made with an error of less than 10 %.

The most obvious unit of energy to employ is the erg. 1 r. of γ-rays or X-rays involves the dissipation of about 90 ergs/g. of tissue. If we are concerned with the energy dissipated in a single cell or smaller entity, an erg becomes an inconveniently large unit, and it will be found more convenient to employ the electron-volt, which is 1.602×10^{-12} erg. Thus 1 r. of X- or γ-rays is about 60 eV./μ^3 of tissue. In view of the fact that it is common physical practice to express energies of radiations in electron-volts, this unit is probably the most convenient to employ.

TABLE 9. Interconversion of units of energy dissipation in tissue

1 eV./10^{-12} g. = 1·602 ergs/g.
1 energy-unit (Gray & Read) = 93·12 ergs/g.
1 gram-roentgen (Mayneord) = 83·78 ergs
1 ionization per 10^{-12} g.:
    α-ray, proton or neutron = 56·07 ergs/g.
    X-ray, γ-ray or electron = 52·07 ergs/g.

As regards the expression of dosage in terms of ionization in tissue, there is some uncertainty in the conversion, since the physical observations are in practice made on ionization of gases, and it is necessary to assume without experimental confirmation that the energy expended in the production of an ionization is the same in tissue as in a gas. The values we take to apply to tissue are 32·5 eV. per ionization by electrons or X- or γ-rays, and 35 eV. per ionization by α-rays, protons or neutrons, these being the values commonly accepted for air.

Gray and Read[1] suggest that when the effects of different radiations are being compared, the doses be expressed in *energy-units*, where by definition 1 energy-unit of any radiation dissipates the same energy in 1 g. of the tissue concerned as 1 r. of γ-rays dissipates in 1 g. of water.

The *gram-roentgen*[2] is the energy dissipated by 1 r. of γ-rays or hard X-rays in 1 g. of air.

In most of what follows we shall express tissue dosage either in ionizations or in eV./10^{-12} g. of tissue. (10^{-12} g. is the weight

[1] Gray, L.H., Mottram, J.C., Read, J. & Spear, F.G. (1940).
[2] Mayneord, W.V. (1940).

of a cubic micron of water.) Table 9 makes clear the relation between the various units, and in Table 2 we have given factors for converting doses measured in roentgens to any of the methods of expressing tissue dosage likely to be employed. It will be realized that although the factors are for convenience given to four figures, the precision with which they are at present known is by no means so great.

Primary ionization, ranges, and energy dissipation

The various ionizing radiations we are considering all dissipate their energy by the passage through the tissue of either electrons, protons or α-particles. These may be produced externally by a radioactive or an artificial source, or they may be liberated in the tissue as when the tissue is irradiated by X-rays or neutrons, or is impregnated with radon. The principal difference from the theoretical point of view is that in the former case, when a film of tissue thin compared with the range of the ionizing particle is used, the efficiency of a particular part of the track of the ionizing particle may be determined, while in the latter case the ionizing particle is completely absorbed in the tissue and one determines an average efficiency of the whole track.

Electrons, protons and α-particles pass through tissue in paths which may be taken as straight for distances of a few microns, though the bending of electron tracks as a result of scattering by atoms is considerable over longer distances. As the particle traverses the tissue it loses energy as a result of collisions with atoms, excitation or ionization of the atom struck resulting. As explained earlier, excitation is probably of small biological importance compared with ionization. As the ionizing particle slows down, the energy dissipation per micron path, and so the number of ionizations per micron path (which we refer to as *ion-density*, or *specific ionization*), increases. In Plate I are shown photographs[1] of electron, proton, and α-ray tracks showing that the ion-density is greater for α-rays and protons than for electrons, and is greater for a slow electron than for a fast electron. Thus the ion-density increases towards the end of an electron track.

In Tables 10–13 are given numerical data[2] of the range, energy

[1] See p. 10, n. 1.
[2] The sources of these and subsequent tabulated data are described in the Appendix.

dissipation, and number of ionizations produced per micron path by the three types of ionizing particle, for various energies of the particle. None of these data are known as accurately as one would wish, and the uncertainty attaching to the number of

TABLE 10. Range and energy dissipation per micron by electrons in tissue of density 1 g./cm.³

Electron energy ekV.	Energy dissipation ekV./μ	Range μ	Electron energy ekV.	Energy dissipation ekV./μ	Range μ
0·1	33·23	0·00301	9	2·436	2·089
0·15	31·59	0·00455	10	2·298	2·517
0·2	28·71	0·00622	11	2·133	2·969
0·3	23·86	0·01005	12	1·993	3·455
0·4	20·84	0·01455	13	1·873	3·972
0·5	18·62	0·01963	14	1·768	4·522
0·6	16·84	0·02528	15	1·674	5·104
0·7	15·40	0·03150	20	1·339	8·464
0·8	14·17	0·03828	25	1·127	12·55
0·9	13·20	0·04559	30	0·9803	17·32
1·0	12·33	0·05344	40	0·7811	28·82
1·2	10·94	0·07068	50	0·6698	42·69
1·4	9·859	0·08996	60	0·5879	58·66
1·6	8·991	0·1112	70	0·5280	76·63
1·8	8·278	0·1344	80	0·4824	96·47
2·0	7·680	0·1595	90	0·4462	118·0
2·5	6·533	0·2304	100	0·4170	141·2
3·0	5·710	0·3124	110	0·3929	166·0
3·5	5·086	0·4054	120	0·3727	192·1
4·0	4·596	0·5089	150	0·3278	278·1
5	3·872	0·7468	180	0·2976	374·3
6	3·359	1·025	210	0·2769	478·9
7	2·975	1·342	240	0·2602	590·7
8	2·676	1·697	300	0·2381	832·0
			360	0·2239	1092
			420	0·2142	1366
			480	0·2073	1651
×ρ		÷ρ		×ρ	÷ρ

For tissue of density ρ g./cm.³ multiply or divide the figures given by ρ as indicated at the foot of each column.

primary ionizations per micron is considerable, since one has really very little information concerning the production of ionization in solids and liquids. The range meant is the distance travelled by a particle of energy E in being brought to rest, and is measured along the curved path.[1] The energy dissipation is

[1] In the case of electrons, the paths of which are by no means straight, this range is not the same as the maximum penetration of a sheet of absorber. According to Williams, E.J. (1930), the range measured along

RANGE AND ENERGY DISSIPATION

TABLE 11. Number of primary ionizations produced by electrons per micron of path in tissue of density 1 g./cm.^{3}

Electron energy ekV.	Primary ions per μ ions/μ	Electron energy ekV.	Primary ions per μ ions/μ	Electron energy ekV.	Primary ions per μ ions/μ
0·1	1697	2·75	95·01	27·5	12·70
0·15	1206	3·25	81·82	35	10·38
0·25	780·1	3·75	71·96	45	8·452
0·35	583·8	4·5	61·09	55	7·206
0·45	469·5	5·5	50·99	65	6·332
0·55	394·2	6·5	43·86	75	5·685
0·65	340·6	7·5	38·54	85	5·186
0·75	300·5	8·5	34·41	95	4·790
0·85	269·2	9·5	31·90	105	4·603
0·95	244·1	10·5	29·21	115	4·202
1·1	214·5	11·5	26·96	135	3·786
1·3	185·1	12·5	25·06	165	3·349
1·5	163·0	13·5	23·43	195	3·048
1·7	145·9	14·5	22·01	225	2·846
1·9	132·1	17·5	18·70	270	2·592
2·25	113·4	22·5	15·06	330	2·382
				390	2·241
				450	2·166
$\times \rho$		$\times \rho$		$\times \rho$	

For tissue density ρ g./cm.^{3} multiply the figures given by ρ as indicated at the foot of each column.

TABLE 12. Range, energy dissipation, and number of primary ionizations produced per micron path in tissue by α-particles

α-particle energy eMV.	Energy dissipation per μ of tissue ekV./μ	Tissue path equivalent to 1 mm. air at 0° & 760 mm. μ	Range in tissue μ	Primary ionizations per μ tissue ions/μ
1	263·9	1·039	5·3	5207
2	176·1	1·070	10·1	2883
3	134·6	1·082	16·8	2031
4	110·1	1·088	25·1	1581
5	93·77	1·093	35·2	1301
6	82·01	1·096	47·0	1109
7	73·10	1·098	60·3	968·0
8	66·09	1·100	75·5	860·5
9	60·41	1·102	91·6	775·4
10	55·71	1·104	108·4	706·4
5·2984 (Po)			38·9	
5·4860 (Rn)			41·1	
5·9981 (Ra A)			47·0	
7·6802 (Ra C′)			70·8	
	$\times \rho$	$\div \rho$	$\div \rho$	$\times \rho$

Above figures apply to tissue of density 1 g./cm.^{3} For density ρ g./cm.^{3} they should be multiplied or divided by ρ as indicated at the foot of each column.

the curved path is some 40 % greater than the extrapolated range deduced from measurements on the penetration of foils by an electron beam.

computed in electron-kilovolts per micron path and refers to a particle of the stated energy E, and is not the average over all the energies between E and 0; this average if desired can be found by dividing the energy E by the range. The number of

TABLE 13. Range, energy dissipation, and number of primary ionizations produced per micron path in tissue by protons

Proton energy eMV.	Energy dissipation per μ of tissue ekV./μ	Tissue path equivalent to 1mm. air at 0° & 760 mm. μ	Range in tissue μ	Primary ionizations per μ tissue ions/μ
1	27·69	1·088	23	398·2
2	16·65	1·100	73	217·0
3	12·20	1·106	147	151·8
4	9·742	1·109	241	117·7
5	8·164	1·112	355	96·56
6	7·058	1·113	486	82·13
7	6·236	1·115	642	71·63
8	5·599	1·116	813	63·62
9	5·089	1·117	1004	57·31
10	4·672	1·118	1211	52·19
	$\times \rho$	$\div \rho$	$\div \rho$	$\times \rho$

Above figures apply to tissue of density 1 g./cm.³ For density ρ g./cm.³ they should be multiplied or divided by ρ as indicated at the foot of each column.

primary ionizations per micron is similarly for the particle of energy E and not an average for all energies between E and 0. By *primary* ionization is meant an ionization caused by a collision of the primary particle (electron, proton or α-particle) not including secondary ionization produced by δ-rays, which we shortly proceed to discuss. The figures given in the tables are for a tissue of density 1 g./cm.³ and need to be multiplied or divided by ρ for a tissue of density ρ as indicated in the tables.

Secondary ionization, clusters, and δ-rays

At each primary ionization an electron is ejected by collision of the primary particle (electron, proton or α-ray) with an atom of the tissue. Often this electron is ejected with insufficient energy to make any ionizing collisions on its own account, and ultimately attaches itself to an atom to form a negative ion, the attachment being, however, probably without any biological significance (see p. 2). If the electron is ejected with rather more energy it may produce a few ionizations on its own account.

CLUSTERS AND δ-RAYS

The range of a slow electron of a hundred volts or so is only a few millimicrons (Table 10), and the *secondary* ionizations produced by this *secondary* electron are therefore produced close to the primary ionization and with it form a cluster of ionizations. In Plate I D the larger droplets are condensed on unresolved clusters of ions. The relative numbers of clusters of different sizes are given in Table 14.[1] α-rays and slow protons produce their

TABLE 14. Frequency of ion-clusters containing various numbers of ionizations

No. of ionizations in the cluster	1	2	3	4	>4	Total
Frequency of cluster of this size	0·43	0·22	0·12	0·10	0·13	1·00

primary ionizations at such close intervals that the successive clusters overlap and lose their individuality, forming a column of ionizations rather than a series of discrete clusters (Plate I A, B).

Occasionally the electron ejected at a primary ionization has an energy of several hundred or even thousand electron-volts. It is then able to travel an appreciable distance and produce a large number of ionizations on its own account. The secondary electron thus forms a separate track branching off the main track, and is called a δ-ray. Plate I A shows δ-rays branching from an α-ray track. Nearly half of the total number of ionizations produced by the primary particle, whether it be electron, proton, or α-ray, are found in δ-rays of energy exceeding 100 eV. The other half of the total number of ionizations are distributed among the isolated primary ionizations and the clusters of two or three ionizations.

Since δ-rays are slow electrons, they have an ion-density greater than that of a fast electron and less than that of an α-particle. It follows that the total range of all the δ-rays branching from an α-ray track actually exceeds the range of the α-ray itself, but the total range of all the δ-rays branching from an electron track is only a small percentage of the total range of the primary electron. In Table 15 are given numerical data of the total range and total number of ionizations produced by δ-rays[2] of energy exceeding 100 eV.

δ-rays turn out to play a rather important part in certain biological actions of radiation (see especially Chapters III and VII),

[1] Experimental observations of Wilson, C.T.R. (1923).
[2] Including tertiary electrons, i.e. δ-rays produced by δ-rays.

TABLE 15. δ-rays exceeding 100 eV. energy produced by electrons, protons and α-particles

Electrons			α-particles			Protons		
Electron energy	δ-ray range per μ range of primary particle	No. of ions produced by δ-rays per primary ionization	α-ray energy	δ-ray range per μ path of α-ray	No. of ions produced by δ-rays per primary ionization	Proton energy	δ-ray range per μ path of proton	No. of ions produced by δ-rays per primary ionization
ekV.	μ		eMV.	μ		eMV.	μ	
0.5	0.064	0.225	1	2.52	0.718	1	0.392	0.905
1.5	0.060	0.504	2	2.01	0.832	2	0.308	0.956
3	0.048	0.597	3	1.73	0.879	3	0.270	0.978
6	0.038	0.672	4	1.56	0.907	4	0.236	0.981
12	0.031	0.734	5	1.44	0.925	6	0.219	1.006
24	0.026	0.780	6	1.35	0.939	8	0.203	1.013
48	0.022	0.802	7	1.29	0.950	10	0.193	1.018
96	0.022	0.856	8	1.23	0.959			
192	0.023	0.891						
384	0.027	0.920						

TABLE 16. δ-ray production by electrons in tissue

The table gives the number of δ-rays of energy exceeding W produced by a primary electron of energy E per micron path in tissue of density 1 g./cm.³ For tissue density ρ g./cm.³, multiply the figures by ρ as indicated at the feet of the columns.

E W ekV.	0.5	1.5	3	6	12 ekV.
0.1	16.50	8.757	4.964	2.662	1.433
0.15	7.382	5.191	3.103	1.712	0.9364
0.25	—	2.475	1.650	0.9609	0.5412
0.35	—	1.390	1.050	0.6448	0.3734
0.45	—	0.8203	0.7283	0.4723	0.2809
0.55	—	0.4668	0.5306	0.3643	0.2226
0.65	—	0.2131	0.3982	0.2910	0.1825
0.75	—	—	0.3041	0.2381	0.1534
0.85	—	—	0.2343	0.1984	0.1314
0.95	—	—	0.1799	0.1676	0.1141
1.1	—	—	0.1167	0.1326	0.09432
1.3	—	—	0.05328	0.09956	0.07532
1.5	—	—	—	0.07602	0.06165
1.7	—	—	—	0.05858	0.05137
1.9	—	—	—	0.04497	0.04339
2.25	—	—	—	0.02695	0.03309
2.75	—	—	—	0.00817	0.02327
3.25	—	—	—	—	0.01664
3.75	—	—	—	—	0.01202
4.25	—	—	—	—	0.008466
5.5	—	—	—	—	0.002120
6.5	—	—	—	—	—
	$\times \rho$	$\times \mu$	$\times \rho$	$\times \rho$	$\times \rho$

δ-RAYS

TABLE 16 (cont.).

W ekV. \ E	24	48	96	192	384 ekV.
0·1	0·7579	0·4011	0·2326	0·1442	0·1012
0·15	0·4996	0·2657	0·1546	0·09600	0·06738
0·25	0·2935	0·1576	0·09215	0·05740	0·04036
0·35	0·2056	0·1113	0·06543	0·04087	0·02880
0·45	0·1570	0·08570	0·05060	0·03169	0·02236
0·55	0·1261	0·06942	0·04117	0·02586	0·01826
0·65	0·1049	0·05817	0·03465	0·02182	0·01543
0·75	0·08938	0·04994	0·02987	0·01885	0·01336
0·85	0·07757	0·04366	0·02622	0·01659	0·01177
0·95	0·06828	0·03871	0·02334	0·01480	0·01052
1·1	0·05757	0·03299	0·02001	0·01274	0·009065
1·3	0·04722	0·02743	0·01677	0·01072	0·007651
1·5	0·03969	0·02338	0·01440	0·009245	0·006614
1·7	0·03399	0·02029	0·01259	0·008118	0·005822
1·9	0·02952	0·01786	0·01117	0·007229	0·005197
2·25	0·02369	0·01467	0·009287	0·006055	0·004370
2·75	0·01804	0·01155	0·007440	0·004898	0·003356
3·25	0·01421	0·009588	0·006168	0·004099	0·002992
3·75	0·01146	0·007862	0·005240	0·003515	0·002579
4·25	0·009409	0·006689	0·004534	0·003069	0·002264
5·5	0·006024	0·004725	0·003340	0·002313	0·001728
6·5	0·004326	0·003724	0·002724	0·001919	0·001448
7·5	0·003145	0·003006	0·002277	0·001632	0·001243
8·5	0·002193	0·002468	0·001938	0·001413	0·001086
9·5	0·001458	0·002052	0·001674	0·001242	0·0009631
10·5	0·0008423	0·001722	0·001462	0·001103	0·0008636
11·5	0·0002794	0·001454	0·001289	0·0009900	0·0007816
12·5	—	0·001233	0·001145	0·0008945	0·0007129
13·5	—	0·001047	0·001024	0·0008140	0·0006544
14·5	—	0·0008887	0·0009205	0·0007449	0·0006042
15·5	—	0·0007517	0·0008314	0·0006850	0·0005595
17·5	—	0·0004227	0·0006858	0·0005865	0·0004883
22·5	—	0·0001077	0·0004429	0·0004195	0·0003648
27·5	—	—	0·0002944	0·0003158	0·0002871
32·5	—	—	0·0001937	0·0002457	0·0002338
35	—	—	0·0001541	0·0002186	0·0002130
45	—	—	0·00003168	0·0001422	0·0001537
55	—	—	—	0·00009512	0·0001167
65	—	—	—	0·00006294	0·00009152
75	—	—	—	0·00003890	0·00007334
85	—	—	—	0·00001929	0·00005962
95	—	—	—	0·00000172	0·00004495
105	—	—	—	—	0·00003678
115	—	—	—	—	0·00003010
125	—	—	—	—	0·00002450
135	—	—	—	—	0·00001971
165	—	—	—	—	0·00000837
195	—	—	—	—	—
	$\times \rho$	$\times \rho$	$\times \rho$	$\times \rho$	$\times \rho$

and it is necessary to have adequate data of their number and energy. Tables 16 and 17 give for various energies of electrons, protons and α-rays the number of δ-rays produced per micron path in tissue of density 1 g./cm.³ If the actual tissue density is ρ g./cm.³, then the number of δ-rays produced per micron path is ρ times greater than in Tables 16 and 17.

TABLE 17A. δ-ray production by α-rays in tissue

Number of δ-rays of energy exceeding W produced by an α-particle per micron path tissue of density 1 g./cm.³ For density ρ g./cm.³ multiply by ρ.

W ekV.	α-particle energy (in eMV.)								
	1	2	3	4	5	6	7	8	10
0·1	416·4	231·4	159·5	121·5	98·15	82·31	70·87	62·22	50·01
0·15	246·6	146·6	102·9	79·08	64·20	54·01	46·61	40·99	33·03
0·25	110·8	78·64	57·59	45·13	37·03	31·38	27·21	24·02	19·44
0·35	52·60	49·53	38·18	30·57	25·39	21·67	18·89	16·74	13·62
0·45	20·26	33·36	27·40	22·49	18·92	16·28	14·27	12·70	10·39
0·55	—	23·07	20·54	17·34	14·80	12·85	11·33	10·12	8·332
0·65	—	15·95	15·80	13·78	11·96	10·48	9·298	8·343	6·907
0·75	—	10·72	12·31	11·17	9·865	8·737	7·805	7·037	5·862
0·85	—	6·729	9·649	9·173	8·267	7·406	6·664	6·038	5·063
0·95	—	3·575	7·546	7·596	7·006	6·354	5·763	5·250	4·432
1·1	—	—	5·109	5·768	5·544	5·136	4·718	4·336	3·701
1·3	—	—	2·735	3·987	4·119	3·949	3·701	3·446	2·989
1·5	—	—	0·993	2·681	3·074	3·078	2·954	2·793	2·466
1·7	—	—	—	1·682	2·275	2·412	2·384	2·293	2·067
1·9	—	—	—	0·894	1·644	1·886	1·933	1·899	1·751
2·25	—	—	—	—	0·810	1·191	1·337	1·378	1·334
2·75	—	—	—	—	—	0·505	0·749	0·863	0·923
3·25	—	—	—	—	—	0·031	0·342	0·507	0·638
3·75	—	—	—	—	—	—	0·044	0·246	0·429
4·25	—	—	—	—	—	—	—	0·046	0·269
5·5	—	—	—	—	—	—	—	—	—
	×ρ	×ρ	×ρ	×ρ	×ρ	×ρ	×ρ	×ρ	×ρ

Number and total range of ionizing particles per unit volume of tissue

It is sometimes necessary to know the number of ionizing particles generated per unit volume of irradiated tissue, i.e. the number of electrons projected (by photoelectric or Compton effect) in unit volume of tissue by a given dose of X- or γ-rays, or the number of protons set into motion per unit volume of tissue by a given dose of neutrons. Table 18 provides the necessary data. The calculations have been made for a 'wet tissue' of the composition assumed on p. 7, and are based on the numerical data given in Tables 2, 3 and 8.

NUMBER OF IONIZING PARTICLES

It is also sometimes necessary to know the number of ionizing particles (electrons, protons or α-rays) which cross each square micron of tissue irradiated by a given dose, or what amounts nearly to the same thing, the total path length in microns of all the ionizing particles which are liberated in $1\mu^3$ of tissue, or which traverse $1\mu^3$ of tissue. This information is also provided in Table 18, the calculation utilizing the range-energy and δ-ray data given in the present chapter.

TABLE 17B. δ-ray production by protons in tissue

Number of δ-rays of energy exceeding W produced by a proton per micron path in tissue of density 1 g./cm.^{3} For density ρ g./cm.^{3} multiply by ρ.

W ekV.	Proton energy (in eMV.)						
	1	2	3	4	6	8	10
0.1	30.58	15.66	10.52	7.921	5.301	3.984	3.191
0.15	19.90	10.32	6.959	5.250	3.520	2.648	2.122
0.25	11.35	6.042	4.110	3.113	2.096	1.580	1.267
0.35	7.685	4.211	2.889	2.197	1.485	1.122	0.9011
0.45	5.650	3.193	2.211	1.689	1.146	0.8673	0.6976
0.55	4.355	2.546	1.779	1.365	0.9304	0.7055	0.5681
0.65	3.458	2.097	1.480	1.141	0.7810	0.5934	0.4784
0.75	2.801	1.769	1.261	0.9764	0.6714	0.5112	0.4126
0.85	2.298	1.517	1.093	0.8507	0.5876	0.4483	0.3624
0.95	1.901	1.319	0.9610	0.7514	0.5214	0.3987	0.3227
1.1	1.441	1.089	0.8076	0.6364	0.4447	0.3412	0.2767
1.3	0.9923	0.8645	0.6582	0.5243	0.3700	0.2852	0.2318
1.5	0.6636	0.7001	0.5486	0.4421	0.3152	0.2441	0.1990
1.7	0.4122	0.5744	0.4648	0.3793	0.2733	0.2127	0.1738
1.9	0.2137	0.4752	0.3986	0.3297	0.2402	0.1879	0.1540
2.25	—	0.3440	0.3112	0.2641	0.1965	0.1551	0.1277
2.75	—	0.2145	0.2248	0.1993	0.1533	0.1227	0.1018
3.25	—	0.1248	0.1651	0.1545	0.1234	0.1003	0.0839
3.75	—	0.0591	0.1212	0.1216	0.1015	0.0838	0.0707
4.25	—	0.0088	0.0877	0.0965	0.0848	0.0713	0.0607
5.5	—	—	0.0306	0.0536	0.0562	0.0498	0.04355
6.5	—	—	0.0007	0.0312	0.0413	0.0386	0.03458
7.5	—	—	—	0.0148	0.0303	0.0304	0.02801
8.5	—	—	—	0.0022	0.0219	0.0241	0.02298
9.5	—	—	—	—	0.0153	0.0192	0.01901
10.5	—	—	—	—	0.0100	0.0151	0.01580
11.5	—	—	—	—	0.0055	0.0118	0.01314
12.5	—	—	—	—	0.0018	0.0090	0.01091
13.5	—	—	—	—	—	0.0067	0.00901
14.5	—	—	—	—	—	0.0046	0.00737
15.5	—	—	—	—	—	0.0028	0.00595
17.5	—	—	—	—	—	—	0.00358
19.5	—	—	—	—	—	—	0.00171
21	—	—	—	—	—	—	0.00053
23	—	—	—	—	—	—	—
	$\times\rho$	$\times\rho$	$\times\rho$	$\times\rho$	$\times\rho$	$\times\rho$	$\times\rho$

TABLE 18. Numbers and ranges of ionizing particles per μ^3 per 1000 r. in wet tissue of unit density

	Wave-length A.	Numbers of electrons[1] projected per μ^3 per 1000 r.	Combined range in μ of all the electrons (including δ-rays) per μ^3 per 1000 r.
γ-rays	0·0152	0·171	173
X-rays	0·0809	1·96	36·0
	0·1618	4·55	30·9
	0·3033	4·31	35·2
	0·4853	3·52	27·4
	1·54	6·70	10·83
	4·15	20·1	5·30
	8·32	37·3	2·86

[1] Recoil plus photoelectrons.

		Deuteron energy eMV.	No. of protons projected per μ^3 per 1000 v-units	Combined range in μ of all the ionizing particles per μ^3 per 1000 v-units		
				Protons	δ-rays	Protons + δ-rays
Neutrons	Li+D	0·9	0·0293	4·05	1·15	5·20
	D+D	0·3	0·0364	2·27	0·85	3·12

	Energy eMV.	Combined range in μ of all the ionizing particles traversing the tissue, per μ^3 per 1000 r.		
		α-rays	δ-rays	α-rays + δ-rays
α-rays (irradiating thin layer in which the α-rays are not completely absorbed)	1	0·27	0·67	0·94
	2	0·39	0·77	1·16
	3	0·50	0·87	1·37
	4	0·61	0·95	1·56
	5	0·71	1·03	1·74
	6	0·81	1·10	1·91
	7	0·91	1·16	2·07
	8	1·00	1·23	2·23

	Range in μ, per μ^3 tissue		
	α-rays	δ-rays	α-rays + δ-rays
α-rays of Rn + Ra A + Ra C' completely absorbed (i.e. radon in solution)			
per 1000 r.	0·56	0·92	1·48
per sec., for 1 curie radon present per cm.³ tissue	5·9	9·7	15·6
for 1 μc. of radon disintegrated per cm.³ tissue	2·8	4·6	7·4

Chapter II

CHEMICAL EFFECTS OF IONIZING RADIATIONS, AND POSSIBLE MECHANISMS OF BIOLOGICAL ACTION

The ionic yield

Experiments have been carried out on the chemical effects of ionizing radiations[1] on pure substances in the gaseous, liquid and solid states, and also on solutions, particularly aqueous solutions. Gaseous reactions have usually been studied using α-rays, radon, either mixed with the gas or else sealed in a thin-walled bulb at the centre of the gas, being the source of the rays. Some experiments have been made with X-rays, cathode-rays and β-rays, but there is much less information than one would like on the relative efficiency of different radiations, and this is at present one of the most urgent requirements. With the doses which it is practicable to give with X-rays, the percentage of the reactants which undergoes chemical change is usually small. In the absence of complications, such as back reactions, the amount of chemical change then increases in direct proportion to the dose, and the yield of the reaction is most conveniently expressed as the number of molecules of a specified substance formed or destroyed per ion-pair produced, a ratio conventionally represented as M/N.

Other methods of representing the yield are sometimes found in the literature. If the energy of the radiation absorbed in the irradiated material has been measured calorimetrically, the yield may be given as gram-molecules reacting per kilogram-calorie. If it has been measured by the complete absorption of a known number of ionizing particles of known energy it may be expressed as the number of molecules reacting per electron-volt of energy absorbed. If radon is used as the source of radiation the yield may be expressed as gram-molecules reacting per millicurie of radon disintegrating, or as $cm.^3$ of gas reacting per millicurie of radon disintegrating. In X-ray experiments on dilute

---

[1] Sometimes referred to as *radiochemical* reactions in distinction from *photochemical* reactions produced by visible and ultra-violet light. General references to this subject are: Lind, S.C. (1928), *The Chemical Effects of α-particles and Electrons*; Allsopp, C.B. (1944).

aqueous solutions the yield is sometimes expressed in terms of micromoles reacting per 1000 r. per litre of solution.

To convert these units into molecules reacting per ion-pair, it is necessary to know the mean energy dissipation per ionization.

TABLE 19. Conversion factors for expressing yields in radiochemical reactions as M/N values (molecules per ion-pair)

	Molecules per ion-pair	
	α-rays	X-rays, β-rays, cathode-rays
1 gram-molecule per kilogram-calorie =	807	749
1 molecule per electron-volt =	35	32·5
1 gram-molecule per millicurie of radon disintegrated =	$5·80 \times 10^4$ [1]	—
1 ml. gas at N.T.P. per millicurie of radon disintegrated =	2·59 [1]	—
1 micro-mole per 1000 r. per litre in dilute aqueous solutions:		
Soft X-rays, 0·3–0·5 A. =	—	0·38
Hard X-rays and γ-rays =	—	0·34

[1] Assuming that the α-rays and β-rays and recoil nuclei from the disintegration of Rn, Ra A, Ra B, Ra C, Ra C′ are all completely absorbed in the reacting system. This will be sufficiently nearly true when the radon is dissolved in a liquid, providing allowance is made for the distribution of the radon between the liquid and any gas space in communication with it. For gas reactions the calculation is more complicated, since a proportion of the α-particles are absorbed in the walls, either of the radon bulb or of the gas container, thus causing the factor to be greater than that given. For the method of calculation in these cases see Lind, S.C. (1928).

In Table 19 this has been taken to have the value valid for air, viz. 35 eV. per ion-pair for α-rays and 32·5 eV. per ion-pair for electrons. If the actual figure is known for the substance being used, say W eV. per ion-pair, then the conversion factor given in the table should be multiplied by $W/35$ for α-rays and $W/32·5$ for X-rays, cathode-rays and β-rays.

Gas reactions [1]

The reactions which are best understood are simple gas reactions in which either the decomposition of a single gas is studied, or the combination of two gases. The interpretation of gaseous reactions is helped when information is available of the types of positive and negative ions formed (from the mass spectrograph), of the ionization and excitation potentials (from electron impact

[1] The mechanism of the induction of chemical change in gases irradiated by ionizing radiations discussed in this section was proposed by Eyring, H., Hirschfelder, J.O. & Taylor, H.S. (1936 a, b).

experiments), and of the probable fate of excited molecules (from absorption spectra and photochemical studies). The following account describes the processes contributing to the total yield in typical gas reactions.

When an α-particle or a fast electron passes through a gas, energy is dissipated partly by ionization and partly by excitation or direct dissociation of the molecules with which it collides, the energy dissipated in ionization usually being of the order of one-half of the total energy. In addition, a small fraction is converted directly into heat by elastic collisions. Excitation often leads to the dissociation of the excited molecule, either spontaneously or when the excited molecule makes a collision with an ordinary molecule. Thus, following excitation, H_2 may dissociate into $H+H$, CO into $C+O$. These products, liberated in atomic form, are then likely to take part in chemical reaction.

The process of ionization of a gas molecule usually does not directly cause dissociation. Thus in CO_2 at low pressure the positive ion CO_2^+ is more copiously formed than CO^+ or C^+, and in hydrogen the ion H_2^+ is more copiously formed than H^+. When the positive ion is neutralized by collision with either an electron or a negative ion, a large amount of energy is set free, and this almost always results in dissociation.[1] The final result of positive ion formation is thus the dissociation of the ionized molecules; the free atoms liberated probably taking part in subsequent chemical reaction.

The electron which is projected from the molecule when ionization occurs may be fast enough to ionize and excite on its own account, and the chemical changes so caused will not differ from those following ionization and excitation by the primary particle. Some electrons will be ejected with insufficient energy for ionization or excitation, and all will eventually be slowed down. In some gases, e.g. pure nitrogen, carbon monoxide, carbon dioxide, neutral gas molecules have very little electron affinity, and the slow electrons therefore remain free until they collide with and neutralize a positive ion. In other gases such as oxygen the electrons form negative ions, e.g. O_2^-. The energy of

[1] If, however, the positive ions are neutralized on a surface, as when an electrically charged plate is introduced into the reaction chamber, then the molecule can get rid of the energy set free at neutralization, and dissociation may be avoided (Smith, C. & Essex, H. 1938).

attachment of the electron to the molecule is low, and the molecule will not usually be dissociated when it gives up the spare electron to a positive ion. If the electron has rather more energy when it makes the attachment collision, an atomic negative ion may be formed by a reaction such as $O_2 + e = O + O^- - 2 \cdot 9 \text{eV}$.

In a given gas, knowledge of the types of positive and negative ion formed, and the energies involved in ionization, in electron attachment and in dissociation enables the number of molecules dissociated per ion-pair to be estimated. Thus it is estimated[1] that in the irradiation of H_2 4 atoms of H in the atomic state are produced per ion-pair as a result of ionization, and probably a further 2 as a result of excitation. Knowledge of the chemical actions likely to ensue between these atoms and whatever molecules are present enables the final result of the reaction to be predicted. In this way[2] the decomposition of HBr into H_2 and Br_2, and the reverse synthesis, have been satisfactorily and quantitatively explained,[3] and also the combination[4] of CO and O_2 to form CO_2, and the decomposition[4] of H_2S in the presence of a large excess of hydrogen. Yields are of the order of one or a few molecules per ion-pair. Reactions such as the combination of H_2 and Cl_2 where very much greater yields are obtained are chain reactions.

From the considerable number of gas reactions which have been quantitatively investigated, we choose for inclusion in Table 20 those for which figures are available for the ionic yield with more than one ionizing radiation. Estimates of the yield of any one reaction with different radiations have usually been made by different authors, and the accuracy is rarely sufficient for a small difference to be established with certainty. It appears from the table that on the whole the ionic yields with different ionizing radiations are equal, and that the ionic yield is several times greater than the ultra-violet quantum yield. The exceptions to this general rule (decomposition of CO_2 and H_2O; production of ozone from oxygen) are all reactions in which a back reaction

[1] Eyring, H., Hirschfelder, J.O. & Taylor, H.S. (1936 a, b).
[2] An alternative mode of description of gaseous reactions is the *cluster theory*, developed particularly by Lind and by Mund. An account of this theory, and relevant references, will be found in Glockler, G. & Lind, S.C. (1939), *The Electrochemistry of Gases*.
[3] Eyring, H., Hirschfelder, J.O. & Taylor, H.S. (1936b).
[4] Hirschfelder, J.O. & Taylor, H.S. (1938).

is liable to occur, and in which therefore the yield obtained is likely to depend upon the experimental conditions.

TABLE 20. Ionic yields (M/N) in gas reactions

Reaction	α-rays	X-rays	Cathode- or β-rays	Ultra-violet light
NH_3 decomposition	1·37[1]	—	1·20[4]	0·2–0·3[18]
	0·80[2]	—	—	—
	1·16[3]	—	—	—
NH_3 synthesis (including hydrazine)	0·2–0·3[5]	—	~0·2[6]	—
	0·28[3]	—	—	—
N_2O decomposition	1·7[2]	—	3·9[7]	1[19]
	4·4[8]	—	—	—
HBr synthesis (great excess H_2)	2·9[10]	2·6[9]	—	—
HI decomposition	~6[11]	8·0[9]	—	2[20]
C_2H_2 polymerization	20[13]	—	26[12]	9·2[21]
O_3 from O_2	2–2·5[14]	—	2·2[15]	2[22]
H_2O vapour decomposition	0·05[17]	1·3[16]	—	—
CO_2 decomposition	0·0[2]	—	0·04[15]	1[23]

1 Smith, C. & Essex, H. (1938).
2 Wourtzel, E. (1919).
3 Jungers, J.C. (1932).
4 Gedye, G.R. & Allibone, T.E. (1930).
5 Lind, S.C. & Bardwell, D.C. (1928).
6 Gedye, G.R. & Allibone, T.E. (1932).
7 Gedye, G.R. (1931).
8 Kolumban, A.D. & Essex, H. (1940).
9 Günther, P. & Leichter, H. (1936).
10 Lind, S.C. & Livingston, R. (1936).
11 Brattain, K.G. (1938).
12 Mund, W. & Jungers, J.C. (1931).
13 Lind, S.C., Bardwell, D.C. & Perry, J.H. (1926).
14 Lind, S.C. & Bardwell, D.C. (1929).
15 Busse, W.F. & Daniels, F. (1928).
16 Günther, P. & Holtzapfel, L. (1939a) (excess of xenon present).
17 Duane, W. & Scheuer, O. (1913).
18 Wiig, E.O. (1935).
19 Noyes, W.A. (1937).
20 Lewis, B. (1928).
21 Lind, S.C. & Livingston, R. (1932).
22 Vaughan, W.E. & Noyes, W.A. (1930).
23 Groth, W. (1937).

Liquids and solids

Radiochemical actions in liquids and solids and solutions are less well understood than in gases, but are of more importance biologically. As regards pure liquids and solids, data are rather meagre. Kailan[1] has determined the ionic yield for a number of organic liquids exposed to β- and γ-rays and finds values ranging from 0·1 to 8, that is to say, much of the same order as for gas reactions. Lind and Ogg[2] state that the ionic yield for decomposition by α-particles is approximately the same in liquid and in gaseous HBr.

Enzymes irradiated in the dry state by X-rays are inactivated. Fig. 1A shows the proportion of ribonuclease remaining active as

1 See Table 11 in Lind, S.C. (1928), *The Chemical Effects of α-particles and Electrons.*
2 Lind, S.C. & Ogg, E.F. (1931).

a function of the dose of radiation.[1] The ionic yield is approximately unity.

It appears, therefore, from the somewhat limited amount of data available, that, as in gas reactions, the typical result of

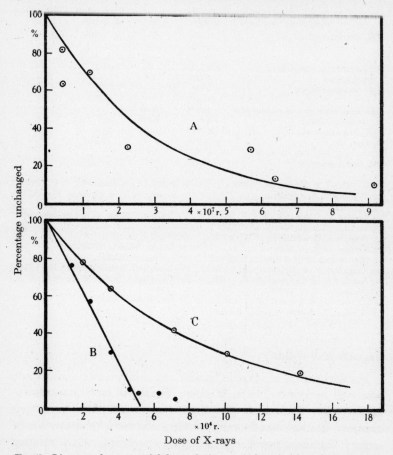

FIG. 1. Linear and exponential dose relations. A, inactivation of ribonuclease (dry; Lea, Smith, Holmes & Markham). B, oxidation of ferrous sulphate ($10^{-3}\ M$; Fricke & Morse). C, conversion of haemoglobin to methaemoglobin ($1·2 \times 10^{-4}\ M$; Fricke & Petersen).

irradiating substances in the solid and liquid states is decomposition at the rate of about 1 molecule per ionization. It is notable that this simple result holds as well for a protein molecule as for a simple inorganic gas molecule.

[1] Lea, D.E., Smith, K.M., Holmes, B. & Markham, R. (1944).

A number of photochemical decompositions have been studied in both the liquid and gaseous states.[1] In some cases (e.g. decomposition of HI) the yield per ultra-violet quantum is approximately the same, but cases are known (e.g. the decomposition of ammonia) in which the yield is much less in the liquid state.

Solutions

As regards reactions in solutions, analogy between photochemical and radiochemical reactions is not very profitable. In studying photochemical reactions, a solvent is usually chosen in which the absorption of ultra-violet light is negligible by comparison with its absorption in the solute, so that the energy of the radiation is primarily liberated in the solute molecules only. In such solutions it is often found that the quantum yield for decomposition of the solute is rather less than for the decomposition of the same substance in the gaseous state. This is plausibly explained[2] either by recombination of the atoms or radicals resulting from the dissociation being facilitated by the caging effect of the solvent molecules, or by collisions between solvent molecules and excited solute molecules removing from the latter sufficient vibrational energy to prevent decomposition.

With ionizing radiations, however, it is not possible to use a non-absorbing solvent, and to a first approximation the relative amounts of energy dissipated in solute and solvent are proportional to the masses of each present. Consequently in a dilute solution the number of molecules of solute directly ionized or excited by the radiation is very small compared with the number of solvent molecules ionized or excited. There is thus the possibility of an indirect effect on the solute, either due to transference of energy from excited or ionized solvent molecules, or due to chemical change occurring in the solvent and the products affecting the solute. Photochemical reactions are known in which a non-absorbing component suffers chemical change as a result of the absorption of the light in an absorbing substance, which may itself be unchanged and merely hands on the energy

[1] Cp. Dickinson, R.G. (1935, 1938).
[2] Franck, J. & Rabinowitsch, E. (1934); Atwood, K. & Rollefson, G.K. (1941).

it absorbs. We shall not be surprised to find that most radiochemical reactions in dilute solution are indirect, chemical change in the solute occurring as the result of ionization and excitation in the solvent molecules. Direct excitation or ionization of the solute, when it occurs, will no doubt also lead to chemical change, but since the number of direct solute ionizations and excitations in a dilute solution is very small compared with the number of ionizations and excitations of solvent molecules, direct action will be negligible unless the probability of an ionized or excited solvent molecule causing change in a solute molecule is small. The evidence is that this probability is quite high when water is the solvent.

Decomposition of water

There are still some obscurities concerning the production of chemical change by ionizing radiations in water itself. There is agreement that in ice at liquid-air temperature the ionic yield is low.[1] It is not clear whether appreciable decomposition occurs in the vapour state (cp. Table 20). As regards the liquid state, it is agreed that the irradiation of ordinary water not specially purified and de-aerated leads to decomposition with the production of H_2 and O_2 gases and the formation of some H_2O_2. With moderate doses of radiation not much oxygen is liberated as is to be expected in view of the H_2O_2 formation. The concentration of H_2O_2 does not, however, increase indefinitely,[2] and with larger doses of radiation the reaction proceeds according to the equation $2H_2O = 2H_2 + O_2$. The ionic yield of the reaction, estimated as the number of molecules of H_2 produced per ion-pair, is, for both α-rays[3] and X-rays, about 1·0. The yield of H_2O_2 during the early stages of the reaction with X-rays is about 1 molecule per ion-pair.[4] If a small quantity of a reducible substance is present in solution, oxygen but no hydrogen will be evolved, while if an

---

1 0·05, the products of decomposition being $2H_2 + O_2$, according to Duane, W. & Scheuer, O. (1913). Decomposition undetectable according to Günther, P. & Holtzapfel, L. (1939b).

2 The decomposition of H_2O_2 by radiation is rapid, and increases with increasing H_2O_2 concentration (Fricke, H. (1935a)).

3 Lanning, F.C. & Lind, S.C. (1938) give 0·87; Nurnberger, C.E. (1934) gives 0·78; Duane, W. & Scheuer, O. (1913) give 1·06.

4 Fricke, H. (1934a) gives 0·8 molecule per ion-pair at acid pH, 0·4 at alkaline pH. Clark, G.L. & Coe, W.S. (1937) give 0·6 at acid pH.

oxidizable substance is present, hydrogen and no oxygen will be evolved.[1]

The behaviour of carefully purified water freed from dissolved oxygen is different from that of water not free from oxygen, and the various experiments which have been reported are difficult to reconcile. It has been stated[2] that no hydrogen and oxygen are produced when gas-free water is irradiated by X-rays. A recent experiment[3] with X-rays, however, reports a yield of about 1 hydrogen molecule per ion-pair. Experiments with α-rays[4] and β-rays[5], using larger doses, also give an evolution of gas, but the yield appears to be lower than 1 molecule per ion-pair.

It seems to be established that the production of H_2O_2 in irradiated water requires the presence of dissolved oxygen. This was reported by Risse,[6] and Fricke[7] found that the yield of H_2O_2 diminished on reducing the oxygen tension from 70 to 4 cm. of mercury, and was undetectable in gas-free water. A trace of iodide or bromide ions in gas-free water catalyses the production of H_2 and H_2O_2 to the extent of 0·2 molecule of each per ion-pair, which is less than the yield in aerated water, while traces of oxidizable or reducible substances lead to the evolution of gas, as already mentioned.

1 Lanning, F.C. & Lind, S.C. (1938).
2 Risse, O. (1929); Fricke, H. & Brownscombe, E.R. (1933b).
3 Günther, P. & Holtzapfel, L. (1939b).
4 Nurnberger, C.E. (1934, 1936a). The yield in the experiment using gas-free water can be calculated from the data given to be 0·06, assuming that the radon is distributed between liquid and gaseous phases in the expected ratio. The distribution of radon is liable to disturbance by the evolution of gas which occurs, leading to there being less radon in the solution than calculated. It seems improbable, however, that this correction could raise the yield to the order of 1 molecule per ion-pair. Nurnberger, however, reports his experiments as consistent with a yield of unity.
5 Kernbaum, M. (1909). This early experiment appears to have been made with de-aerated water. If all the β-rays reached the water the yield can be calculated from the data given to be 0·015 molecule per ion-pair. A considerable proportion of the β-ray energy would be absorbed in the wall of the vessel, and some gas would remain dissolved in the solution. It seems improbable however that these corrections could raise the yield to the order of 1 molecule per ion-pair.
6 Risse, O. (1929).
7 Fricke, H. & Brownscombe, E.R. (1933b); Fricke, H. (1934a); Fricke, H. & Hart, E.J. (1935d).

The reported production of H_2O_2 in gas-free water irradiated by α-rays[1] or β-rays[2] is not necessarily inconsistent with the X-ray results, since in these experiments higher doses were used and sufficient oxygen was probably liberated to serve in the production of H_2O_2.

We draw the following conclusions which, however, are to be considered surmises liable to correction when the experimental contradictions are resolved. Pure gas-free water is decomposable by ionizing radiations, though the ionic yield is low and may be due to residual impurities. In the absence of dissolved substances no appreciable amount of H_2O_2 is produced. By the use of considerable doses, which are usual in experiments using α- and β-rays, but not in experiments using X-rays, sufficient oxygen accumulates in solution to lead to the production of H_2O_2. If oxygen is present in the water at the start of the experiment, then production of H_2O_2, with an ionic yield of the order unity, and evolution of hydrogen and eventually of oxygen also, begin immediately.

Indirect action in aqueous solution

A number of authors have studied the chemical changes produced by X-rays and radioactive radiations in dilute aqueous solutions. Among the principal researches may be mentioned those of Fricke[3] using inorganic and simple organic compounds, and of Dale[4] using enzymes. Inorganic reactions which have been studied are mainly oxidations and reductions. The change induced in a simple organic compound is usually oxidation to CO_2, and is accompanied by an evolution of hydrogen. The actual chemical change occurring when enzymes are irradiated has not been studied; the effect observed is the loss of enzyme activity.

All these actions of radiation on dilute aqueous solutions are *indirect actions*, that is to say, most of the molecules of solute which react have not been excited or ionized directly by the

[1] Nurnberger, C. E. (1936*b*).
[2] Kernbaum, M. (1909).
[3] Fricke, H. & Petersen, B.W. (1927); Fricke, H. & Morse, S. (1927, 1929); Fricke, H. & Brownscombe, E.R. (1933*a*, *b*); Fricke, H. (1934*a*, *b*, 1935*a*, *b*, 1938); Fricke, H. & Hart, E.J. (1934, 1935*a*, *b*, *c*, *d*, 1936); Fricke, H., Hart, E.J. & Smith, H.P. (1938).
[4] Dale, W.M. (1940, 1942, 1943*a*, *b*); Dale, W.M., Meredith, W.J. & Tweedie, M.C.K. (1943).

radiation, but their reaction follows excitation or ionization of the solvent molecules.[1] This conclusion is based principally on the results of experiments in which a given substance is irradiated

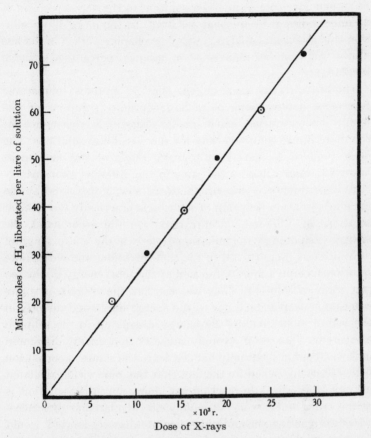

FIG. 2. The independence of absolute yield upon concentration of solute: ● 10^{-4} M formic acid, ⊙ 10^{-1} M formic acid (Fricke, Hart & Smith).

in solutions of different concentration. In such experiments it is found that the weight of solute reacting as a result of a given number of roentgens given to the solution is, over a wide range of concentrations, independent of the concentration of the solution. Thus Fig. 2[2] shows that 25 micromoles of hydrogen gas

[1] Solute molecules which have been excited or ionized directly by the radiation probably react also, but these usually constitute a minute fraction of the total number reacting.

[2] Fricke, H., Hart, E.J. & Smith, H.P. (1938).

are liberated per litre of a solution of formic acid given 10,000 r. of X-rays, irrespective of whether the solution contains 10^{-4} or 10^{-1} gram-molecules of formic acid per litre. This emission of hydrogen corresponds in either solution to the decomposition of 25 micromoles of formic acid per litre. In the more dilute solution this represents a large percentage change (25%); in the less dilute solution it represents a minute percentage change (0·025%).

Similarly, it has been shown[1] that to produce comparable *percentage* inactivations in carboxypeptidase solutions of different concentrations, much smaller doses of X-rays suffice for the more dilute solutions than for the less dilute solutions. The dose required to inactivate a given *weight* of the enzyme is, however, approximately the same in the different solutions.

In experiments of this sort the total energy dissipated by the ionizing radiation per gram of solution is practically the same in solutions of different concentration. On the other hand, the energy dissipated by the radiation directly in the solute per gram of solution is proportional to the concentration, and with dilute solutions is only a minute fraction of the total energy dissipated per gram of solution. Thus we see that the weight of solute reacting is proportional not to the energy dissipated directly in the solute alone, but to the energy dissipated in the solution altogether. This result strongly suggests that energy dissipated in the solvent is eventually handed on to the solute, a conclusion which is strengthened by the fact that the ionic yield calculated as molecules-of-solute-reacting-per-ionization-in-the-solution is of the order unity, while an ionic yield calculated as molecules-of-solute-reacting-per-molecule-of-solute-directly-ionized would be very much greater than unity (and would depend upon the concentration).

Evidently the ionization or excitation of a water molecule by an ionizing particle causes the production of an intermediary body of finite life, capable of causing reaction in many solutes. The nature of this intermediary was for some years obscure, and it is usually referred to in the literature as *activated water*. The probable nature of activated water is discussed later.

By the addition to a solution of a substance capable of reacting with activated water and so of competing with the

[1] Dale, W.M. (1940).

original solute for the activated water, we should expect the amount of the first solute reacting per unit dose to be diminished. This expectation has been confirmed, which gives further support to the view that the chemical changes in question involve an intermediary.[1]

Providing that there is no reverse reaction, and that the products of the reaction do not compete for the activated water, the quantity of the solute remaining unchanged diminishes linearly with the dose, as illustrated in Fig. 1 B.[2] If, however, there is a protective agent present which secures most of the activated water, or if the products of the reaction themselves compete for activated water, then the gradient of the curve becomes less steep as the percentage of unchanged solute diminishes, owing to the solute securing a continually smaller proportion of the total amount of activated water available.[3] A curve of this type is shown in Fig. 1 C.[4]

Ionic yields in aqueous solution

In Table 21 are collected values obtained for the ionic yield, i.e. the number of molecules of solute reacting per ion-pair in the solution, for a number of reactions investigated in dilute aqueous solution. Many of the reactions are oxidation or reduction reactions, and in these cases the number of equivalents reacting per ion-pair has been listed as well as the number of molecules. Some of the reactions could be caused by H_2O_2. But the ionic yields are not consistent with this being the whole explanation of oxidation and reduction in irradiated solutions, and some reactions (e.g. reduction of KIO_3) take place which do not occur with

[1] Fricke, H., Hart, E.J. & Smith, H.P. (1938); Dale, W.M. (1942, 1943*a*).
[2] Fricke, H. & Morse, S. (1929). The reaction studied was the oxidation of $FeSO_4$ (10^{-3} M in $0 \cdot 8$ N H_2SO_4). The departure of the points from the curve when the oxidation is 90 % complete is probably due to the back reaction also occurring under the irradiation.
[3] Cp. Dale, W.M., Meredith, W.J. & Tweedie, M.C.K. (1943).
[4] Fricke, H. & Petersen, B.W. (1927) irradiating haemoglobin. The curve is exponential. Approximately exponential curves have also been obtained by Dale, W.M. (1940, 1942), irradiating enzymes. Probably this is the usual result when a large organic molecule such as a protein is irradiated, since the products of the reaction will doubtless be usually capable of further reaction, and so will compete for the activated water.

H_2O_2. As pointed out by Allsopp,[1] the assumption that H_2O_2 is involved would not avoid the necessity for invoking the hypothesis of activated water, since the formation of H_2O_2 itself requires this hypothesis.

TABLE 21. Ionic yields in dilute aqueous solution

Reaction	Radiation	Yield per ion-pair Molecules	Yield per ion-pair Equivalents	Reference
HBr decomposition	α-rays	1·2	—	Lanning, F.C. & Lind, S.C. (1938)
HI decomposition	,,	1·9	—	,, ,,
$KMnO_4$ reduction	,,	0·5–0·9	2·5–4·5	,, ,,
$KMnO_4$ reduction	X-rays	0·74	3·7	Clark, G.L. & Coe, W.S. (1937)
$Ce(SO_4)_2$ reduction	,,	2·0	2·0	,, ,,
KIO_3 reduction	,,	0·1	0·62	,, ,,
Nitrate reduction	,,	0·08[1]	0·16[1]	Clark, G.L. & Pickett, L.W. (1930)
Nitrite oxidation	,,	0·2	0·4	Fricke, H. & Hart, E.J. (1935b)
Ferrocyanide oxidation	,,	0·4	0·4	Fricke, H. & Hart, E.J. (1935c)
Selenite oxidation	,,	0·2	0·4	,, ,,
Arsenite oxidation	,,	0·2	0·4	,, ,,
$FeSO_4$ oxidation[3]	,,	3[2]	3	Fricke, H. & Hart, E.J. (1935a)
$FeSO_4$ oxidation[6]	,,	~10	~10	Gray, L.H. & Weigert, F. (unpublished)
$FeSO_4$ oxidation[3]	α-rays	0·8–1·7	0·8–1·7	Nurnberger, C.E. (1934)
$Fe_2(SO_4)_3$ reduction[6]	X-rays	~0·5	~1·0	Gray, L.H. & Weigert, F. (unpublished)
$K_2Cr_2O_7$ reduction[3]	X-rays	0·2	1·2	Fricke, H. & Brownscombe, E.R. (1933a)
Oxyhaemoglobin to methaemoglobin	,,	0·6	—	Fricke, H. & Petersen, B.W. (1927)
Various organic acids to H_2 and CO_2. H_2 yield	,,	0·4–1·2	—	Fricke, H., Hart, E.J. & Smith, H.P. (1938)
$H_2O + CO$ to H_2, CO_2, HCHO	,,	1·5	—	,, ,, ,,
d-amino-acid-oxidase inactivation:				
Prosthetic group	,,	0·1	—	Calculated from Dale, W.M. (1942)
Specific protein	,,	0·14	—	,, ,,
Tyrosine decomposition	,,	0·1	—	Stenström, W. & Lohmann, A. (1928)
Tyrosine decomposition	α-rays	0·0035	—	Nurnberger, C.E. (1937)
Glutathione	X-rays	0·4–1·7	—	Kinsey, V.E. (1935)
Ascorbic acid	,,	0·7	—	Anderson, R.S. & Harrison, B. (1943)
Ribonuclease	,,	0·03	—	Lea, D.E. & Holmes, B. (unpublished)

1 Recalculated. The authors report a yield of 0·2–0·3 molecule per ion-pair, but this appears to be based on a misconception of the relation between energy absorption and dose in roentgens.
2 In oxygen-free solutions. The yield is about doubled in oxygenated solution.
3 In 0·8 N sulphuric acid.
4 In Dale's preparation approximately 20% of the protein was enzyme protein. The calculation of ionic yield has been made on the assumption that the 80% of non-enzyme protein has equal affinity for activated water as the enzyme protein. Dale (in a private communication) suggests that the affinity might be lower. In that event the ionic yield for the specific protein could be as low as 0·02.
5 This is the figure given by Nurnberger. Radon was used as the source of α-particles, and about 5% of the total ionization would thus be due to β-rays. Accepting Stenström and Lohmann's figure for X-rays as also applying to β-rays, it follows that the whole of Nurnberger's yield can be accounted for by the β-rays, with no decomposition at all produced by the α-rays.
6 Ferrous or ferric ammonium sulphate in $10^{-3} N$ H_2SO_4. The reactions are complicated; the yields given are the initial yields.

There are very few data bearing on the relative efficiencies of different radiations in promoting reactions in solution, and work on this subject is urgently needed. The conversion of oxyhaemo-

[1] Allsopp, C.B. (1944).

globin to methaemoglobin is stated to be independent of X-ray wave-length from 0·25 to 0·76 A. The ionic yields obtained with α-particles and X-rays in the reduction of $KMnO_4$ agree. The results for the irradiation of $FeSO_4$ suggest that the ionic yield is lower with α-particles than with X-rays, but this is a complicated reaction in which the yield depends upon the pH, the $FeSO_4$ concentration, the dose, and the degree of oxygenation of the solution, and no reliance can be placed upon a comparison made by different authors under different conditions. The most striking difference is that reported for the decomposition of tyrosine by α-particles and X-rays, also in the direction of α-rays being less efficient.

Inspection of the table shows that almost all the ionic yields lie between 0·1 and 2·0. It appears probable that activated water is formed at the rate of about 1 molecule per ion-pair, and that reaction yields smaller than this are due to a proportion of collisions between reactant molecules and activated water leading to deactivation without reaction, or to other causes which are discussed later.

Chemical mechanism of the indirect action in aqueous solution

Weiss[1] has suggested that activated water consists of free H and OH radicals. To convert a water molecule into H and OH radicals requires 5 eV. (i.e. 115 kilocalories per gram-molecule), and two mechanisms may be suggested to account for this conversion by an ionizing radiation, following ionization or excitation of water molecules respectively.

If an electron is ejected from an H_2O molecule, the molecule may split into a hydrogen ion and a hydroxyl radical $H_2O^+ \rightarrow H^+ + OH$, thus producing a hydroxyl radical at the site of the ionization. The electron which is ejected will travel a distance determined by the energy with which it was ejected, and eventually become attached either to a hydrogen ion or to a water molecule, in either event giving rise to a hydrogen radical $H^+ + e \rightarrow H$, or $H_2O + e \rightarrow H + OH^-$. The H radical will thus be produced at an appreciable distance from the OH radical.

It is possible that some of the H_2O molecules which are excited by the radiation decompose directly into H and OH radicals. The radicals in this case will be produced close together.

[1] Weiss, J. (1944). That H atoms and OH radicals are involved was suggested tentatively by Risse, O. (1929) and Fricke, H. (1938).

The peculiarities noted earlier concerning the decomposition of water by ionizing radiations can largely be understood in terms of the production of H and OH radicals. The low yield in pure gas-free water is due to the back reaction $H + OH \rightarrow H_2O$. For decomposition, it is necessary for collisions to occur between two hydrogen radicals or two hydroxyl radicals, $H + H \rightarrow H_2$ or $OH + OH \rightarrow H_2O + O$ followed by $O + O \rightarrow O_2$. Proximity will favour the $H + OH$ combination in the case of radicals produced by the decomposition of an excited molecule, though not so much in the case of radicals produced following an ionization.

If dissolved oxygen is present the reaction is $H + O_2 \rightarrow HO_2$, followed by $2HO_2 \rightarrow H_2O_2 + O_2$. The removal of the hydrogen radicals by the oxygen reduces the rate of recombination of H and OH. The OH radicals which accumulate combine to give oxygen as before.

The oxidation of inorganic ions by OH radicals, or their reduction by H radicals, explains most of the inorganic reactions which have been studied. Thus $Fe^{++} + OH \rightarrow Fe^{+++} + OH^-$; $Ce^{++++} + H \rightarrow Ce^{+++} + H^+$. Since oxidations remove OH radicals, the H radicals accumulate in the solution and combine with the evolution of gaseous hydrogen. Similarly, reductions are accompanied by the evolution of gaseous oxygen, resulting from the combination of OH radicals, which accumulate in the solution when the H radicals are removed.

The oxidation of simple organic compounds, accompanied by an emission of hydrogen, is explained in a similar fashion.

The OH radical is highly reactive as an electron acceptor ($OH + e \rightarrow OH^- + 3 \cdot 7$ eV.), and it is this which probably accounts for the fact that almost all organic compounds are decomposed by irradiation in aqueous solutions. Conversely almost any organic compound, if present in sufficient concentration, is able to act as a protective agent, by reducing the concentration of OH radicals in the solution by the above reaction. The H radical reacts with solutes which are oxidizing agents, but is probably less reactive than the OH radical with most organic molecules.

Spatial distribution and recombination of the active radicals

In typical chemical reactions induced by ionizing radiations in aqueous solution, the ionic yield is independent of the concentration over a wide range of concentrations. It has been shown, however, with some solutes, and is probably true in general, that at sufficiently low concentration the ionic yield is no longer constant but diminishes with diminution of concentration.[1] The explanation is presumably that in sufficiently concentrated solutions the H and OH radicals react with the solute before they have time to collide with each other, while in sufficiently dilute solutions the H and OH radicals collide with each other and combine before they have time to react with the solute molecules.[2]

A quantitative treatment requires a knowledge of the spatial distribution of the radicals, which distribution is far from uniform. When a solution is irradiated, ionizing particles pass through it (fast electrons in an X-ray experiment, α-rays in an α-ray experiment), and the H and OH radicals are produced along the paths of these ionizing particles. The OH radicals (resulting from the positive ions) are initially localized along the path of the ionizing particle, the H radicals (resulting from the attachment to an H_2O molecule or an H^+ ion of the electron ejected at ionization) are produced at a distance away depending on the distance travelled by the ejected electron before it is attached. The initial distribution of the OH and H radicals is thus the same as the initial distribution of positive and negative ions.[3] This latter distribution can be studied in gases by the Wilson chamber method.[4] It can also be studied in gases and insulating liquids by a less direct method depending on the comparison of observed ionization currents with calculations of the

[1] Stenström, W. & Lohmann, A. (1928) using tyrosine; Kinsey, V.E. (1935) using glutathione; Fricke, H., Hart, E.J. & Smith, H.P. (1938) using formic acid, oxalic acid, formaldehyde and methyl alcohol; Lanning, F.C. & Lind, S.C. (1938) using $KMnO_4$.

[2] We are assuming that there is nothing in the diluting fluid with which the active radicals can react. Precautions should be taken in experiments of this sort to use gas-free water free from traces of organic impurity capable of acting as protective agents. It is not certain that precautions were adequate in all the experiments we discuss.

[3] We are neglecting in this treatment any H and OH radicals produced by excitation rather than ionization.

[4] Klemperer, O. (1927) photographed α-ray tracks in hydrogen gas at a pressure of one-tenth of an atmosphere, and found that the radius of the column of negative ions was equivalent to 0·002 cm. of air at atmospheric pressure.

proportion of ions which escape ionic recombination, these calculations involving the initial distribution of the ions.[1] It is concluded that the number (n) of ions per cm.^{3} of each sign at a distance r from the axis of the ionizing particle can be represented initially[2] by the formula

$$n = \frac{N_0}{\pi b^2} e^{-r^2/b^2}, \qquad (\text{II-1})$$

N_0 is the number of ions of each sign produced by the ionizing particle per cm. path, and b is a measure of the radius of the column of ions. b has been found to have the values $1{\cdot}79 \times 10^{-3}$ cm. in air (density $0{\cdot}0012$) and $2{\cdot}34 \times 10^{-6}$ cm. in hexane (density $0{\cdot}677$). We shall take the value $b = 1{\cdot}5 \times 10^{-6}$ for water (density 1). We shall suppose formula (1) to represent also the initial distribution of OH and H radicals.

The radicals rapidly diffuse away from the path of the ionizing particle, and after time t it can be shown[3] that the number of radicals of each sort per cm.^{3} at a distance r from the path of the ionizing particle is

$$n = \frac{N}{\pi (4Dt + b^2)} e^{-r^2/(4Dt + b^2)}, \qquad (\text{II-2})$$

where D is the diffusion constant, which for the purpose of our calculation we shall assume to be 2×10^{-5} cm.^{2} sec.$^{-1}$ at room temperature for both H and OH radicals, though it may be higher for the former.[4] The radius of the column is now $\sqrt{(4Dt + b^2)}$ instead of b. N is the total number of ions per cm. length of path of the ionizing particle, and progressively diminishes from its initial value N_0 as a result of recombination of the radicals or their reaction with any solute present.

[1] Jaffé, G. (1913). Cp. also Kara-Michailova, E. & Lea, D.E. (1940).

[2] Strictly, not initially, but after the very short time interval needed for the positive ions which are to begin with closer to the axis of the ionizing particle than the negative ions to diffuse to a comparable distance.

[3] Here and elsewhere in this section the treatment of Jaffé, G. (1913) is followed. Jaffé worked out his calculations for the diffusion and recombination of ions, but the mathematics apply equally well to the diffusion and recombination of radicals, providing the appropriate values of the constants are used.

[4] The diffusion constant of deuterium hydroxide in water is 2×10^{-5} cm.^{2} sec.$^{-1}$ at 15° (Orr, W.J.C. & Butler, J.A.V. 1935).

RECOMBINATION OF RADICALS

After a short time (a fraction of a second) diffusion has broadened the columns sufficiently for adjacent columns to overlap. While the radicals are diffusing, their number is diminishing owing to combination with each other and reaction with any solute which is present. If, by the time adjacent columns overlap, most of the radicals in a column have disappeared, then each column may be considered an isolated entity, and the concentration of radicals we have to consider in calculating reaction rates is the concentration given by equations (1) or (2). But if, by the time the columns overlap, only a small proportion of the radicals have disappeared, then the initial localization of the radicals in columns can be neglected and we can consider the reaction to occur in the liquid as a whole, and take as the concentration of radicals not the values in the column given by equations (1) or (2), but the much lower average values obtained by dividing the total number of radicals by the total volume of the solution. It is necessary therefore to determine whether reaction occurs mainly before the columns mingle, or mainly after the columns mingle.

If the dose-rate is I roentgens per second, approximately $2 \times 10^{12} It$ ion-pairs (and therefore pairs of radicals) will be produced per cm.³ in t seconds. There being N_0 ion-pairs per cm. path, $2 \times 10^{12} It/N_0$ ionizing particles will cross each square cm. in t seconds. Now in t seconds the radius of the column is $\sqrt{(4Dt+b^2)}$, and its area is therefore $\pi(4Dt+b^2)$. Thus for the columns to overlap in t seconds we must have

$$\pi(4Dt+b^2) \times 2 \times 10^{12} It/N_0 = 1.$$

In practice $4Dt \gg b^2$, and we deduce that

$$t = 10^{-6} \times \sqrt{(N_0/8\pi DI)} \quad \text{and} \quad \sqrt{(4Dt+b^2)} = 10^{-3} \times (2N_0 D/\pi I)^{\frac{1}{4}}.$$

These expressions give respectively the time required for the columns to overlap, and the radius of a column when adjacent columns overlap.

To work out typical numerical values we insert in these formulae $I = 10$ r. per second, $D = 2 \times 10^{-5}$ cm.² sec.⁻¹, $N_0 = 3 \times 10^7$ ionizations per cm. for α-rays or $N_0 = 6 \times 10^5$ ionizations per cm. for X-rays. We deduce that in X-ray experiments adjacent columns overlap after 0·01 second when the column radius is 0·001 cm., and that in α-ray experiments the columns overlap after 0·08 second when the column radius is 0·0025 cm.

In the absence of a solute, the radicals disappear by the reaction $H + OH \to H_2O$. The number of pairs of radicals disappearing per cm.[3] per second will, according to the mass action law, be αn^2, where n is the number of radicals of each kind per cm.[3] and α is a constant the value of which can be calculated on the kinetic theory assuming (provisionally) that every collision between H and OH is effective. The value used[1] is $\alpha = 4 \times 10^{-10}$.

The number N of pairs of radicals which remain uncombined after a time t, during which the column diffuses to a radius $\sqrt{(4Dt + b^2)}$, is related to the original number N_0 by the formula[2]

$$N_0/N = 1 + \frac{\alpha N_0}{8\pi D} \log \frac{4Dt + b^2}{b^2}. \qquad (II\text{-}3)$$

Using the values of α, b, N_0, and D already quoted, we find that by the time the columns overlap the proportion of radicals remaining uncombined is 0·3 % in an α-ray column and 13 % in an X-ray column.

If a solute is present the radicals will disappear still more quickly. We conclude therefore that the reaction takes place independently in the paths of the individual ionizing particles, and that the radicals produced by different ionizing particles do not mix appreciably.

We can calculate from formula (3) the time required for half of the radicals to combine, the times obtained being $1·2 \times 10^{-9}$ second in an α-ray experiment and 2×10^{-7} second in an X-ray experiment. These times will be reduced if a solute is present.

Kinetics of indirect actions in aqueous solution

Fig. 3 shows the calculated number[3] of collisions per second undergone by a (specified) H or OH radical with solute molecules, the solute being of molecular weight M and present in

[1] $6 \times 10^{20} \alpha$ is the reaction constant in gram-molecular units of the reaction $H + OH \to H_2O$. α is calculated from formulae given by Moelwyn-Hughes, E.A. (1933), *Kinetics of Reactions in Solution*.

[2] Jaffé, G. (1913).

[3] The calculation can be made only approximately, and assumes that the diameter of a molecule of molecular weight M is $1·33 \times 10^{-8} M^{\frac{1}{3}}$ cm. Cp. Moelwyn-Hughes, E.A. (1933), *Kinetics of Reactions in Solution*; Hinshelwood, C.N. (1940), *Kinetics of Chemical Change*.

concentration either of 1 gram-molecule per litre (curve A and left-hand ordinate scale) or 1 g. per litre (curve B and right-hand ordinate scale). For equal molar concentrations the collision

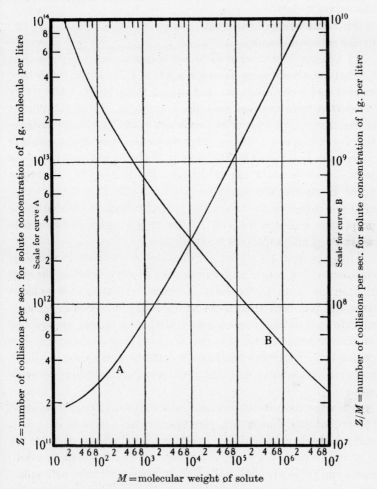

FIG. 3. Number of collisions per second made by a (specified) active radical with solute molecules, the solute being of molecular weight M and present in a concentration of: A, 1 gram-molecule per litre; B, 1 g. per litre.

rate with different solutes increases with increasing molecular weight, but for equal concentrations in grams per litre the collision rate decreases with increasing molecular weight. We shall call Z the collision frequency for a concentration of 1 gram-molecule per litre.

A collision of an active radical and a molecule of a solute capable of reacting can result in one of the following alternatives:

(a) Elimination of the active radical and chemical change in the solute molecule.

(b) Elimination of the active radical without chemical change in the solute molecule.

(c) Change in neither radical nor solute.

That not all collisions between a solute molecule and an active radical which eliminate the latter lead to reaction in the solute is inferred from the fact that the ionic yield in many reactions is less than unity under circumstances when radical recombination is not suspected. Where large molecules such as enzymes are concerned, it is possible that change in the solute molecule may occur which is not made evident by the method of estimation used. In most of the reactions listed in Table 21 it appears that of those collisions in which the active radical is eliminated, a proportion between 0·1 and 1·0 result in change in the solute molecule. This proportion we designate by P.

That the probability of elimination of an active radical at collision with a solute molecule is zero or very small in the case of many solutes, particularly inorganic substances, is evident from the fact that these substances could be present in Dale's experiments in high concentration without exerting appreciable protective action on the enzyme. We have no guarantee that this probability, which we shall call p, attains unity even with the most reactive solutes, but unity is at any rate its highest possible value.

When the concentration of solute is c gram-molecules per litre, the probability that in time dt a (particular) active radical shall suffer a collision in which it is eliminated is $cZp\,dt$. If there are n active radicals per cm.^{3}, $ncZp\,dt$ active radicals will be eliminated in time dt, and $ncZpP\,dt$ solute molecules will suffer chemical change in time dt.[1] If two solutes are competing, the concentrations being c_1 and c_2, the number of molecules of solute 1 reacting will be $nc_1Z_1p_1P_1/(c_1Z_1p_1+c_2Z_2p_2)$, and of solute 2,

$$nc_2Z_2p_2P_2/(c_1Z_1p_1+c_2Z_2p_2). \qquad \text{(II-4)}$$

[1] In these formulae we are neglecting the probability that an active radical shall be eliminated by combination with another active radical. This is justifiable providing c is sufficiently high. We discuss later the case of low solute concentration.

Competition between two solutes; protective action

The concentration c_2 of solute 2, at which the reaction of solute 1 is reduced to 50% of its value in the absence of solute 2, is, from equation (4), given as

$$c_2 = c_1 \frac{Z_1 p_1}{Z_2 p_2}, \quad \text{whence} \quad p_2/p_1 = c_1 Z_1 / c_2 Z_2. \tag{II-5}$$

Fricke, Hart and Smith[1] have irradiated mixtures of formic acid with caproic acid, methyl alcohol, formaldehyde, oxalic acid or acetone, plotting the yields of H_2 and CO_2 as a function of the relative concentrations of the formic acid and the other solute. From their curves we can read off the ratio c_1/c_2 of the concentrations of the formic acid and the other solute at which the activated water appears to be equally shared between the two solutes. The values of Z for the different solutes being taken from Fig. 3, we are able to deduce that the values of p for the six solutes caproic acid : formic acid : methyl alcohol : formaldehyde : oxalic acid : acetone are in the ratios 1·6 : 1·0 : 0·44 : 0·22 : 0·067 : 0·024 respectively. The experiment is an illustration of one method by which relative (but not absolute) values of p may be determined. A variant of this method is to compare the efficiencies of different substances in protecting the same solute.

Dale[2] lists the concentrations (c_2) of a number of solutes which suffice to reduce to about 50% the number of molecules of alloxazin-adenine-dinucleotide reacting with a given dose, the dinucleotide itself being present in a concentration of $c_1 = 6·1 \times 10^{-7}$ gram-molecule per litre. With the aid of equation (5) and Fig. 3 we can deduce the ratios p_2/p_1 from these experimental values, p_1 referring to the dinucleotide. The values of p_2/p_1 so obtained are given in Table 22, column 2.

Dale also gives similar though less extensive data for the protection of the specific protein component of d-amino-acid-oxidase, and the values of p_2/p_1, where p_1 now refers to this protein, are given in column 3.[3] Comparison of columns 2 and 3

[1] Fricke, H., Hart, E.J. & Smith, H.P. (1938).
[2] Dale, W.M. (1942).
[3] Dale's preparation contained 20 μg. of active protein to 105 μg. total protein; we have assumed that from the point of view of competing for activated water all the protein and not merely the enzyme protein is to

of Table 22 enables us to determine the ratio of the values of p for the dinucleotide and the protein. The ratio inferred from the glucose data appears to be anomalous. The others agree fairly

TABLE 22. Relative deactivation efficiencies of different solutes

Solute 2 (1)	p_2/p_1, where p_1 is value for		Inferred ratio of p values dinucleotide protein (4)	Relative affinities for activated water of solutions containing	
	Dinucleotide (2)	Specific protein (3)		1 g.-mol. per litre Zp/p_1 (5)	1 g. per litre Zp/p_1 (6)
Glycine	1.8×10^{-4}	8.5×10^{-4}	4.8	4.6×10^7	6.2×1
Na oxalate	1.5×10^{-4}	—	—	4.6×10^7	3.4×1
Na nitrate	1.8×10^{-4}	—	—	4.6×10^7	5.5×1
Leucylglycine	1.3×10^{-3}	2.1×10^{-3}	1.5	4.6×10^8	2.3×1
Alanine	1.7×10^{-3}	2.6×10^{-3}	1.5	4.6×10^8	5.2×1
K ferricyanide	1.0×10^{-3}	—	—	4.6×10^8	1.4×1
K ferrocyanide	1.0×10^{-3}	—	—	4.6×10^8	1.3×1
Na hippurate	1.3×10^{-2}	—	—	4.6×10^9	2.3×1
Glucose	4.3×10^{-2}	2.1×10^{-4}	0.005	1.5×10^{10}	8.1×1
Sucrose	3.2×10^{-2}	—	—	1.5×10^{10}	4.3×1
K thiocyanate	5.2×10^{-2}	—	—	1.5×10^{10}	1.5×1
Na formate	5.9×10^{-2}	—	—	1.5×10^{10}	2.2×1
Fructose	1.4×10^{-1}	—	—	4.6×10^{10}	2.6×1
Na nitrite	1.8×10^{-2}	—	—	4.6×10^{10}	6.7×1
Na nucleate	5.8×10^{-2}	—	—	4.6×10^{10}	4.2×1
Specific protein of d-amino-acid-oxidase	0.4	—	—	3.8×10^{12}	5.4×1
Alloxazin-adenine-di-nucleotide	1.0	—	—	7.6×10^{11}	8.3×1

well and suggest that the value of p is about 2.5 times as great for the dinucleotide as for the protein, i.e. the probability of an active radical being deactivated on collision is somewhat greater for the dinucleotide than for the protein. The large size of the protein molecules more than compensates for the smaller value of p, and taking the values of Z from Fig. 3, Zp is found to be five times greater for the protein than for the dinucleotide. Thus when irradiating a mixture containing approximately equal molar concentrations, as in one of Dale's experiments,[1] the protein

be allowed for. If, however, the affinity for the active radicals of the non-enzyme protein is smaller than that of the enzyme protein, the figures in the third and fourth columns of Table 22 will need to be reduced by some factor between one and five, and the value of p for the enzyme protein might thus become equal to, or slightly in excess of, that for the dinucleotide.

[1] The molar concentration of the dinucleotide was 2.2×10^{-7} gram-molecule per litre, of the active protein 1.1×10^{-7} gram-molecule per litre, or of the total protein about five times this.

should almost completely protect the dinucleotide, and not be noticeably protected by it. This was in fact found by Dale. If, however, the concentration of the dinucleotide were to be increased ten times or more without changing that of the protein, some protection of the protein by the dinucleotide should become noticeable according to the present calculations.

As regards the absolute values of p, these experiments give no information. Since the dinucleotide has the largest value of p for all the substances investigated by Dale, and since this cannot exceed unity, the figures in the second column may be taken to be maximum values of p for the corresponding solutes. We shall give evidence later that the actual values are not very different from these maximum values.

Reduced yield in dilute solutions, and with densely ionizing radiations

Providing that the solute concentration is sufficiently high for nearly all the active radicals to be eliminated by collisions with solute molecules rather than by collisions with each other, the ionic yield will be independent of the solute concentration. If, however, the solute concentration is so low that an appreciable proportion of the total number of active radicals combine with one another rather than react with solute molecules, then the ionic yield will fall. The determining factor is evidently the ratio of the number of solute molecules per cm.^{3} to the number of active radicals per cm.^{3} Since the latter number is much higher in an α-ray track than in the track of a fast electron, it is evident that the diminished ionic yield due to combination of radicals will become appreciable in α-ray experiments at higher concentrations than in X-ray experiments. This effect probably accounts for the ionic yields in certain of the reactions listed in Table 21 being less for α-rays than for X-rays.

It is possible to make calculations of the rate at which active radicals disappear by the two processes of combination of radicals and reaction with the solute,[1] during the diffusion of the radicals away from the axis of the column. In Fig. 4A the results of these calculations are given in the form of graphs showing the proportion of radicals which are eliminated by collision with the solute as a function of cZp. cZp is a measure of the concentra-

[1] By an extension of the calculation given on pp. 49–52.

tion of the solute, and the graphs show, as anticipated, that the proportion of the active radicals which react with the solute diminishes with diminishing solute concentration, and (for concentrations such that $cZp < 10^{10}$) is less for α-rays than for X-rays.

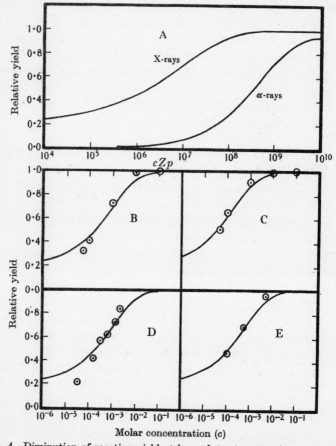

FIG. 4. Diminution of reaction yield at low solute concentrations. A, calculated curves for X-rays and α-rays. B-E, comparison of calculated curves and experimental results (X-rays): B, oxalic acid (Fricke, Hart & Smith); C, methyl alcohol (Fricke, Hart & Smith); D, glutathione (Kinsey); E, tyrosine (Stenström & Lohmann).

In Figs. 4 B, C, D, E a comparison is made between the results of experiments[1] in which the ionic yield has been studied as a function of the solute concentration, and the theoretical curve

[1] Stenström, W. & Lohmann, A. (1928) using tyrosine; Kinsey, V.E. (1935) using glutathione; Fricke, H., Hart, E.J. & Smith, H.P. (1938) using oxalic acid and methyl alcohol.

for X-rays. It is seen that the theory satisfactorily fits the experimental results. In fitting the theory to the experiments, p is treated as an arbitrary constant. Experiments of this sort thus serve to determine p, and values of p deduced in this manner are given in Table 23. The values of p for methyl alcohol and oxalic acid are in the same ratio (7 : 1) as was deduced on p. 55 from a different experiment.

TABLE 23. Values of p deduced from the diminution of reaction yield at low solute concentrations

Substance	p
Oxalic acid	0·037
Methyl alcohol	0·25
Glutathione	0·018
Tyrosine	0·037

The absolute values of p given in Table 23 are subject to considerable numerical uncertainty, owing to the fact that any errors in the values assumed for N_0, α, D and b will result in rather serious displacement of the scale of abscissae in the X-ray curve of Fig. 4A. Also, we have assumed throughout that a collision between an H and an OH radical will always result in combination. If this is not so, the values of p in Table 23 will be too high.

It is unfortunate that no substance occurs in both Tables 22 and 23. Comparison of the values of p in Table 23 with the values of p_2/p_1 for chemically related substances in Table 22 suggests that the former values are probably somewhat too high. They also suggest that p_1 in Table 22 cannot be much less than unity, so that the figures in column 2 of this table can be taken as values of p for the corresponding substances.

Having deduced the value of p for tyrosine from consideration of the manner in which the X-ray ionic yield varies with solute concentration (Fig. 4E), it is possible to predict the ionic yield with α-rays by making use of the α-ray curve of Fig. 4A. In this way it is calculated that the ionic yield in a $2 \cdot 2 \times 10^{-4}$ M solution of tyrosine should be 18 times smaller to α-rays than to X-rays. Experimentally it has been found to be about 30 times smaller.[1]

[1] Nurnberger, C.E. (1937) found that the ionic yield in a $2 \cdot 2 \times 10^{-4} M$ solution of tyrosine irradiated by α-rays was 0·0029. Stenström, W. & Lohmann, A. (1928), using X-rays, found ionic yields of 0·17 in a $5 \cdot 5 \times 10^{-3}$ M solution, and 0·08 in a $1 \cdot 1 \times 10^{-4}$ M solution. Nurnberger's explanation was in principle the same as we have given.

The low ionic yield with α-rays is thus fairly satisfactorily accounted for by the present calculation. It is highly desirable that the interpretation offered should be tested by irradiating more concentrated solutions with α-rays, when (according to Fig. 4A) the ionic yield should increase to values approaching those in X-ray experiments.

It is of interest to see whether, with solute concentrations of the order found in living cells, any appreciable difference between the ionic yields of X-rays and α-rays is to be anticipated on account of the mechanism we have been discussing. By inspection of Table 22, column 6, it appears likely that there will be proteins, sugars and other cell constituents having values of Zp/M as high as 10^8. If such constituents are present in a concentration of 10 g. per litre, $cZp = 10^9$. Fig. 4A shows that for this value of cZp the ionic yield with α-rays is two-thirds as great as with X-rays. It appears not impossible therefore that chemical changes induced in cells by the indirect action via the water may have smaller yields in α-ray experiments than in X-ray experiments.

Direct and indirect actions of radiation

We have mentioned that one of the characteristics of the indirect action is that, in the absence of complications, the yield is directly proportional to the dose. Fig. 1B, for example, shows how the percentage of ferrous sulphate decomposed increases with the dose; the curve is linear from 0 to 90 % decomposition. A back reaction probably accounts for the reaction not proceeding to completion. Back reaction between the products of the primary action is one cause for a reaction curve being non-linear. Another cause is the presence of protective agents. If we are estimating the proportion of a given solute 1 which survives irradiation with various doses in the presence of an excess of solute 2, then we see, from equation (4), that the number of gram-molecules of solute 1 reacting for a dose increment corresponding to the production of dn gram-molecules of activated water is

$$-dc_1 = dn\, c_1 Z_1 p_1 P_1 / (c_1 Z_1 p_1 + c_2 Z_2 p_2). \tag{II-6}$$

With a sufficiently large excess of solute 2 the term in c_1 in the bracket can be neglected, and so we get the result that the *fraction* of solute 1 reacting is proportional to the dose increment dn and is independent of the solute concentration c_1, i.e.

$$-dc_1/c_1 = dn\ Z_1 p_1 P_1 / c_2 Z_2 p_2, \qquad (\text{II-7})$$

or
$$c_1 \propto e^{-n/n_0},$$

where
$$n_0 = c_2 Z_2 p_2 / Z_1 p_1 P_1. \qquad (\text{II-8})$$

In other words, the concentration of solute c_1 diminishes from its initial value as an exponential function of the dose of radiation, instead of as a linear function.

It is not always necessary for there to be an excess of a second solute for the curve to be of this type, for, as pointed out by Fricke,[1] and in greater detail by Dale, Meredith and Tweedie,[2] it may happen that the products of the reaction are also capable of deactivating activated water, and if they are of approximately the same efficiency as the original solute, c_1 in the bracket term of equation (6) will need to be replaced by its *constant* initial value, and so the bracket will remain constant without the necessity for the solute 2 to be in large excess, or even to be present at all. Fig. 1c is an example of an exponential curve explained in this way.

When, as in the present case, the amount of solute surviving a given dose diminishes as an exponential function of the dose, it is convenient to indicate the rate of reaction by specifying the dose which reduces the amount surviving to a fraction $e^{-1} = 37\%$ of the initial amount, i.e. by the dose referred to as n_0 in equation (8). As can readily be seen from equation (8) this 37% dose[3] increases approximately in proportion to the concentration of the protecting solute 2, and at large concentrations of protective solute would become very large. However, in addition to reacting with activated water which may be termed the *indirect* action of the radiation, there is little doubt that a molecule of solute will undergo chemical change if itself *directly* ionized by the radiation. Normally in dilute solutions a dose which suffices to cause a considerable proportion of a solute to react with the active radicals will only suffice to ionize directly a negligible proportion of the solute molecules. In the presence

[1] Fricke, H. (1934*b*).
[2] Dale, W.M., Meredith, W.J. & Tweedie, M.C.K. (1943).
[3] Alternatively referred to as the *inactivation dose* or *mean lethal dose* where these terms are appropriate.

of a protective agent however, reducing the efficiency of the indirect action via the water, it is clear that the relative importance of the direct action increases.

From the foregoing theory the relation between the 37 % dose in solution, when both direct and indirect effects occur, and the 37 % dose which would be obtained when direct action only occurred, can readily be determined. In a solution containing only the one solute, the reaction products of which are supposed to have an affinity for activated water equal to that of the unchanged solute, we obtain

$$\frac{37\% \text{ dose for direct action only}}{37\% \text{ dose in solution}} = 1 + \frac{\gamma}{\Gamma}\left(\frac{\text{water content}}{\text{solid content}}\right). \quad \text{(II-9)}$$

γ is the ionic yield for indirect action, i.e. the number of solute molecules reacting, per ionization in the solvent, under conditions in which only the solvent contributes to the deactivation, while Γ is the ionic yield for direct action, i.e. the number of molecules reacting per ionization directly produced in the solute, under conditions in which ionizations not directly in the solute do not appreciably contribute to the yield. The term (water content/solid content) simply means the ratio of the weight of water to the weight of solute in a given quantity of solution, e.g. is 90/10 for a solution containing 10 % by weight of solute.

If the solid content is low enough for the ionic yield of the indirect action to be appreciably reduced by combination of the radicals, γ must be reduced by multiplying by the factor read from Fig. 4A.

If the solution contains the solute being investigated at low concentration, and in addition contains a protective agent in much higher concentration, then we obtain

$$\frac{37\% \text{ dose for direct action only}}{37\% \text{ dose in solution}}$$

$$= 1 + \frac{\gamma}{\Gamma}\left(\frac{\text{water content}}{\text{solid content}}\right)$$

$$\times \left(\frac{\text{deactivating efficiency per unit mass of solute}}{\text{deactivating efficiency per unit mass of protective agent}}\right).$$

$$\text{(II-10)}$$

By deactivating efficiency per unit mass we mean Zp/M, relative values of which may be read off from column 6 of Table 22 for the substances investigated by Dale. Solid content includes the protective agent. γ/Γ refers to the solute being investigated.

We see from equation (9) that as the concentration of the solute being investigated is increased, the 37 % dose in solution increases, and tends as a limiting value to the value for direct action only, which is presumably the value to be expected when the solute is irradiated dry.[1]

In the event of the ionic yield for the indirect effect being considerably less than for the direct effect, i.e. $\gamma/\Gamma \ll 1$, then the limit will be approached for concentrations of solute a good deal less than 100 %.

Broda has recently published figures for the decomposition of ammonium persulphate in glycerine, the persulphate concentration varying from 0·04 to 0·13 g./cm.³ The 37 % dose appears to be independent of concentration over this range, and it is inferred that there is no indirect action in glycerine solutions such as occurs in aqueous solutions. It is unwise, however, to make this deduction on the basis of observations which do not extend to low concentrations of the solute, and Broda's experiments do not prove more than that the ionic yield for the indirect action is less than one-tenth of the ionic yield for the direct action.

As the concentration is diminished, the 37 % dose diminishes, but owing to recombination of radicals it does not do so indefinitely. If the solute is one for which the ionic yield for the indirect action (γ) is considerably less than for the direct action (Γ), there may not be a great deal of difference between the 37 % dose for the direct action (i.e. the 37 % dose when the solute is irradiated dry, or in concentrated solution, or in the presence of sufficient protective agent), and the 37 % dose in dilute solution. This happens with virus proteins, which appear to

[1] Broda, E. (1943), irradiating ammonium persulphate, finds that the yield dry is much less than in concentrated solution in glycerine. The doses of radiation he used are not stated in roentgens, but, making an estimate from the data given, it appears that the ionic yield in solution is of the order of 25, suggesting a chain reaction of some sort. The negative result when the persulphate is irradiated dry probably means that the chain reaction cannot occur in the dry state. The results do not rule out an ionic yield of the order of unity in the dry state.

have γ much less than unity, and will be discussed further in Chapter IV.

From equation (10), we see that the 37 % dose in solution may approach the 37 % dose for direct effect even in a solution of quite low solid content, in the event of the protective agent having a deactivating efficiency per unit mass much higher than that of the solute being investigated (or of γ/Γ being much less than unity).

POSSIBLE MODES OF BIOLOGICAL ACTION OF RADIATIONS

We shall take it for granted that the biological effects of ionizing radiations are due in some way to the chemical changes induced by the radiations. We are immediately faced with the problem of explaining why marked biological effects are produced by doses of radiation which produce only a small degree of chemical change. Marked biological effects in different materials are produced by doses ranging from about 50 r. to about 5×10^5 r. The number of ionizations produced in a cubic micron of tissue by a dose of 5×10^5 r. is about 10^6, and judging from the results of chemical experiments the number of molecules reacting will be of this order. The number of atoms in $1\mu^3$ is, however, about 10^{11}. Thus even the very large dose of 5×10^5 r. produces a rather small percentage chemical change, and the dose of 50 r. seems quite negligible from this point of view. There are several ways, however, in which a small overall percentage chemical change may be imagined to be effective, and these we now proceed to discuss.

Cell poisons

The products of decomposition of proteins and other cell constituents by radiations have not been much investigated, but it is quite possible that they may be injurious in quite low concentration. It is possible that there are some biological effects due to this cause. There is not much one can say, however, about their mechanism, and they will not be discussed in this book.

Activated water reactions

As we have seen, by the use of a sufficiently dilute solution a large percentage of chemical change in the solute can be accomplished by a moderate dose of radiation. Thus dilute solutions of enzymes may be largely inactivated by doses of a few thousand roentgens which would have practically no effect upon a concentrated solution or a dry preparation. Enzymes are present in cells in low concentration, and since their destruction would produce marked effects it is natural to suspect that enzyme destruction is of importance. This argument has been developed by Dale.[1] The low concentration of the enzyme in the cell does not in itself necessarily lead to a large percentage destruction by moderate doses, since other cell constituents present in larger amounts will protect it. In fact, referring to equation (10) we see that the dose required to inactivate a given percentage of an enzyme present in concentration small compared with that of the other cell solutes is independent of the enzyme concentration, and is determined principally by the value of Zp/M. As may be seen from column 6 of Table 22, enzymes have the largest Zp/M values of any substances investigated by Dale. It is quite possible, however, that other proteins are equally effective in this respect. It seems therefore that while the sensitivity of an enzyme in a cell will be higher than that of the enzyme in concentrated solution, it may not be as much higher as is sometimes supposed.[2]

Direct action on large molecules

As will be explained in greater detail in Chapter III, the dose required to produce chemical change in a given proportion of the molecules of a substance by *direct* action is inversely proportional to the molecular weight, supposing that the ionic yield (number of molecules affected per ion-pair) is constant. As a rough working rule, a dose of 10^6 r. will produce chemical change in half the molecules of a substance of molecular weight 10^6 (if the ionic yield is about unity). Now the smallest viruses appear simply to be proteins of very high molecular weight, and in view of their high molecular weight the dose required to produce chemical change in a given proportion of the virus molecules is much smaller than is required to produce chemical change in a comparable proportion of a chemical of lower molecular weight. We shall find in Chapter IV that the inactivation of these small viruses can be explained adequately on the view that

[1] Dale, W.M. (1940, 1942).   [2] Forssberg, A. (1945, 1946).

chemical change in one molecule occurs whenever one or more ionizations are produced in it.

There are some effects of radiation on higher cells which are believed to be due to the direct ionization of large molecules in the cell. These are reactions which may be included under the general heading of gene mutation, and will be discussed in Chapters v and ix.

Localization of ionization

While the overall dose of radiation may be such that only a minute proportion of chemical change occurs in the solution as a whole, in the immediate neighbourhood of the path of an ionizing particle, particularly a densely ionizing particle such as an α-ray, practically every solute molecule may be affected. This will be no less true for direct actions not depending on active radicals. Thus, while the overall chemical change in the cell may be small, it may be high locally in the particular structures through which the ionizing particle passes. If effects in these structures are microscopically observable, or if the structures are sufficiently vital for changes in them to affect the cell as a whole, then a biological effect will be recorded. The best example of this type of action so far studied is the breakage of chromosomes by radiation, discussed in Chapters vi and vii, in which a chromosome is broken by the passage through it of an ionizing particle, providing the latter is densely ionizing and can produce (in the case of *Tradescantia*) something like 20 ionizations in its passage through the chromosome thread of diameter 0.1μ. These breaks are microscopically visible.

The target theory

When the biological effect observed is due to the production of ionization in some particular molecules, as in the induction of gene mutations, or is due to the passage of an ionizing particle through some particular structure, as in the induction of chromosome breakage, it is possible to calculate the size of the molecule or structure involved from a knowledge of the proportion of the organisms irradiated which are affected by a given dose of radiation. It is further possible to predict the variation of ionic efficiency of different radiations in producing effects of this sort. The interpretation of biological effects of radiation along these lines has become known as the *target theory* or *Treffertheorie*. When making calculations in general terms one often speaks of the molecule or structure in which ionization has to be produced as the 'target', and the production of ionization in it as a 'hit'. This mechanistic approach has been found unplausible by some workers in this field, but the successes of the theory in explaining in particular the different ionic efficiencies of different radiations make it evident that the model is not too crude to represent the facts adequately in the cases which are discussed in this book.

Spread of the effect of an ionization

When a chromosome is broken by radiation, a phenomenon briefly referred to above, and which is discussed in detail in Chapter VII, the evidence is that the passage of a densely ionizing particle, e.g. a proton or slow electron, anywhere through the chromosome thread causes a break. Now the chromosome thread being (in the case of *Tradescantia*) of a diameter of about 0.1μ, it must be made up of a very large number of chain molecules, and only a small fraction of these chains will be broken by the direct ionization or excitation of bonding electrons by impact of the ionizing particle. Some spread of the effects of ionization or excitation must therefore occur. Transference of energy from one part of a molecule to another is a process known to occur, and capable of interpretation on current quantum-mechanical theory. A cruder representation is to regard the column of ionization produced by a densely ionizing particle as a line source of heat, and to consider effects produced at a finite distance from it as due to temperature rise calculable in terms of thermal conductivity and specific heat. Jordan[1] has developed this point of view in a revival of the old 'point-heat' theory of Dessauer.[2]

Still a third mechanism, suggested by Gray,[3] takes account of the fact that ionization produced in the water inside and immediately outside the chromosome leads to the production of active radicals, and that the active radicals are capable of producing chemical changes, including presumably the disruption of bonds, at a finite distance from the place where ionization occurred. As we saw on p. 50, the H radicals are produced at

[1] Jordan, P. (1938c).   [2] Dessauer, F. (1923).
[3] Gray, L.H. (unpublished).

distances of the order of 15 mμ from the path of the particle. If the H radicals are effective in causing the change studied, then effects may be found at this distance from the path of the ionizing particle, especially in the case of a densely ionizing radiation such as α-radiation.

If only the OH radicals are effective, then this will not be so, since the OH radicals are produced much nearer to the path of the ionizing particle, and in a cell, where there is an appreciable concentration of protein in solution, the distance diffused by a radical before it is eliminated by collision with a solute molecule is only of the order of 2 or 3 mμ.[1]

[1] If the solute concentration is such that elimination of active radicals by radical combination can be neglected by comparison with elimination by solute reaction, the number of radicals diminishes according to the formula e^{-cZpt}. Taking $cZp \sim 10^9$ in the cell (cp. p. 60) it follows that the active radicals persist for a time of the order of 10^{-9} sec., which suffices only for them to diffuse a distance of 2 or 3 mμ.

Chapter III

THE TARGET THEORY

The target theory has been briefly introduced in Chapter II. Biological effects of radiation to which this theory is applicable are those in which the effect studied is due to the production of ionization by the radiation in, or in the immediate vicinity[1] of, some particular molecule or structure. Thus the production of gene mutation by ionization of the gene molecule, or of chromosome breakage following the passage of an ionizing particle through the chromosome, are actions of radiation to which the target theory is applicable. There are many actions of radiation on living organisms which are not to be interpreted on the target theory. Thus, if the effect observed in a given cell is found to be due to changes in the surrounding tissue, or in the blood circulation, the target theory will not be applicable to this effect. Or if the effect observed in some cell structure is due to change in its chemical environment as a result of ionizations produced in the cell fluids, then the theory will not be helpful. It is clear that a large number of possible modes of action of radiation lie outside the scope of the theory, and some writers[2] have questioned whether there exist in fact any actions to which the theory is applicable. It is the opinion of the present author, the basis of which will be made clear in the sequel, that the validity of the target theory is as certain as a scientific theory ever is in a rapidly developing subject, in the case of the inactivation of small viruses by radiation, and the production of certain chromosome aberrations in higher cells. We shall apply it also to the killing of larger viruses and bacteria, and the production of gene mutations, where we regard its validity as highly probable. Very possibly there are other actions of radiation which are correctly interpretable in terms of the target theory, but we do not consider that sufficient experimental evidence is yet available to give the theory more than the status of a working hypothesis in these cases. Much needless controversy has arisen in the past

[1] The qualification 'or in the immediate vicinity' is made to take account of the possibility of the spread, perhaps over distances of the order of a few millimicrons, of the effect of an ionization (as discussed on p. 67).

[2] E.g. Scott, C.M. (1937), *The Biological Actions of X- and γ-rays*.

when biological effects of radiation have been interpreted on the target theory with inadequate evidence, and we shall for this reason exclude from discussion instances where the application of the target theory has been based only on the determination of the shape of a survival curve.

Three investigations should form a part of any attempt to determine whether a given biological action of radiation is of the target-theory type. One of these is the determination of the manner in which the number of organisms or cells affected increases with the dose of radiation. When a lethal action is being studied, this is essentially the same as determining the shape of the survival curve, i.e. the curve obtained by plotting the proportion of organisms surviving against the dose. The second is the determination of the manner in which the effect of a given dose depends upon the intensity at which it is administered. The third is an investigation of the relative effectiveness of different types or wave-lengths of radiation. This last is of particular importance, and it is from its success in explaining the manner in which the effect of a given dose varies with wave-length and type of radiation that the target theory derives its principal value. We shall in this book be concerned to a considerable extent with this application of the theory, and it will be found that considerable progress both on the experimental and theoretical sides has been made.

One class of action to which we apply the target theory is the class in which the biological effect is believed due to a single ionization. We interpret in this manner the inactivation of viruses (Chapter IV), the production of gene mutations (Chapter V), and the killing of bacteria (Chapter IX). From a radiochemical standpoint we may group these actions as those in which the biological effect is due to change produced in a single molecule by the ionization of that molecule. From a biological standpoint we may consider all these actions to be of the nature of gene mutations.

A second class of action to which we apply the target theory is the production of certain chromosome aberrations in higher cells by radiation. The aberrations follow the breakage of chromosomes caused by the passage through the chromosomes of ionizing particles. A single ionization in the chromosome has very small probability of causing breakage (in the case of

Tradescantia, discussed in Chapter VII), but the passage of a single ionizing particle suffices, providing that it is densely ionizing and so produces a sufficient number of ionizations in the chromosome in its passage through it.

A third class of action to which the target theory has been applied is one in which a large number of ionizations must be produced within the target, and therefore a number of ionizing particles must pass through it. The more densely ionizing is the ionizing particle, i.e. the more ionizations per micron it produces, the fewer ionizing particles will be required, and this should be shown up by a change in the shape of the survival curve with different wave-lengths or types of radiation. Examples of actions of radiation which have been interpreted in terms of the existence of a target through which several ionizing particles must pass, are the killing of bacteria,[1] of bean seeds,[2] of yeasts,[3] of Protozoa,[4] and the inhibition of division in tissue cells.[5] In the opinion of the present author the validity of the application of the target theory in the manner proposed is less well established in these cases, and we shall not discuss in detail the 'multi-hit' target theory in which the biological effect is supposed due to the cumulative effect of several ionizing particles separately passing through the target, but shall confine our attention to the actions which are attributable to a single ionization or a single ionizing particle.

Recognition of the single-ionization type of action

Apart from the question whether the action studied can, on biological grounds, plausibly be believed to be due to change occurring in a single molecule, discussion of which we defer until later, there are a number of lines of evidence provided by the

[1] Lacassagne, A. & Holweck, F. (1929a), using '*Pyocyanique S*'. Lea, D.E., Haines, R.B. & Coulson, C.A. (unpublished) have repeated these experiments on a number of bacteria including the strain of *Pyocyaneus* used by Lacassagne & Holweck, but have failed to confirm their experimental results, and instead obtain data consistent with the interpretation that a single ionization is responsible for the lethal effect observed.

[2] Glocker, R. (1932).

[3] Glocker, R. (1932); Lacassagne, A. & Holweck, F. (1930).

[4] Crowther, J.A. (1926).

[5] Spear, F.G., Gray, L.H. & Read, J. (1938). The authors suggest that either a single proton, or a large number of electrons, are required to pass through the target, the total ionization required being almost the same.

radiation experiments themselves which should be investigated before concluding that a particular action studied is caused by a single ionization. The following results are to be expected for this type of action:

(a) The survival curve is exponential.

(b) The effect of a given dose is independent of the intensity at which it is given, or of the manner in which it is fractionated.

(c) For the same degree of effect, the dose required with different radiations increases in the order γ-rays, hard X-rays, soft X-rays, neutrons, α-rays. Often the difference is not detectable between γ-rays and the different wave-lengths of X-rays, but becomes noticeable with neutrons and α-rays. It is probable also that the effect of a given dose will be independent of the temperature.

Shape of survival curve

It is convenient, while developing the theory in general terms, to speak of the region within which ionization has to be produced to obtain the mutation, killing, or other effect studied as the *target*. To ionize the target it is necessary for an ionizing particle (electron, proton, etc.) to pass through it, and the passage of an ionizing particle through the target producing ionization in it may be spoken of as a *hit*. The type of action we are studying is caused by a single hit. Now it is evident that the number of hits is simply proportional to the dose of radiation given. If the dose given is such that only a small proportion of the targets are hit, no distinction need be made between the total number of hits and the number of targets hit. The number of targets hit is then proportional to the dose, and a straight line is obtained by plotting (as in Fig. 5A) the yield of the reaction against the dose. If the dose used is larger so that the number of targets hit is a considerable proportion of the whole number, cases will occur of several hits being obtained in a single target. The number of targets hit will thus be less than the total number of hits. Although the total number of hits increases in strict proportionality to the dose, the number of targets hit increases more slowly, so that the yield plotted against the dose gives a curve which is convex upwards (curve B, Fig. 5) tending asymptotically to 100% at large doses. If one is following, say, the killing of bacteria or of other single-celled organisms, the

SHAPE OF SURVIVAL CURVE

numbers killed by successive increments of dose are not equal, but each increment of dose kills the same *proportion* of the number of organisms which have survived until then. The

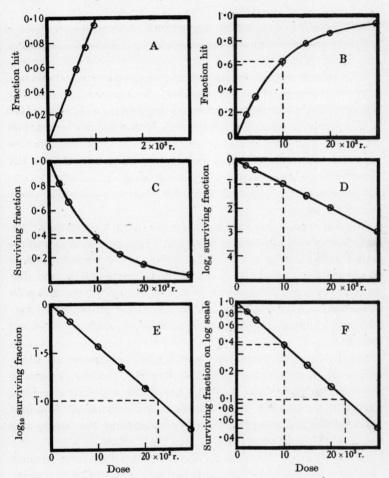

FIG. 5. Methods of plotting experimental data for the single-ionization type of action.

number of viable organisms falls off in a geometrical progression, or what is merely another way of saying the same thing, *the survival curve is exponential* (curve C, Fig. 5).

For, if n_0 is the initial number of organisms, and n the number which survive a dose D, then the proportion of organisms not so far hit which will be hit by an increment dD in the dose, will be given by the formula $-dn/n = dD/D_0$, where D_0 is the dose required to score an average of one hit per organism. This formula integrates to

$$\log_e (n/n_0) = -D/D_0, \quad \text{or} \quad n = n_0\, e^{-D/D_0}. \qquad \text{(III-1)}$$

The logarithms are natural logarithms to a base e, a table of which is given (Table 24), together with a table of the exponential function e^{-D/D_0}.

The easiest way of testing whether an experimental survival curve is exponential is to plot against the dose not the surviving fraction, but its logarithm, when a straight line will be obtained if the survival curve is exponential. Either natural logarithms to base e (Fig. 5D), or logarithms to base 10 (Fig. 5E), may be used for this test, or the surviving fractions may be plotted directly on to logarithmic graph paper (Fig. 5F). Natural logarithms are to be preferred since they facilitate the subsequent calculations.

Having established that the survival curve is exponential we can most concisely report the sensitivity of the material to radiation by stating what we may define as the *mean lethal dose*, or *inactivation dose*, or 37% *dose*. By this we mean D_0, the dose which corresponds to an average of one hit per target. It can be obtained from the experimental results in the following alternative ways, depending on which method of plotting has been adopted:

(A) If the experiment has been limited to dosages at which only a small proportion (e.g. less than 10%) of the organisms have shown the effect studied, and if the results have been exhibited by plotting the proportion affected against dose, one extrapolates the straight-line curve best fitting the results and reads off the dose corresponding to 100% effect.

(B) If the number of organisms affected has been plotted against dose, and the experiment has not been limited to a small proportion of organisms affected, the dose required is that affecting 63·2% of the organisms, because $1 - e^{-1} = 0·632$.

(C) If the number of organisms unaffected, i.e. surviving the dose, has been plotted against dose, one reads off the dose giving 36·8% survival, because $e^{-1} = 0·368$. (Hence the name, 37% dose.)

(D) If the natural logarithm of the surviving fraction has been plotted against dose, one draws the straight line best fitting the experimental points and reads off the dose giving the natural logarithm $= -1$.

(E) If logarithms to base 10 have been used, one reads off the dose giving $\log_{10} n/n_0 = -1$, and divides it by 2·3, because $\log_e 10 = 2\cdot 3$.

TABLE 24. Natural logarithms and exponentials

n/n_0	$\log_e n/n_0$		D/D_0	e^{-D/D_0}	D/D_0	e^{-D/D_0}
0·95	$\bar{1}$·949 =	−0·051	0·02	0·980	2·0	0·1353
0·90	$\bar{1}$·895	−0·105	0·04	0·961	2·1	0·1225
0·85	$\bar{1}$·837	−0·163	0·06	0·942	2·2	0·1108
0·80	$\bar{1}$·777	−0·223	0·08	0·923	2·3	0·1003
0·75	$\bar{1}$·712	−0·288	0·10	0·905	2·4	0·0907
0·70	$\bar{1}$·643	−0·357	0·12	0·887	2·5	0·0821
0·65	$\bar{1}$·569	−0·431	0·14	0·869	2·6	0·0743
0·60	$\bar{1}$·489	−0·511	0·16	0·852	2·7	0·0672
0·55	$\bar{1}$·402	−0·598	0·18	0·835	2·8	0·0608
0·50	$\bar{1}$·307	−0·693	0·20	0·819	2·9	0·0550
0·45	$\bar{1}$·201	−0·799	0·25	0·779	3·0	0·0498
0·40	$\bar{1}$·084	−0·916	0·30	0·741	3·1	0·0450
0·35	$\bar{2}$·950	−1·050	0·35	0·705	3·2	0·0408
0·30	$\bar{2}$·796	−1·204	0·40	0·670	3·3	0·0369
0·25	$\bar{2}$·614	−1·386	0·45	0·638	3·4	0·0334
0·20	$\bar{2}$·391	−1·609	0·50	0·607	3·5	0·0302
0·18	$\bar{2}$·285	−1·715	0·55	0·577	3·6	0·0273
0·16	$\bar{2}$·167	−1·833	0·60	0·549	3·7	0·0247
0·14	$\bar{2}$·034	−1·966	0·65	0·522	3·8	0·0224
0·12	$\bar{3}$·880	−2·120	0·70	0·497	3·9	0·0202
0·10	$\bar{3}$·697	−2·303	0·75	0·472	4·0	0·0183
0·08	$\bar{3}$·474	−2·526	0·80	0·449	4·1	0·0166
0·06	$\bar{3}$·187	−2·813	0·85	0·427	4·2	0·0150
0·05	$\bar{3}$·004	−2·996	0·90	0·407	4·3	0·0136
0·04	$\bar{4}$·781	−3·219	0·95	0·387	4·4	0·0123
0·03	$\bar{4}$·493	−3·507	1·0	0·368	4·5	0·0111
0·025	$\bar{4}$·311	−3·689	1·1	0·333	4·6	0·0101
0·020	$\bar{4}$·088	−3·912	1·2	0·301	4·7	0·0091
0·015	$\bar{5}$·800	−4·200	1·3	0·273	4·8	0·0082
0·010	$\bar{5}$·395	−4·605	1·4	0·247	4·9	0·0074
0·008	$\bar{5}$·172	−4·828	1·5	0·223	5·0	0·0067
0·007	$\bar{5}$·038	−4·962	1·6	0·202	5·1	0·0061
0·006	$\bar{6}$·884	−5·116	1·7	0·183	5·2	0·0055
0·005	$\bar{6}$·702	−5·298	1·8	0·165	5·3	0·0050
0·004	$\bar{6}$·479	−5·521	1·9	0·150	5·4	0·0045
0·003	$\bar{6}$·191	−5·809				
0·002	$\bar{7}$·785	−6·215				
0·001	$\bar{7}$·092	−6·908				

(F) If logarithmic graph paper has been used, one reads off the dose reducing the surviving fraction to 0·368, or else takes the dose reducing the surviving fraction to 0·1 and divides by 2·3.

These various methods of plotting radiation data are illustrated in Fig. 5, the (hypothetical) data on which they are based being given in Table 25. In the particular example chosen, the 37 % dose is 10×10^3 r. This, then, is the dose needed to produce sufficient ionization in the tissue to obtain an average of one hit per target. The deduction of the 37 % dose is the first stage in calculating the size of the target from the experimental data.

TABLE 25. *Example of the method of working up experimental data for the single-hit type of action*

Dose ($\times 10^3$ r.)	0	0·2	0·4	0·6	0·8	1·0	2
Fraction killed	0·00	0·02	0·04	0·06	0·08	0·10	0·18
Fraction surviving	1·00	0·98	0·96	0·94	0·92	0·90	0·82
$\log_e$ surviving fraction	0·00	$\bar{1}$·98	$\bar{1}$·96	$\bar{1}$·94	$\bar{1}$·92	$\bar{1}$·90	$\bar{1}$·80
Dose ($\times 10^3$ r.)	3	4	5	10	15	20	30
Fraction killed	0·26	0·33	0·39	0·63	0·78	0·86	0·95
Fraction surviving	0·74	0·67	0·61	0·37	0·22	0·14	0·05
$\log_e$ surviving fraction	$\bar{1}$·70	$\bar{1}$·60	$\bar{1}$·50	$\bar{1}$·00	$\bar{2}$·50	$\bar{2}$·00	$\bar{3}$·00

The dose of radiation that suffices to affect an appreciable proportion of the organisms irradiated will usually correspond to the passage of a large number of ionizing particles through each organism. Thus, when irradiating bacteria with α-particles, to kill 50 % of the organisms a dose corresponding to the passage

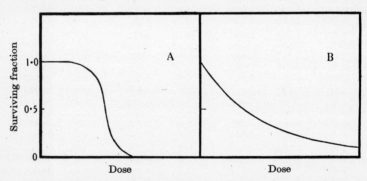

FIG. 6. Survival curves expected for: A, cumulative type of action; B, single ionization, or single ionizing particle, type of action.

through each bacterium of about 100 α-particles is required. On the target theory the lethal action is due to a single one of these hundred α-particles, which happens to go through the sensitive region or target. A superficially more natural explanation would be that the death is the cumulative effect of the chemical change produced in the organism by the hundred α-particles. If the effect were cumulative, however, one would expect a shape of survival curve such as Fig. 6A, in which a dose equal to the mean lethal dose resulted in the survival of 50% of the organisms, while a certain proportion of the organisms were killed by somewhat smaller doses or survived somewhat larger doses, depending on individual resistance. This *sigmoid* shape of survival curve is in contrast to the *exponential* shape Fig. 6B expected when the action is due to a *single-unit action* (i.e. a single ionization or a single ionizing particle), and the experimental realization of an exponential curve instead of a sigmoid curve is an argument suggesting that the effect being studied is due to a single ionization or a single ionizing particle rather than the cumulative effect of many ionizing particles.

When a curve definitely of the type of Fig. 6A is obtained, it does of course rule out the possibility of the cause of death being a single-unit action, but it is not equally certain that an exponential survival curve such as Fig. 6B rules out a cumulative action. For such a curve could be obtained with a cumulative action if the resistance of individual organisms to the radiation varied very widely. There may be found in the literature discussion *ad nauseam* of whether the exponential survival curve obtained, e.g. with bacteria, proves the disinfection to be of the single-unit action type, opponents of this view preferring to ascribe the exponential curve to an extremely skew distribution of resistance. As long as the argument is based only on the shape of the survival curve, the conclusion must be largely subjective, since it depends on whether one regards the *a priori* improbability of the target theory to be greater or less than the *a priori* improbability of the distribution of resistance to radiation among the organisms being of the extremely skew type required. Under these circumstances additional criteria rather than further discussions are called for. One remark, however, we shall add before leaving the subject of shape of survival curve.

The argument for the target theory interpretation clearly increases in force the more exactly the survival curve is exponential, since while the distribution of resistance may be skew, there seems no reason why it should approximate closely to a simple mathematical function. Hence the value of determining the survival curve as accurately as possible. The establishing of a small systematic departure from the exponential curve does not, however, necessarily exclude the target theory, unless certain complicating factors capable of distorting an exponential survival curve have been excluded. Thus if one is irradiating, say, bacteria, and some clumping occurs, then the survival curve will be thereby made slightly sigmoid, since several hits, each on a separate bacterium, will be needed to make the clump incapable of generating a colony. Or if the organisms irradiated are not quite uniform, the more sensitive will be killed more rapidly than the more resistant, and the curve obtained by plotting the logarithm of the surviving fraction against dose will be concave upwards. Or if attempts to extend the experiment to very low survival are made by the use of large doses, then some additional lethal action may become important, such as the production of poisons in the medium, which will make the logarithmic survival curve convex upwards. Very often these complications do not occur. When they do, they naturally weaken the force of the survival curve argument as an indication of the applicability of the target theory to the particular case, but a slight departure from exponential shape should not be regarded as ruling out a single-unit action interpretation when other indications support this interpretation, until the possible existence of complicating factors has been investigated. There are several examples in the literature of the bactericidal action of radiations, in which sigmoid survival curves have been reported, but experiments in which special care has been taken to avoid disturbing factors such as clumping have yielded exponential survival curves.

Time-intensity factor

The fact that the effect in one organism, when it occurs, is due to one ionization, and is not the cumulative effect of several ionizations, means that the manner in which the ionizations are distributed in time has no effect. For each ionization produced in the irradiated tissue has a certain probability of producing the effect, but this probability is uninfluenced by the time elapsed since the last ionization occurred or before the next one will occur. Hence, in the class of action of radiation we are considering, the effect of a given dose is independent of the intensity of radiation and the manner in which it is fractionated. Naturally this statement supposes that the material irradiated remains of constant sensitivity during the period over which the irradiation is spread.

On the alternative view that the effect observed is not due to a single ionization or single ionizing particle, but to the cumulative action of a large number of ionizing particles, we might expect to find that the effect of a given dose depended on the intensity. There are numerous instances in which a biological effect of radiation, not of the single-unit action type, is found to require a larger dose when the radiation is administered at low intensity than when administered at high intensity. The natural explanation is that the organism is capable of recovery from the effects of radiation provided this is administered sufficiently slowly.

The experiment of spreading a given dose over various lengths of time should always be made when investigating an action of radiation suspected to be of the single-unit action type, since to find that the effect of a given dose did indeed vary with the duration over which it was spread would provide a strong argument against the action being of this type. But to find the effect of a given dose to be independent of this duration cannot be regarded as conclusive evidence for a single-unit action, since even with cumulative actions there is usually found a range of durations over which the effect of a given dose is independent of the duration.

Dependence on type of radiation

If the ionizations produced in irradiated tissue were spatially distributed at random, the yield in a reaction, in which a single ionization in the target sufficed to produce the effect, would depend only on the number of ionizations produced per unit volume in the tissue, and would not depend on the wave-length or type of radiation used. Actually the ionizations produced by X-rays or radioactive radiations are localized along the paths of ionizing particles, and ionization can only be produced in the target if it is traversed by an ionizing particle. If the latter is densely ionizing, i.e. produces a large number of ionizations per micron path, then it is found with targets of the size existing in practice that several ionizations are produced when the ionizing particle traverses the target. Since one suffices to produce the effect being studied, the additional ionizations which contribute to the dose, but not to the biological effect, reduce the yield *per ionization*. Thus densely ionizing radiations are less effective in this type of action for equal ionization in the tissue. This is the principal means by which actions due to a single ionization can be recognized, since in some others in which many ionizations are required (e.g. chromosome breakage, Chapter VII) exactly the reverse is true, the densely ionizing radiations being more effective per ionization than the less densely ionizing radiations.

In the single-ionization type of action no appreciable difference is to be expected, however, between γ-rays and different wave-lengths of hard X-rays with targets of the size range commonly found (4–40 mμ diameter). This is true despite the fact that the difference of ion density between the ionizing particles (electrons) produced by a γ-ray of 0·01 A. and an X-ray of 0·1 A. is considerable. The reason is that along the track of the ionizing particle successive primary ionizations are separated by distances greater than the target diameter, so that more than one primary ionization rarely falls in the target. (Many of the primary ionizations are the centres of small clusters of secondary ionization, so that it frequently happens that two or three ionizations composing a single cluster fall in a target. The mean number of secondary ionizations per primary ionization is, however, approximately the same for different wave-lengths of γ-rays and hard X-rays, so that this complication does not introduce any difference in efficiency per ionization.)

Thus the yield of the reaction studied for a given number of ionizations per unit volume of irradiated tissue is practically independent of wave-length over the range of wave-lengths most easily accessible, viz. γ-rays and X-rays down to say 50 kV. It is usually only when soft X-rays of wave-lengths exceeding 1 A. are used, or neutrons or α-rays, that a significant variation is found.

Relation between target size and inactivation dose

Various biological actions of radiations have been interpreted on the target theory for a number of years, though it is only fairly recently that adequate tests for the validity of such interpretations have been applied. The usual aim is to deduce from the observations the size of the target, in the hope of identifying it with some known cell structure.

In the case of actions in which a single ionization in the target suffices, knowledge of the size and shape of the target should

CALCULATION OF TARGET SIZE

enable the 37 % dose to be predicted, since we know sufficiently well the number of ionizations produced per unit volume in tissue by 1 r. of any radiation, and the spatial distribution of the ionizations (numerical information on these points is to be found in the tables of Chapter I). The converse problem of deducing the size of the target from the measured value of the 37 % dose can also be solved if we assume the target to have some simple shape, e.g. spherical. If data are available for the 37 % doses with several different radiations, then the constancy of the estimates of target size obtained from the several radiations affords a highly desirable check of whether the single-ionization theory is applicable to the particular action being studied, and whether the target is sufficiently nearly spherical to be assumed so in the calculation.

As will have been realized from the account given in Chapter I, the spatial distribution of ionization in irradiated tissue is complicated, and an exact calculation is laborious. If the ionizations were produced in the tissue singly and at random, the calculations would be much easier. One approximate method of calculation, which we describe as Method I,[1] is to treat the problem as if the ionizations were in fact produced singly and at random, and we describe below the results given by this method and indicate the conditions under which it may be approximately valid. Another approximate method which has been widely used (Method II)[2] is to take account of the fact that ionizations are localized along the paths of ionizing particles (electrons, protons or α-particles) so that ionization can only be produced in the target when an ionizing particle passes through it, by postulating that the effect occurs *whenever* an ionizing particle passes through the target. Still another method (Method IV)[3] takes account of the fact that when tissue is irradiated by X-rays the ionizing particles (electrons) are liberated by the absorption of X-ray quanta, and regards a 'hit' as the absorption of a quantum in the target.

Each of these methods is valid for certain sizes of target and certain radiations, but each of them leads to serious error when

[1] Crowther, J.A. (1924).
[2] Glocker, R. (1932); Mayneord, W.V. (1934); Lea, D.E., Haines, R.B. & Coulson, C.A. (1936).
[3] Lacassagne, A. & Holweck, F. (1929a); Wyckoff, R.W.G. (1930a,b).

applied outside this range. A method[1] of calculation may be devised (Method III) not subject to this limitation, but valid for any size of target and ionizing particles of any ion-density.

As explained in an earlier chapter, less than half of the total ionization is produced by the primary ionizing particle (electron, proton or α-particle), the rest being secondary ionization produced by secondary electrons (δ-rays) ejected by the primary ionizing particle from some of the atoms it ionizes. All the methods so far catalogued either neglect the distinction between primary and secondary ionizations completely, or else take it into account crudely by considering the secondary ionizations produced by a δ-ray to form a compact cluster.[2] This procedure, while valid for the majority of secondary electrons which are of low energy, fails to allow adequately for the appreciable proportion of the total ionization contained in the tracks of more energetic δ-rays of longer range. A rather laborious calculation can be made to take these δ-rays into account. The details of this calculation are relegated to the Appendix, but in Figs. 8–11 of this chapter the results of the calculation are presented in a form enabling experimental data of inactivation doses to be immediately converted to target size.

We now outline briefly the various methods we have referred to for calculating the target size from the 37% dose.

Method I. If the ionizations were produced in the tissue singly and at random, the dose needed to produce an average of 1 ionization per target of volume v (i.e. the 37% dose) would be that dose which produces $1/v$ ionizations per unit volume in tissue. This procedure for calculating the 37% dose would be valid if ionizations were distributed at random in the tissue. Actually they are localized along the paths of ionizing particles. If no ionizing particle passes through the target then no effect is achieved. If an ionizing particle does pass through, then several ionizations may be produced (in the event of the target dimensions being several times greater than the separation of consecutive ionizations), and hence the dose required for a given biological effect will be proportionately greater than on this method of calculation. The error will be great for large targets,

[1] Lea, D.E. (1940a).
[2] Lea, D.E., Haines, R.B. & Coulson, C.A. (1936); Jordan, P. (1938b); Lea, D.E. (1940a).

and will be in the direction of underestimating the target size, but the method will be correct for targets smaller than the separation of consecutive ionizations in the track of the ionizing particle.

Method II. A method which has also been commonly adopted is to suppose that a 'hit' is obtained whenever the ionizing particle traverses the target. This method will clearly be correct if the target is sufficiently much larger than the mean separation of ionizations along the path of the ionizing particles to make it improbable that the ionizing particles should traverse the target without leaving any ionization in it. This method of calculation will, however, be in error if the target is small, so that it is possible for the ionizing particle to traverse the target without leaving any ionization in it. In such cases, application of this incorrect theory will underestimate the target size.

It is clear that of the two procedures, Method I is valid for very small targets and Method II for very large targets. Attempts have been made to apply one or other of these methods of calculation to predict the variation of biological effect with wave-length of X-rays for targets of given size. The weakness of these attempts lies in the fact that it turns out that the targets of biological interest are often of a size intermediate between the small size for which the first method is valid, and the large size for which the second is valid. Thus when the inactivation of small viruses is being considered, the first method is approximately valid for γ-rays and hard X-rays, while the second method is approximately valid for very soft X-rays. For medium wave-length X-rays the relation between the size of the target and the separation of successive ionizations is such that neither method of calculation is satisfactory. It is evident that any attempt to base a theory of wave-length variation upon one or other of the two methods of calculation would be inaccurate.

Method III. The associated volume method. The present author[1] has given a method of calculation which should be satisfactory over the important intermediate range of target size, while tending to the same results as Method I for small target size, and to the same result as Method II at large target size. The method is illustrated by Fig. 7A, which represents diagrammatically the distribution of ionization produced by an electron

[1] Lea, D.E. (1940 *a*, *b*).

over a small portion of its path. The solid dots are ionizations, some clusters of secondary ionization being indicated. Suppose that to get an effect, e.g. a gene mutation, it is necessary for an ionization to be produced within a spherical target of radius r.

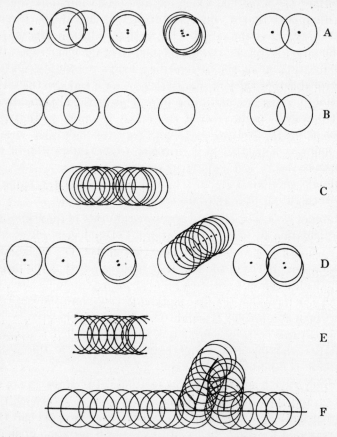

Fig. 7. Illustrating the 'associated volume' method of calculation.

What is the probability of this happening when a dose equivalent to the production of N ionizations per cubic micron is given? The calculation may be effected by the following construction, indicated in Fig. 7A. Round each ionization as centre a sphere of radius r is imagined. These spheres, which in general overlap to a greater or lesser degree, will occupy a nett volume which may be described as the *associated volume*. For mutation to occur, the centre of the target must lie within this associated volume; this is the necessary and sufficient condition for an ionization to be produced in the target. On account of the overlapping the associated volume will be less than $N.\tfrac{4}{3}\pi r^3$ in $1\mu^3$ of tissue. The overlapping is most in the case of the spheres associated with the separate ionizations of a cluster. These ionizations, in a cluster of up to five or six ionizations, are only separated by about $1\,\mathrm{m}\mu.(10^{-7}\,\mathrm{cm.})$, and since the diameters of targets found in examples studied to date range from about 4 to about $40\,\mathrm{m}\mu$, the overlapping in small clusters is practically complete, so that the volume associated with a cluster is only $\tfrac{4}{3}\pi r^3$. The calculation is simplified by assuming this to be true of all clusters, so that the picture simplifies to Fig. 7 B, in which a sphere is shown for each *primary* ionization only. Since the ratio of the total number of ionizations, N, to the number of primary ionizations, n, is approximately $3:1$, there will be $n=N/3$ spheres per unit volume. On account of the overlapping of the spheres belonging to consecutive primary ionizations the associated volume will be less than $n.\tfrac{4}{3}\pi r^3$ and will be, say,

$$\frac{1}{F} n . \tfrac{4}{3}\pi r^3 = \frac{1}{F}\frac{N}{3}\tfrac{4}{3}\pi r^3 \qquad (\text{III-2})$$

per cubic micron. Expression (2) is the sum of the associated volumes of all the ionizing particles in $1\mu^3$ of tissue, and is equal to the mean number of hits per target. The individual associated volumes partly overlap, and the probability p that there will be at least one hit per target, which is also equal to the proportion of targets hit, is given by the following expression:

$$p = 1 - \exp\left(-\frac{1}{F} n . \tfrac{4}{3}\pi r^3\right). \qquad (\text{III-3})$$

F may be calculated [1] in terms of the diameter of the target $2r$ and the mean separation (L) of consecutive primary ionizations in the path of the ionizing particle, and is found to be

$$F = \frac{2\xi}{3}\bigg/\{1 - 2(1-e^{-\xi})/\xi^2 + 2e^{-\xi}/\xi\}, \qquad (\text{III-4})$$

where $\xi = 2r/L$. This function F is tabulated against ξ in Table 26.

We can readily show that for large targets (i.e. $\xi \gg 1$), when the overlapping spheres fuse into a cylinder, (2) reduces to $nL\pi r^2$,

[1] See Appendix, also Lea, D.E. (1940a).

i.e. πr^2 times the total length of track per cubic micron. This is the probability of obtaining a hit on the assumption that a hit is obtained whenever the ionizing particle passes through the target area πr^2. Our method thus reduces, for large targets, to

TABLE 26. The overlapping factor F for values of ξ from 0 to 10 at intervals of 0·1

	0·0	0·1	0·2	0·3	0·4	0·5	0·6	0·7	0·8	0·9
0	1·000	1·038	1·077	1·116	1·156	1·197	1·239	1·282	1·325	1·369
1	1·414	1·460	1·506	1·552	1·599	1·648	1·696	1·745	1·795	1·846
2	1·897	1·948	2·000	2·053	2·106	2·159	2·213	2·267	2·322	2·377
3	2·433	2·489	2·545	2·602	2·659	2·717	2·774	2·832	2·891	2·949
4	3·008	3·067	3·127	3·187	3·246	3·307	3·367	3·428	3·488	3·549
5	3·610	3·672	3·733	3·795	3·857	3·919	3·981	4·043	4·106	4·168
6	4·231	4·294	4·357	4·420	4·483	4·546	4·609	4·673	4·736	4·800
7	4·864	4·928	4·991	5·055	5·120	5·184	5·248	5·312	5·376	5·440
8	5·505	5·569	5·634	5·699	5·763	5·828	5·892	5·957	6·022	6·087
9	6·152	6·217	6·282	6·347	6·412	6·477	6·542	6·607	6·672	6·737
10	6·803									

For values of ξ outside the range covered by the table, the following approximations may be used:

For large ξ: $\quad F = \frac{2}{3}(\xi + 2/\xi)$.

For small ξ: $\quad F = 1 + \frac{3}{8 \cdot 5}\xi + \frac{13}{320}\xi^2$.

what we have described as Method II. For targets so small that the spheres do not overlap, (2) tends to the value $n \cdot \frac{4}{3}\pi r^3$ which is the same result as given by Method I, except for the modification that we take the *cluster* rather than the individual ionization as the effective unit, thus counting $n = N/3$ clusters per cubic micron instead of N ionizations.

We illustrate the three methods of calculation in Table 27 in which experimental data[1] on the inactivation of S 13 Dysentery bacteriophage is analysed in terms of the target theory. The experiments give the inactivation doses for three radiations of widely differing ion-density, namely, α-rays, X-rays (1·5 A.) and γ-rays. These inactivation doses increase in the order of increasing ion-density, the survival curves are exponential, and the effect of a given dose is found to be independent of the intensity at which it is administered. These considerations suggest that we are dealing with a single-ionization type of action, and one proceeds to calculate the size of the target, supposing it to be spherical. Each of the three radiations gives an independent estimate of the target size. In Table 27 the three methods of

[1] Lea, D.E. & Salaman, M.H. (unpublished).

calculation listed above as I, II and III have been used. For the reasons stated Method I is likely to underestimate seriously the target size deduced from the α-ray data, and Method II similarly to underestimate the target size deduced from the

TABLE 27. Target diameters deduced from experimental data on the inactivation of a bacteriophage by radiations: comparison of four methods of calculation

Radiation	γ-rays	X-rays (1·5A.)	α-rays (4eMV.)
Inactivation dose ($\times 10^5$ r.)	5·8	9·9	35
Target diameter in mμ:			
Method I	11·2	10·0	5·9
Method II	3·8	11·3	24·4
Method III	16·3	16·2	24·4
Improved method	15·5	15·9	16·3

γ-ray data. The figures given in the table bear out these anticipations, and Method III which should be free from these errors gives a more consistent set of estimates. The principal remaining discrepancy is in the high value obtained with α-rays. An improved method of calculation discussed later in the chapter, which takes account of δ-rays, removes this discrepancy, as the last line in the table shows.

Method IV. If the target is rather large, and X-rays of very long wave-length giving very short photoelectron tracks are used, the photoelectron has an associated volume which is represented in Fig 7c. In an extreme case this evidently is little bigger than a sphere $\frac{4}{3}\pi r^3$. Thus for very large targets and very short photoelectron tracks the whole photoelectron tracks act as a unit. This is the justification for considering a 'hit' to be the absorption of an X-ray quantum within the sensitive volume, which is the assumption made in Method IV. In practice, it is doubtful if there are any cases of direct action of radiation, of the type where a single ionization suffices, where the target is so large that the associated volume is practically a sphere. The type of geometry illustrated in Fig. 7c, where the photoelectron range is greater, but not many times greater, than the target diameter, is, however, known, and the associated volume in this case is clearly $\frac{4}{3}\pi r^3(1+n/F)$ if there are n primary ionizations in the photoelectron track, thus giving a slightly greater yield per ionization than would be obtained if the ionizing particle produced the same number of ionizations per micron path but was not of restricted length. Clearly the associated volume method tends to the same limit as the single-quantum method in cases (if any) where the latter is applicable.

Secondary ionization

We have not so far taken account of secondary ionization except to point out that *small* clusters of secondary ionization will not add appreciably to the effect of primary ionization with targets of the size found in practice. This will not be true, however, for the more energetic secondary electrons or δ-rays, which produce tracks of appreciable length. Their importance is greatest with α-rays, since with this radiation the total length of δ-ray tracks is greater than the length of the α-ray track itself (Plate I A and Table 15, p. 28). Neglect of the secondary ionization will result in an estimate of target size being obtained in excess of the true size of target.

The proposal was made by Mohler and Taylor[1] and followed up by other authors,[2] that secondary ionization should be taken into account by regarding the α-ray track not as a geometrical line but as a column of radius b. If any part of this cylinder was inside the target of radius r a hit was considered to be scored. The effective target area thus became $\pi(r+b)^2$ in place of πr^2. The value of b was to be determined either from Wilson chamber photographs[3] or from calculations based on the recombination of ions in an α-particle column in air.[4]

On further examination this method is seen to be untenable. The size of column found by either of these methods is determined mainly by the distance an electron too slow to ionize travels before attachment to a neutral atom to form a negative ion. As we have given reasons for believing that the negative ions make no appreciable contribution to the biological effect, their spatial distribution is not relevant to the problem.

There is obviously no difficulty in principle in making an exact allowance for the δ-rays by the method of associated volume. The method in this case is illustrated by Fig. 7 D. Here the usual construction of describing about each ionization as centre a sphere of radius r is applied also to the δ-ray track. A knowledge

[1] Mohler, F.L. & Taylor, L.S. (1934).
[2] Lea, D.E., Haines, R.B. & Coulson, C.A. (1936); Jordan, P. (1938b).
[3] Klemperer, O. (1927).
[4] Jaffé, G. (1913); cp. Chapter II, p. 50.

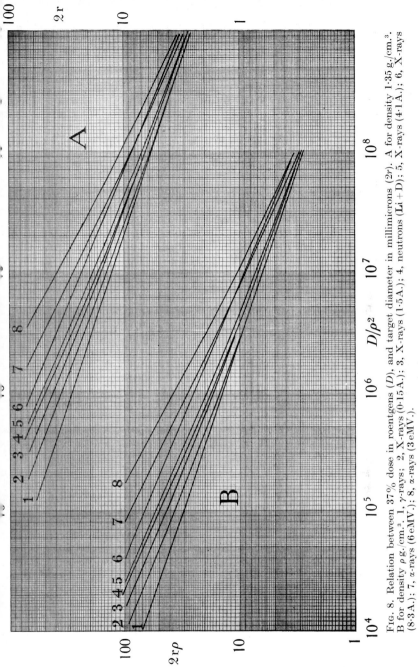

Fig. 8. Relation between 37% dose in roentgens (D), and target diameter in millimicrons ($2r$). A for density 1·35 g./cm.³. B for density ρ g./cm.³. 1, γ-rays; 2, X-rays (0·15 A.); 3, X-rays (1·5 A.); 4, neutrons (Li + D); 5, X-rays (4·1 A.); 6, X-rays (8·3 A.); 7, α-rays (6 eMV.); 8, α-rays (3 eMV.).

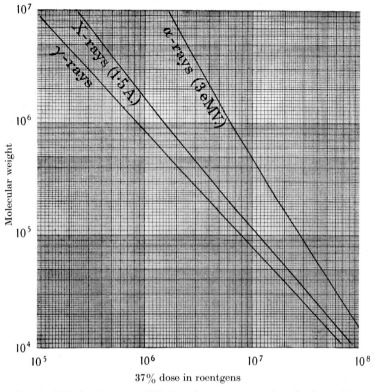

Fig. 9. Relation between 37% dose in roentgens and molecular weight (density 1·35 g./cm.³).

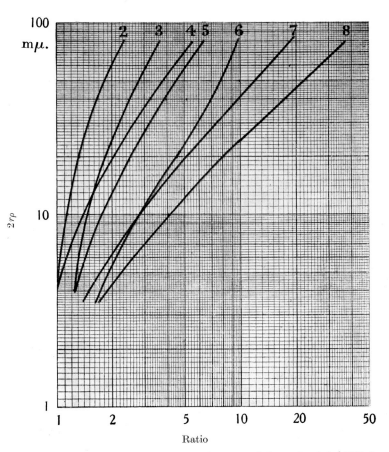

Fig. 10. Dependence upon the target diameter of the ratio of the 37% dose with a given radiation to the 37% dose with γ-rays. 2, X-rays (0·15A.); 3, X-rays (1·5A.); 4, neutrons (Li+D); 5, X-rays (4·1A.); 6, X-rays (8·3A.); 7, α-rays (6eMV.); 8, α-rays (3eMV.).

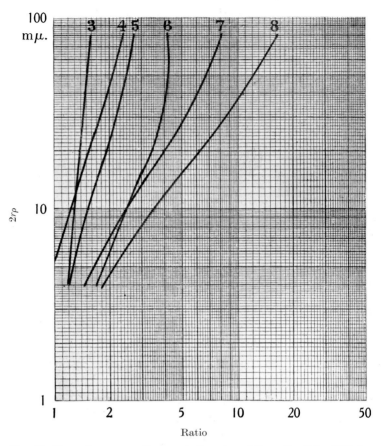

Fig. 11. Dependence upon the target diameter of the ratio of the 37% dose with a given radiation to the 37% dose with X-rays (0·15 A.). 3, X-rays (1·5 A.); 4, neutrons (Li+D); 5, X-rays (4·1 A.); 6, X-rays (8·3 A.); 7, α-rays (6 eMV.); 8, α-rays (3 eMV.).

of the number and energies of the δ-rays is required, this being provided by the tables of Chapter I. With the aid of this more exact method of calculation we have computed the apparent target diameters to α-rays which would be given by the simple method of calculation which neglects consideration of the secondary ionization, for each of a number of (true) sizes of target. The results are given in Table 28, and as expected the 'apparent' target diameters are greater than the true target

TABLE 28. Apparent target diameters to α-rays of 5 eMV.

True target diameter (mμ)	4	10	20	40	80
Apparent diameter	5·6	15·1	30·9	62·8	126·2
Apparent − true	1·6	5·1	10·9	22·8	46·2

diameter. The difference between apparent and true target diameter is also tabulated, and is seen to be by no means constant, revealing the inadequacy of the notion that an α-ray track is to be regarded as a column of ionization of definite diameter b. Rather than the *difference* of the apparent and true target diameters being constant, the *ratio* is seen to be more nearly constant. In particular, there is no justification for the idea[1] that with large targets the apparent target diameter is approximately correct, while with small targets it is made up mainly of the column size and is greatly in excess of the true target size.

The calculation of the relation between target size and 37 % dose for any given radiation, using the associated volume method and taking δ-rays into account, is tedious to perform and tedious to describe. We content ourselves therefore by giving in Figs. 8 and 9 the results of the calculation in the form of curves for several different radiations, relating the 37 % dose to target diameter or molecular weight. An outline of the method of calculation will be found in the Appendix. The result depends on what density is assumed for the material of which the target is composed. Figs. 8A and 9 are calculated for a density of 1·35 g./cm.^{3}, this being about the density of dry protein.[2]

[1] Mohler, F.L. & Taylor, L.S. (1934).
[2] If it is desired to assume some other value for the density, Fig. 8B is used. In this figure D/ρ^2 is plotted against $2r\rho$, D being the 37 % dose in roentgens, $2r$ the target diameter in mμ, and ρ the density in g./cm.^{3} These curves are valid for any value of ρ.

As will appear in Chapter IX, it is sometimes necessary to deduce the target size not from the absolute values of the 37% doses for any one radiation, but from the *ratio* of the values of the 37% dose for two different radiations. To facilitate this type of calculation, Figs. 10 and 11 are provided. If one has experimentally determined the ratio of the 37% doses for some densely ionizing radiation (soft X-rays, neutrons or α-rays) compared to γ-rays, then Fig. 10 enables one to read off the target diameter corresponding to this ratio. If the density of the target material is 1 g./cm.^{3}, the abscissae in Fig. 10 are simply the diameters $2r$ in millimicrons. If the density is ρ g./cm.^{3}, the abscissae are values of $2r\rho$. Fig. 11 gives similar curves for experiments in which X-rays (0·15 A.) have been used in place of γ-rays.

The multi-target theory

The assumption that there is a single spherical target, one or more ionizations within which cause the effect being studied, is plausible as an interpretation of experiments on the inactivation of small viruses, and is found to represent satisfactorily the dependence of inactivation dose on ion-density of radiation. There is one class of action of radiation in which we should expect to find a multiplicity of targets. This is the production of lethal mutations. When *Drosophila* males are irradiated (see Chapter V) and suitable breeding tests subsequently carried out, the production of sex-linked lethal mutations in the irradiated sperm can be established. There are, on genetical grounds, known to be numerous genes residing in the X-chromosome of the irradiated sperm, capable of showing lethal mutation, and a model according to which there are N targets each of radius r, ionization *in any one* of which causes the effect studied, is clearly required in this case. The modification this introduces into the theory is simply that the 37% dose is N times smaller than if there were only one target of radius r.

Apart from its plausibility on biological grounds, evidence indicating that a multi-target model is required is given by the radiation experiments themselves. For, if in the lethal mutation experiments we attempt to interpret the experimental data on the single-target model, we find that consistent estimates for the target size are not obtained with radiations of different ion-density. The estimates obtained with densely ionizing radiations (α-rays or neutrons) exceed the estimates obtained with less densely ionizing radiations (γ-rays or X-rays).

There are some actions of radiation, namely, the killing of large viruses and bacteria (discussed in Chapter IX), in which on biological grounds we have not much guide whether to expect a single or multi-target model to be required. The fact that inconsistent estimates of target size are obtained when experiments with different radiations are used, in the sense of the estimate of size being larger for α-rays than for γ-rays, strongly suggests the necessity for the multi-target theory, with which a consistent representation of the numerical data can be obtained.[1] In this manner radiation experiments are able to contribute to our knowledge of the nature of viruses and bacteria.

α-rays and γ-rays

The sizes of target which are found in practice for actions of radiation interpreted on the single-ionization target theory range from about 4 to 40 mμ in diameter. When a target of this size is traversed by an electron having an energy of several hundred electron-kilovolts, such as is produced by γ-rays or hard X-rays, it will not often happen that more than one cluster of ionization will be left in the target. Thus to a first approximation we can regard the action as due to a single cluster, and the 37% dose is that dose which produces an average of one cluster in a volume equal to the target volume. Thus *γ-ray experiments determine the total volume of the sensitive region*. This remains true even if the target is not spherical or if there is a multiplicity of targets, any one of which ionized causes the effect.

If an α-particle traverses a target of the size mentioned, a large number of ionizations are inevitably produced in the target. Hence we can regard every passage of an α-particle through a target as effective, and the 37% dose is that dose which corresponds to an average of one α-particle crossing each area equal to the area of the sensitive region. Thus *α-ray experiments determine the total cross-sectional area of the sensitive region*,

[1] A consistent representation could also be obtained on the assumption of a single target in the form of a filament of width small compared with its length, or of a lamina of thickness small compared with its other dimensions.

this statement again remaining true for non-spherical targets or for multiple targets. In the case of non-spherical targets we mean by cross-sectional area the average area presented to the ionizing particles, allowing for the presumably random orientation of the targets relative to the ionizing particles.

These italicized statements are of course only approximations, and it is intended that not these statements but the curves of Figs. 8–11 should be relied upon in interpreting experiments using α-rays and γ-rays as well as radiations of intermediate iondensity. The italicized statements are, however, of value in that they bring out the physical principles underlying, for example, the use of α-ray and γ-ray experiments to determine both the number and the size of the targets when the multi-target model is being used.

For, suppose there are N spherical targets, each of radius r, and suppose that the α-ray experiment has determined their total cross-sectional area to be A, and the γ-ray experiment has determined their total volume to be V. Then

$$N.\pi r^2 = A, \quad N.\tfrac{4}{3}\pi r^3 = V, \qquad \text{(III-5)}$$

whence, solving these equations,

$$2r = 3V/2A, \quad N = 16A^3/9\pi V^2. \qquad \text{(III-6)}$$

Since the formula for N involves the square and cube of V and A, it is strongly sensitive to experimental error in the determination of the 37 % doses.

The assumptions of the single-ionization target theory[1]

Calculations of the relation between target size and 37 % dose for any given type of radiation involve approximations and simplifications necessarily introduced to make the calculations manageable. Some of these simplifications concern the spatial distribution of ionization, the calculation being made on the assumption of a spatial distribution which is less complex than the actual distribution. The method of calculation outlined in the Appendix, which leads to the curves of Figs. 8–11 relating target size and 37 % dose, has, at the expense of a considerable increase in the labour of computation, avoided most of the approxima-

[1] These considerations are to some extent suggested by an article of Fano, U. (1942).

tions entering into previous calculations. In particular, account is taken of secondary ionization, and of the fact that the number of ionizations per micron path of an electron increases towards the end of the path. It is believed that this calculation is fairly adequate as regards the manner in which it deals with the spatial distribution of ionization, though liable of course to revision when better numerical physical data become available.

The calculation is based on the assumption that an effect to which the single-ionization target theory applies is produced when one or more ionizations occur inside a spherical target of definite radius r, and is not produced when ionizations occur outside the target. While the notion of a sensitive region is of course basic in the target theory, the actual conditions may be more complicated than those allowed for in the calculation, in any of the following respects:

(a) The target may not be spherical.

(b) When an ionization is produced inside the target, the probability of the effect studied ensuing may be less than unity.

(c) There may not be a sharp demarcation between the inside and the outside of the target in the manner envisaged by the simple theory, according to which the probability of an ionization causing the effect is unity if the ionization occurs just inside the target, and zero if it occurs just outside. Instead, there may be a diminution of the probability as we move out of the target without any discontinuity marking a precise boundary.

It is not possible to make a quantitative calculation taking these three factors into account, since we lack the necessary numerical data. It is, however, possible to see qualitatively in what sense the deductions of the simple theory will be in error by neglect of these factors. Further, by examining the extent to which the simple theory is capable of giving a consistent interpretation of experimental results, it is possible to form an opinion of the extent to which the deductions of the simple theory will be in error by neglect of these factors. We shall find that the error is not very great, so that the method of calculation we have outlined, while admittedly a simplified model, gives results which are essentially correct.

There are, of course, precedents in all sciences for a model which is a simplification of the true state of affairs giving a satisfactory representation of the essential experimental facts. The assumption made in the kinetic theory of gases that molecules are hard spheres, and the concept in genetics that phenotypical characters are determined by single Mendelian characters, are familiar examples.

We proceed to examine the effects of factors (a), (b) and (c) separately.

(a) *Non-spherical target*. The effect of shape of target can be illustrated by taking the extreme case where the target, instead of having a spherical shape, is in the form of a long thin filament, i.e. a cylinder of length $2b$ and diameter $2a$, the ratio $f = 2b/2a$ of length to breadth much exceeding unity. Instead of attempting the general solution for a radiation of any ion-density, it will suffice for our purposes to take two extremes, viz. an ionizing particle producing its ionizations (or ion-clusters) at such wide separations that two clusters practically never fall in the target, and an ionizing particle so densely ionizing that ionization is certainly left in the target whenever the ionizing particle passes through it. We shall for the present purpose neglect δ-rays. The experiment with the densely ionizing radiation will in effect measure the *area* presented to the radiation by the target. For an ionizing particle parallel to the axis of the cylindrical target this is πa^2; for a particle perpendicular to the axis it is $4ab$. For randomly orientated cylindrical targets of length much exceeding the breadth we can readily show that the mean area presented is πab. If now we interpret experiments made on a material in which the sensitive volume is a long filament on the incorrect assumption that it is a sphere (which presents an area πr^2), we shall deduce for the radius of the sphere a value r where $\pi r^2 = \pi ab$, i.e. $r = af^{\frac{1}{2}}$.

Experiments with the radiation of low ion-density will in effect measure the volume of the target, i.e. $2\pi a^2 b$. If we incorrectly assume the target to be spherical, we shall deduce for it a radius r, where $\frac{4}{3}\pi r^3 = 2\pi a^2 b$, i.e. $r = a(3f/2)^{\frac{1}{3}}$. This is smaller than the previous estimate in the ratio $1 : 0.87 f^{\frac{1}{6}}$. Thus the incorrect assumption of a spherical target when the target is actually filamentous will lead to the estimate of target radius deduced from experiments made with different radiation being inconsistent. The target radius deduced from the experiments made with a densely ionizing radiation (e.g. α-rays) will exceed that deduced from a radiation of low ion-density (e.g. γ-rays)

ASSUMPTIONS OF THE TARGET THEORY 95

in the ratio of $0.87 f^{\frac{1}{3}} : 1$. For a filamentous target 10 times as long as broad this is $1.28 : 1$, for a filamentous target 100 times as long as broad it is $1.87 : 1$. It is evident that *small* deviations from a spherical shape will not prevent the calculation made on the assumption of a spherical target giving a fair representation of the dependence of the efficiency of different radiations on the target size and ion-density of the radiations.

In certain applications of the target theory, particularly the production of lethal mutations (Chapter v), we consider the cell to have a large number of spherical targets and use the experimental data to calculate the number and size of these targets. We could fit the experimental values of 37 % dose for different radiations equally well on the assumption of a single *filamentous* target, or a small number of filamentous targets. On the one interpretation we use the experiments to determine the number of genes capable of showing lethal mutation, on the other we determine the length of chromosome they occupy. These are of course related quantities.

To conclude: if we have reason to believe that a particular action of radiation is of the single-ionization type, and yet find that when we work out the size of the target from experiments made with different radiations we fail to get consistent results, but the estimates increase in the order of increasing ion-density, then a possible explanation is that the target is filamentous, or else that there are a number of spherical targets, ionization in any one of which causes the effect studied. Examples will be found in Chapter IX illustrating this point.

(b) *Probability less than unity*. We imagine the target to be spherical and of definite radius r. Only ionizations produced within the target can cause the effect studied, but an ionization within the target is not certain of causing the effect but has a certain probability less than unity of doing so. Clusters of different size will have different probabilities. Thus if the probability of a single ionization causing the effect is 0.5, the probability of a cluster of two ionizations doing so will be 0.75, of three ionizations will be 0.875, and so on. Call p the average probability of causing the effect for one cluster in the target. When a radiation is used which produces clusters of ionization widely separated along the track of the ionizing particle, e.g. γ-rays, radiation experiments normally give us the target volume. Here they will give us $p \cdot \frac{4}{3}\pi R^3$, where $\frac{4}{3}\pi R^3$ is the target volume. If we incorrectly neglect the factor p we shall calculate an estimate of the target radius r which will be connected to the real value R by the relation $p \cdot \frac{4}{3}\pi R^3 = \frac{4}{3}\pi r^3$, or $r = Rp^{\frac{1}{3}}$.

A sufficiently densely ionizing particle passing through the target will be certain to cause the effect. Hence experiments with such radiations, e.g. α-rays, will yield a correct estimate for the target radius.

The existence of the probability p less than unity will thus lead to estimates of the target size being greater for α-rays than for γ-rays, the α-ray value being more nearly correct.

For a radiation of intermediate ion-density, for which it is necessary to use the method of calculation involving the overlapping function F (compare p. 85), we find that the introduction of the probability p leads to the result that the 37 % dose, i.e. the dose required for an average of one effective hit per target, is $\dfrac{3F(\xi)}{p \cdot \frac{4}{3}\pi r^3}$, where $F(\xi)$ is the function of ξ defined by equation (4) but $\xi = 2rp/L$ instead of $2r/L$. It follows that if we are investigating an action of radiation of the lethal mutation type, in which we have a large number of spherical targets, and determine their size not from the 37 % dose determined for any one radiation, but from the ratio of the doses of two radiations of different ion-density, then our neglect of the factor p will lead to our estimate of the target diameter being too small in the ratio $p : 1$, and our estimate of the number of targets too large in the ratio $1 : p^2$.

(c) *No sharp boundary to the target.*[1] Instead of the probability of an ion-cluster causing the effect being unity inside the target and zero outside, we suppose that the probability diminishes as we move outwards from the target without any discontinuity. In Fig. 12 we plot the probability p of an ion-cluster being effective as a function of its distance x from the centre O of the target. The normal hypothesis is that the probability is unity up to radius r and zero beyond, as indicated by the square-cornered curve A. Our present hypothesis can be represented by the smooth curve B. The curve we have chosen for illustration is the normal error curve, because it has the right general shape and is amenable to calculation. With a radiation which produces its

[1] Refer to the Appendix for fuller details.

ion-clusters at wide separations (e.g. γ-rays) the experiment normally determines the total volume $V = \frac{4}{3}\pi r^3$ of the target. On the present model it will determine $\int p\,dv$, i.e. the volume integral of the probability. We have drawn the two curves A and B of Fig. 12 so that $\int p\,dv = \frac{4}{3}\pi r^3$, i.e. so that if in a particular instance there is no sharp boundary to the target, and the curve B represents the variation of probability p with distance from

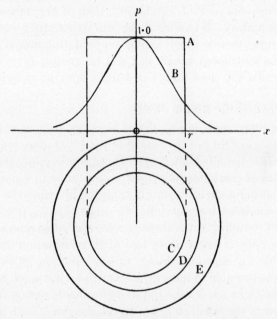

Fig. 12. Target with indefinite boundary.

the centre, then the radius r marked by the square-cornered curve A is the radius of target which will be deduced when the experiment with γ-rays is interpreted on the assumption of a target of definite radius.

When more densely ionizing radiations are used, and the experimental data interpreted in terms of the usual theory which supposes a definite boundary for the target, the radius calculated for the target increases as the ion-density of the radiation increases. Thus the circles C, D, E in Fig. 12 represent the sizes of target to be expected for γ-rays (i.e. radius r), and two more densely ionizing radiations. D is calculated for an ion-density such that an average of five ion-clusters are produced in path length $2r$, and E for an ion-density five times greater.

The more densely ionizing radiation of the two last mentioned is, per ionization, the less effective. According to the simple theory with a target of fixed radius r, it is 4·63 times less effective. On the model represented by the probability curve B of Fig. 12 it is 2·75 times less effective. Thus the absence of a definite boundary to the target will make the efficiency per ionization fall off less rapidly with increase of ion-density than if the target has a definite boundary. If in such a case the target size is calculated on the simple theory which assumes a definite boundary to the target, the calculated target size will be greater for a densely ionizing radiation than for a less densely ionizing radiation.

The validity of the simple model

That a curve such as Fig. 12 B is likely to exist in practice is due to the possibility of the spread of the effect of an ionization, such as was discussed on p. 67. If we knew more about the mechanism of this spreading effect we would be in a position to decide whether a curve deviating considerably from the square-cornered curve, such as curve B, or a curve differing from A only by a slight rounding of the corners, were needed to represent this spreading effect. Again, if we had more information about the ionic yields for chemical change in large protein molecules we would be better able to judge whether values of p much less than unity were to be expected,[1] and so cause the deviation from the simple theory discussed on p. 95. In the absence of such information the following consideration is helpful. All three of the complicating factors we have considered act in the same sense, namely, that when target sizes are calculated with different radiations, the estimate obtained from α-ray data exceeds that from γ-ray data. Or, if a lethal mutation experiment is being interpreted on the multi-target theory, the number of targets will be overestimated. Now cases are known where consistent estimates of target size are obtained from α-ray and γ-ray data. These cases are the inactivation of the small viruses discussed in Chapter IV. Since all the complicating factors act in the same direction, we cannot explain this agreement with the simple

[1] The very limited information at present available suggests that p is approximately unity even for large protein molecules (p. 38).

theory as fortuitous, and due to the different complicating factors happening to cancel each other out. *The complicating factors are evidently not of sufficient importance to invalidate the simple theory.*

A number of other cases are known which, when interpreted in terms of a single spherical target, yield estimates of target size which are greater for α-rays than for γ-rays. This discrepancy may be due either to a multiplicity of targets, or to any of the three complicating factors we have been discussing. Among these cases is the production of lethal mutations in *Drosophila* sperm. Here we know from genetical evidence that there are numerous genes capable of showing a lethal mutation. When we estimate their number on the multi-target theory, neglecting complications, the estimate obtained is, on genetical grounds, entirely reasonable. There is no suggestion from these data that neglect of the complicating factors has led to a gross over-estimate of the number of targets. We are encouraged therefore to believe that the complicating factors which, if they were of sufficient numerical importance, would cause the estimate of the number of targets on the multi-target theory to be exaggerated, do not in fact cause gross error.

Chapter IV

THE INACTIVATION OF VIRUSES BY RADIATION

The viruses[1]

The viruses are agents responsible for a number of infectious diseases in animals and plants. In addition, the bacteriophages, which attack bacteria, are also considered to be viruses. The viruses are distinguished from bacteria and other microbes by their small size and their purely parasitic nature. Nearly all the viruses are too small to be seen by the ordinary microscope, and they pass through the filters of unglazed porcelain or kieselguhr commonly used to sterilize liquids by filtering off bacteria and other small organisms. They will only multiply on or in living cells, and have not so far been made to multiply in a non-living medium in the manner in which bacteria can be grown in broths of suitable composition, or the cells of higher organisms cultured outside the tissue to which they normally belong.

The viruses may be described as occupying an intermediate position between the obviously living and the obviously non-living, since they have some properties which we ordinarily think of as peculiar to life, and others which are in marked contrast to living organisms. Their capacity for reproduction is the characteristic in which they resemble living organisms. The fact that some of the smallest viruses are crystallizable is the most striking difference. Plate II D, for example, shows crystals of the virus responsible for the 'bushy stunt' disease of tomatoes.[2] Tobacco necrosis virus has also been crystallized, while tobacco mosaic virus has been obtained in a pseudo-crystalline state (differing from a true crystal in that the molecules, which are rod-shaped, are arranged regularly in two dimensions but without regularity in the third dimension, which is the direction of the axes of the rods). The crystals are obtained from concentrated and purified suspensions of the virus. It is possible that

[1] This brief introduction is intended for the convenience of the reader who is not a virus specialist. It may be supplemented by reading Smith, K.M. (1940), *The Virus, Life's Enemy*; Bawden, F.C. (1943), *Plant Viruses and Virus Diseases*; the chapter on viruses in Northrop, J.H. (1939), *Crystalline Enzymes*; and the chapter on bacteriophage by Delbrück, M. in Nord, F.F. & Werkman, C.H. (1941–3), *Advances in Enzymology*.

[2] Smith, K.M. & Markham, R.

other small viruses will be obtainable crystalline when methods have been devised for producing sufficiently pure and concentrated suspensions, but it is very improbable that this will happen with the larger viruses.

The crystalline viruses are, chemically, nucleoprotein. Nucleoprotein is a component of living cells, being the substance of which chromosomes are constituted. The chromosomes, which are threads of nucleoprotein, constitute the genetical mechanism of the cell. It is because the sperm and the egg each contribute their quota of chromosomes to the fertilized egg that the organism which develops from it inherits characters from both male and female parent. In the chromosomes are located the genes, the physical entities corresponding to the Mendelian characters. Changes in individual genes, or in their number or arrangement in the chromosome, cause changes in the behaviour of the cell, and it is the genes which largely determine the potentialities for development and behaviour of the cell. At cell division the chromosomes split longitudinally, each gene exactly reproducing itself. The two halves of each chromosome separate, one going into each daughter cell. It is by this mechanism that each daughter cell obtains exactly the same complement of genes as the mother cell. Evidently a gene has the capacity for synthesizing from the cell fluids an exact replica of itself.

Now the viruses have this property of synthesizing exact replicas of themselves from the cell fluids of the appropriate host plant or animal. The smallest, crystallizable, viruses are molecules of nucleoprotein, as are presumably the genes. There is thus a strong analogy, chemically and in behaviour, between these viruses and genes, and if we regard them as living things, they are not to be thought of as minute cells so much as naked genes. (This does not apply to the largest viruses which probably are minute cells.)

Radiation experiments are consistent with this view of the nature of the crystallizable viruses. The smallest viruses can be inactivated by a single ionization almost anywhere in the virus particle. There are some actions of radiation on higher cells which can also be produced by a single ionization, providing that it occurs in the right part of the cell. The best established of these actions is the induction of gene mutation.

Larger viruses show some internal structure when examined under the electron microscope, as shown in the case of vaccinia

Plate IV D,[1] and probably resemble higher cells in that
...tically important nucleoprotein comprises only a part
...irus.

The sizes of the viruses

Knowledge of the sizes of the viruses is of some importance in the interpretation of experiments on their inactivation by radiation. The methods available for the estimation of the sizes of viruses[2] are, in addition to ultra-violet light photography, which is only possible with the largest viruses, sedimentation on high-speed centrifuges, diffusion, filtration through graded collodion membranes, electron micrography, and X-ray diffraction. Determination of the density is necessary in interpreting sedimentation results, and viscosity measurements are sometimes helpful.

There are a few viruses for which consistent estimates of size have been obtained by several different methods, and the sizes of these can probably be considered known to an accuracy of 10 % or better. These are the viruses of tomato bushy stunt, tobacco mosaic, vaccinia, and the Shope rabbit papilloma. All four of these viruses appear to be hydrated in solution, i.e. to take up water with resulting increase of size and diminution of density. When dried, the water is lost without permanent loss of infectivity, the reduction of diameter on drying being from 15 to 30 %. It is a natural presumption that hydration of viruses in solution is a general phenomenon.

In the case of a number of viruses which have been used in radiation experiments, the sizes have so far only been estimated by filtration. Filtration through Elford's 'gradocol' membranes[3] is capable of yielding consistent estimates of the average pore diameter (a.p.d.) of the filter which just stops all the virus. The principal source of uncertainty attaching to this method is the ratio of virus diameter to a.p.d. Elford has recommended the use of the ratio $\frac{1}{3}-\frac{1}{2}$ for a.p.d.'s of 10–100 mμ, and $\frac{1}{2}-\frac{3}{4}$ for a.p.d.'s of 100–500 mμ. It appears, however,[4] that these factors are too

[1] Green, R.H., Anderson, T.F. & Smadel, J.E. (1942).

[2] For a review of the methods see Markham, R., Smith, K.M. & Lea, D.E. (1942, 1944).

[3] See an article by Elford, W.J. in Doerr, R. & Hallauer, C. (1938), *Handbuch der Virusforschung*.

[4] Markham, R., Smith, K.M. & Lea, D.E. (1942, 1944).

A. Vaccinia virus titrated on rabbit.

B. Bacteriophage plaques.

C. Local lesions (tobacco necrosis virus).

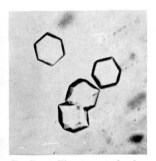

D. Crystalline tomato bushy stunt virus.

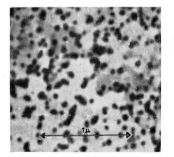

E. Electron micrograph of tobacco necrosis virus.

PLATE II. Viruses

low, and a factor of 0·83 is suggested by Table 29 in which the filtration end-point is given for several viruses whose hydrated diameters are reliably determined by other methods.

When making use of the published results of filtration experiments to determine the sizes of viruses, we have recalculated the sizes on the basis of the factor 0·83 relating the diameter of the hydrated virus to the a.p.d. of the filter membrane.

TABLE 29. The filtration of viruses
(Diameters in millimicrons)

Virus	Hydrated diameter	a.p.d.	Ratio
Phage S 13	18	25	0·72
Bushy stunt	37	40	0·92
Shope papilloma	73	70	1·04
Phage *Staph*. K	73	110	0·66
Vaccinia	200	250	0·80
		Mean ratio	0·83

A number of phages have been investigated by Schlesinger by the sedimentation method,[1] and the size calculated on the assumption of a phage density of 1·10 or 1·12. We have recalculated these sizes using instead the density of 1·22 now believed to be more probably correct.[1]

For the purposes of the interpretation of the radiation experiments, the dry size is more significant than the wet size. This is obvious when the experiment has been made on the dry virus. It is also so when the experiment has been made in solution, since there are indications (see p. 111) that the ionic yields for inactivation by ionization of the virus protein itself is considerably greater than the ionic yield for indirect inactivation following ionization of water. When the size has been determined by a method, such as filtration, which gives the size of the hydrated particle, it is therefore necessary to deduce from this the size of the unhydrated particle. The phages have, when hydrated, a density of 1·22. We have arbitrarily assumed that when dry they have a density of 1·35, this being a typical figure for the dry density of such plant and animal viruses as have been obtained sufficiently pure for the determination of density.

It will be realized that there is some uncertainty attaching to the sizes of most of the viruses. In the list of sizes included in

[1] Elford, W.J. in Doerr, R. & Hallauer, C. (1938).

Table 33 we have prefixed by the sign ~ the sizes which are regarded as least reliable, either owing to their being based on a single method of size determination, or on two methods which have not agreed well.

The estimation of virus activity

For experiments such as the investigation of the inactivation of viruses by radiations, a quantitative method of assessing virus activity is needed. Since viruses cannot be grown on artificial media, the test of activity necessarily involves inoculating the appropriate sensitive organism with the virus under test, and the methods of test are thus different for plant viruses, for animal viruses, and for bacteriophages. The quantitative assessment is most accurate in the case of the bacteriophages. The method used is as follows. A plate of a solid nutrient medium suitable for the growth of the bacterium concerned is inoculated with a drop of bacterial suspension containing a few million organisms, together with a measured drop of the phage suspension, and the liquid uniformly spread over the surface of the nutrient medium with a glass spreader. On incubation of the plate, the heavy bacterial inoculum results in a continuous film of bacterial growth on the surface of the plate, except for clear areas caused by the multiplication of the individual phage particles causing lysis of the bacteria in their immediate vicinity. These clear areas or 'plaques' are easily countable with the naked eye, and are shown in Plate II B. The number of plaques is proportional to the strength of the phage suspension (providing an excessive phage inoculum is not used, in which case the plaques will run together), and is not far short of the number of individual phage particles put on to the plate.[1]

As in any counting method the precision obtainable is limited by the number of plaques counted. The estimation of a phage concentration based on a count of n plaques is subject, on this account alone, to a standard deviation of $100/n^{\frac{1}{2}}$ %. In practice, further uncertainties are introduced by errors, e.g. in the dilution of the phage suspension to a concentration suitable for inoculation, and by variation in the size of drops delivered by the standardized dropping pipette used to measure the inoculum, so that the standard deviation obtained in practice is likely to be

[1] Ellis, E.L. & Delbrück, M. (1938) suggest that the ratio is about 1:2.

about twice the theoretical minimum value. When the assessment is made on the average count of three or four plates each containing about a hundred plaques, an accuracy of about 10 % is commonly obtained.

The purely statistical error, determined by $n^{\frac{1}{2}}$, may be reduced in methods[1] in which a culture of bacteria in a liquid nutrient medium is inoculated with a quantity of phage suspension containing from 10^4 to 10^6 phage particles. The lysis of the bacterial culture is followed by observation of its diminishing opacity, and the time required for a standard culture to be lysed to a given degree is determined. Exact standardization of all conditions is of course imperative in this method, and it is in less general use than the plaque counting method.

Plant viruses are most easily assessed by the local lesion technique. In this method a drop of the virus suspension is rubbed on to a leaf of the susceptible plant with a glass spatula. After a few days local lesions appear (as in Plate II c),[2] these marking the points where virus has entered the leaf. The method is not universally applicable, since with some viruses and some host plants the plant may become infected without the appearance of local lesions at the points of entry. It appears that a local lesion can be caused by the entry of a single virus particle, but determination of the weight of a pure virus preparation required to be rubbed on to a leaf to give one local lesion makes it evident that the number of lesions obtained on a leaf is only a minute fraction of the number of virus particles rubbed on to the leaf. The number of local lesions obtained with a given quantity of virus depends on factors such as age of plant, size of leaf, technique of the operator, and whether or not carborundum powder or sand has been used to increase, by abrasion, the number of possible points of entry on the leaf.

It will be realized, therefore, that while the lesion count is a measure of the activity of a virus preparation, the assessment is liable to considerable experimental error. This is reduced to a minimum by use of suitable procedures. One such is the Latin square method, in which, for example, if the activities of five preparations are to be compared with the aid of five plants each bearing five leaves, the inoculations are arranged so that each virus preparation is inoculated once on to each plant and once in

[1] Krueger, A.P. (1930). [2] Smith, K.M. & Markham, R.

each leaf position. A rather simpler procedure, when determining, for example, the activity remaining after several different exposures of radiation, is to inoculate half of every leaf with the unirradiated virus, and the other half with the irradiated virus. This method depends on the fact that the two halves of a leaf are more nearly alike in sensitivity than different leaves. By the use of about ten half-leaves to each virus preparation, an accuracy of about 20 % in the estimation of relative virus concentration may be expected.

The estimation of animal virus activity may be exemplified by describing the assessment of vaccinia virus. When a suspension of this virus is introduced into the skin of a rabbit by means of a hypodermic needle, a red swelling appears in the course of a few days at the site of the inoculation. To compare the activities of a series of virus preparations, the hair is clipped from the back and flanks of a rabbit and the skin marked out into squares, say six rows of eight squares each. Each row serves to test one virus preparation. A series of eight tenfold or threefold dilutions of the suspension are made and a measured volume of each dilution injected. The most dilute inoculum will contain no virus and give rise to no lesion, the higher concentrations will give lesions. The highest dilution at which a lesion appears is a measure of the strength of the virus preparation. Plate II A shows the flank of a rabbit carrying two rows of squares.[1] The highest concentration in each row is inoculated at the head end of the animal, and produces the largest lesion. The end-point in this particular experiment was reached at the fifth dilution of the series.

A single virus particle suffices to produce a lesion,[2] and the degree of variation of the end-point obtained when the same virus preparation is inoculated in several independent dilution series is no greater than that statistically inevitable.[3]

Technique of virus irradiation experiments

If the virus is to be irradiated wet, a tube of the suspension is exposed to the radiation and samples taken at intervals and the activity assessed by the appropriate methods, as outlined in the previous section. A control preparation is maintained under conditions as comparable as possible, apart from being irradiated,

[1] Salaman, M.H.   [2] Parker, R.F. (1938).
[3] Lea, D.E. & Salaman, M.H. (1942).

and is also sampled at the beginning and end of the experiment to check the possibility of spontaneous inactivation occurring. In this way a curve is obtained (e.g. Fig. 13) of virus activity against dose, and from this may be read off the dose required for

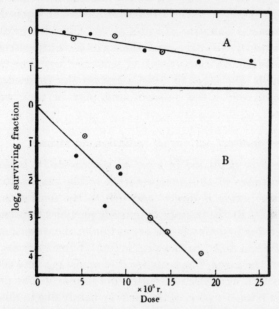

FIG. 13. The inactivation of plant viruses by radiations. ● Virus irradiated dry; ⊙ virus irradiated in aqueous suspension. A, X-rays (1·5A.) on tobacco necrosis virus; B, γ-rays on tomato bushy stunt virus (Lea & Smith).

any percentage reduction of activity. It is strongly to be recommended that this procedure be adopted, rather than the determination of the dose which completely inactivates a virus preparation.

Experiments on suspensions are liable to be complicated by an indirect inactivation following ionization in the water rather than in the virus particles, and if it is desired to study the direct effect, irradiation of the virus dry is to be recommended. This procedure also has the merit of enabling soft X-rays and α-particles and other easily absorbed radiations to be used. Fortunately, many of the viruses can be dried without loss of activity. The addition of ½% of lactose to the solution before drying may be advantageous in the case of a virus liable to loss of activity on drying.

A method of working with dry viruses[1] which has been found satisfactory is the following. A drop of molten 5% gelatin is spread over a cover-slip and dried off on a hot plate. A drop of a virus suspension is put on to the cover-slip with a calibrated dropping pipette and allowed to dry, leaving a circular deposit 6–8 mm. in diameter. After irradiation the cover-slip is dropped into a measured quantity of water at 40° when the gelatin film almost instantly dissolves, and the virus is resuspended. Providing the original virus suspension contained not more than say 1% of solids, the layer of dried virus on the cover-slip has a thickness of only a few microns and may be used with soft radiations.

Direct and indirect actions of radiation on viruses

If a virus is irradiated in aqueous suspension, it is clearly of first importance in the interpretation of the results to decide whether the action is direct, and due to the ionization of the virus particles by the passage of ionizing particles through them, or indirect due to ionization or excitation of the water molecules leading, for example, to the production of free radicals, which then affect the virus. The tests for distinguishing between these types of action have been discussed in Chapter II. The principal test is that if inactivation of the virus is mainly due to the formation of 'activated water', the inactivation dose[2] should increase with increasing concentration of the virus, and should also be increased by adding protective agents to the solution, which are capable of competing with the virus for the activated water. If the action is a direct action on the virus without the intermediary of activated water, the 37% dose should be independent of the concentration of the virus or of the addition of protective agents.

Luria and Exner[3] found that when bacteriophages were irradiated in aqueous or saline suspension the rate of inactivation was more rapid than in broth suspension. Addition of gelatin to the aqueous suspension reduced the inactivation rate to the value found in broth. The indication is that in aqueous suspen-

[1] Lea, D.E. & Salaman, M.H. (1942).
[2] Defined as the dose reducing activity to 37% of the initial activity (see p. 74).
[3] Luria, S.E. & Exner, F.M. (1941).

sion the inactivation is partly indirect, presumably by the intermediary of the same activated water which is responsible for the chemical effects of radiation in solution, but that broth or gelatin protect the virus by competing for the activated water.

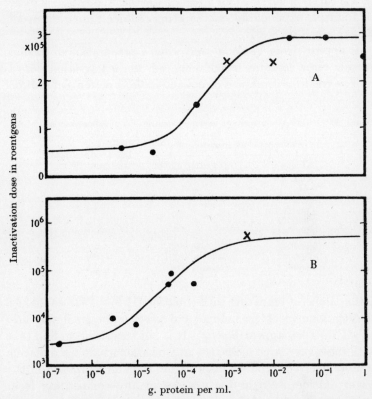

Fig. 14. Dependence of inactivation dose on protein concentration. ● Mainly virus protein; × mainly extraneous protein. A, Tobacco mosaic virus (Lea, Smith, Holmes & Markham); B, rabbit papilloma virus (Friedewald & Anderson).

Somewhat similar results have been obtained[1] with a purified preparation of tobacco mosaic virus. The preparation was irradiated in several different concentrations and also dry. The inactivation dose was the same dry and in concentrated solution, but was less in dilute solution. The addition of gelatin to the dilute solution resulted in the inactivation dose being raised almost to the value for the dry preparation. The results are exhibited in

[1] Lea, D.E., Smith, K.M., Holmes, B. & Markham, R. (1944).

Table 30 and in Fig. 14 A. The conclusion is that the indirect action can be largely inhibited by use of a sufficient concentration of protein, whether virus protein or gelatin. In dilute solution, however, the indirect action predominates. The shape of the survival curve is not a test of whether the action is direct or indirect, exponential curves being obtained in either event, as explained on p. 61.

It appears from Table 30 that when a preparation of tobacco mosaic virus contains 0·00022 mg./ml. the inactivation dose is half as great for combined direct and indirect action as for direct

TABLE 30. *Inactivation doses of various preparations of tobacco mosaic virus*[1]

Concentration in g./ml. of		Inactivation dose
Virus	Protective agent	(in units 10^5 r.)
Solid	—	2·5
0·14	—	2·9
0·022	—	2·9
0·00022	—	1·5
0·000022	—	0·5
0·0000044	—	0·6
0·000022	Glucose 0·05	0·5
0·000022	Gelatin 0·001	2·4
0·000022	Gelatin 0·01	2·4

action alone. Comparing with equation (II-9) (p. 62) we deduce that the ionic yields for indirect and direct action are in the ratio of $\gamma/\Gamma = 1/4000$ approximately. It is not surprising to find that with these very large molecules an ionization inside the molecule is more likely to produce inactivation than an ionization in the water outside, even in the absence of any competition from protective agents.

Less purified virus suspensions, containing foreign proteins and other impurities capable of exerting a protective action, are found to be inactivated at the same rate in solution and dry. The curves of Fig. 13, obtained with plant-virus preparations,[2] illustrate this point. The explanation is that at any dilution at which virus activity is high enough for the purpose of estimation, the total concentration of virus plus protective agent is

[1] Lea, D.E., Smith, K.M., Holmes, B. & Markham, R. (1944).
[2] Lea, D.E. & Smith, K.M. (1940, 1942). Gowen, J.W. (1939) also finds that the rate of inactivation of plant viruses is the same wet and dry.

sufficient to make the indirect action of smaller importance than the direct action.

Fig. 14 B shows how the inactivation dose of the Shope rabbit papilloma virus irradiated by X-rays depends on the concentration of the solution which is irradiated.[1] The general shape of the curve resembles that obtained with tobacco mosaic virus (Fig. 14 A), and the explanation is presumably the same.

There is some indication with both viruses that the inactivation dose does not continue to diminish indefinitely with diminution of concentration, though the experimental evidence is at present not conclusive on this point. This may be due to the majority of the active radicals in a very dilute solution recombining instead of reacting with the solute, as discussed in Chapter II (p. 57). On the other hand, since specially purified water was not used in these experiments, the effect may be due to the unintentional presence in the water of a minute quantity of an impurity having a protective effect.

The curves in Fig. 14 are theoretical curves calculated according to equation (II-9) (p. 62) on the provisional assumption that the water contained a concentration of impurity having a deactivating efficiency equivalent to $6 \cdot 1 \times 10^{-5}$ g./ml. of virus (Fig. 14 A), or $2 \cdot 5 \times 10^{-6}$ g./ml. of virus (Fig. 14 B). The ratios of efficiencies for indirect and direct action needed to make the calculated curves fit the observations are $\gamma/\Gamma = 2 \cdot 6 \times 10^{-4}$ (tobacco mosaic virus), and $\gamma/\Gamma = 4 \cdot 6 \times 10^{-4}$ (rabbit papilloma virus). With both viruses the ionic yield is evidently very much smaller for the indirect than for the direct action.

EVIDENCE THAT VIRUS INACTIVATION IS DUE TO A SINGLE IONIZATION

The remaining discussion of this chapter will be concerned with the direct action of radiation on viruses, and we shall therefore confine ourselves to experiments in which the virus was either irradiated dry or, if in solution, in the presence of a sufficient concentration of protein, whether virus protein or extraneous protein, for the indirect effect to be unimportant. This actually includes the majority of experiments, since it is only by careful purification of the viruses that the experiments can be carried

[1] Friedewald, W.F. & Anderson, R.S. (1940, 1941), as worked up by Lea, D.E., Smith, K.M., Holmes, B. & Markham, R. (1944).

out in a sufficiently low protein concentration for the indirect effect to show up.

All detailed investigations which have been made of the inactivation of viruses by ionizing radiations point to the inactivation of a single virus particle being causable by a single ionization. The tests by which actions of this type may be recognized have been listed in Chapter III (p. 72), and we proceed now to give the experimental results of the application of these tests to the viruses.

Exponential survival curves

In Figs. 13 and 15–17 we give a selection of survival curves for various plant viruses, animal viruses, and bacteriophages, irradi-

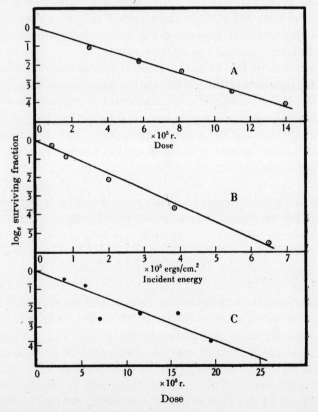

Fig. 15. Exponential survival curves of irradiated plant viruses (Lea & Smith). A, γ-rays on potato virus X; B, ultra-violet light (2536 A.) on tobacco necrosis virus; C, γ-rays on tobacco mosaic virus.

ated by various radiations.[1] Additional examples may be found in the papers from which these curves are taken. The logarithm of the surviving fraction has been plotted against dose, so that an exponential survival is shown by the points lying on a straight

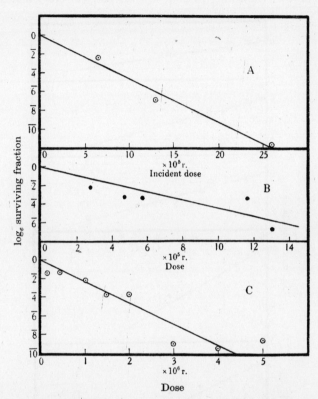

Fig. 16. Exponential survival curves of irradiated animal viruses. A, X-rays (1·5A.) on vaccinia virus (Gowen & Lucas); B, α-rays on vaccinia virus (Lea & Salaman); C, X-rays on the Shope rabbit papilloma virus (Syverton *et al.*).

line within the error of the experiment. The error in the assessment of virus activity is sometimes rather large, but no *systematic* deviations from exponential survival, as distinct from *random* variations, are suggested by the curves.

[1] Gowen, J.W. & Lucas, A.M. (1939); Lea, D.E. & Smith, K.M. (1940, 1942); Wollman, E. & Lacassagne, A. (1940); Wollman, E., Holweck, F. & Luria, S.E. (1940); Syverton, J.T., Berry, G.P. & Warren, S.L. (1941); Luria, S.E. & Exner, F.M. (1941); Lea, D.E. & Salaman, M.H. (1942 and unpublished).

Independence of inactivation dose on intensity

Only a few experiments have so far been made to test whether the effect of a given dose of radiation is independent of whether it is spread over a long time at low intensity or concentrated in

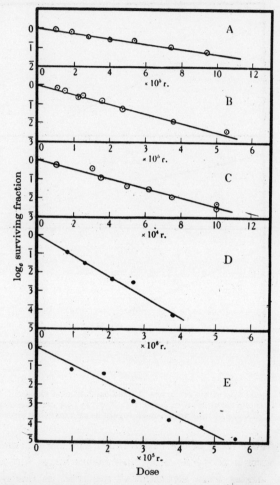

Fig. 17. Exponential survival curves of irradiated bacteriophages. A, X-rays (0·7A.) on phage S 13 (Wollman & Lacassagne); B, α-rays and β-rays on phage C 16 (Wollman, Holweck & Luria); C, X-rays (hard) on phage C 16 (Luria & Exner); D, α-rays on phage C 36 (Lea & Salaman); E, X-rays (1·5A.) on phage *Staph.* K (Lea & Salaman).

a short time at high intensity. The results of these experiments are collected in Table 31. They are considered to show that there

TABLE 31. Independence of inactivation dose on intensity

Virus	Radiation	Intensity	Inactivation dose	Intensity ratio	Reference
...cco necrosis	Ultra-violet light (2536 A.)	6.7×10^4 (erg cm.$^{-2}$ sec.$^{-1}$) 5.6×10^2 (erg cm.$^{-2}$ sec.$^{-1}$)	1.3×10^4 (erg cm.$^{-2}$) 1.9×10^4 (erg cm.$^{-2}$)	120 : 1	1
...inia	X-rays (1.5 A.)	1.0×10^4 (r./min.) 8.32×10^2 (r./min.)	1.04×10^5 r. 0.98×10^5 r.	12 : 1	2
... K phage	X-rays (1.5 A.)	1.34×10^4 (r./min.) 4.72×10^2 (r./min.)	0.935×10^5 r. 0.822×10^5 r.	28 : 1	3
C36 phage	X-rays (1.5 A.)	1.54×10^4 (r./min.) 8.13×10^2 (r./min.)	5.08×10^5 r. 4.17×10^5 r.	19 : 1	3

1 Lea, D.E. & Smith, K.M. (1940).   2 Lea, D.E. & Salaman, M.H. (1942).
3 Lea, D.E. & Salaman, M.H. (unpublished).

TABLE 32. Dependence of inactivation dose on ion-density of radiation

(Ion-density increases from left to right in the table. All doses in units of 10^6 r.)

Virus	Radiation			Reference
	X-rays 0.7 A.	X-rays 1.5 A.	X-rays 2.1 A.	
Strains of tobacco mosaic virus:				
Mosaic	0.07	0.25	0.50	1
Aucuba	0.09	0.15	0.45	
A derivative	0.11	0.22	0.40	

	γ-rays	X-rays 1.5 A.	X-rays 8.3 A.	α-rays $\sim$ 4eMV.	
Tobacco mosaic virus	0.37	0.43	1.49	1.90	2
Tobacco necrosis	0.67	0.94	5.15	—	
Tomato bushy stunt	0.45	0.62	3.10	2.56	
Vaccinia virus	0.080	0.104	—	0.211	3
Dysentery phage S 13	0.58	0.99	—	3.50	4
Coli phage C 36	0.21	0.43	—	0.94	
Staph. phage K	0.079	0.109	—	0.45	

	X-rays 0.15 A.	X-rays 0.7 A.	α-rays	
Dysentery phage C 16	0.039	0.045	0.30	5

1 Gowen, J.W. (1940); virus irradiated in solution.
2 Lea, D.E. & Smith, K.M. (1942); virus irradiated dry.
3 Lea, D.E. & Salaman, M.H. (1942); virus irradiated dry.
4 Lea, D.E. & Salaman, M.H. (unpublished); virus irradiated dry.
5 Wollman, E., Holweck, F. & Luria, S.E. (1940); irradiations made in solution.

is no variation of inactivation dose with intensity outside the error of the experiments.

Dependence of inactivation dose on ion-density of radiation

In Table 32 are tabulated the results of experiments designed to determine whether the inactivation dose is different for different wave-lengths and types of radiation. It is advisable in making experiments of this sort to carry out the treatment with the different radiations on samples prepared from the same batch of virus treated and estimated as far as possible identically. It is not advisable to attempt to compare the inactivation doses for different radiations when these have been determined by different authors.

Table 32 shows unmistakably that the inactivation doses increase in the order γ-rays X rays, and α-rays, and that when different wave-lengths of soft X-rays are employed, the inactivation doses increase with increase of wave-length. No appreciable difference is found if attention is confined to different wave-lengths of hard X-rays.[1]

This general increase of inactivation dose with increase of ion-density of the radiation is to be expected for actions of radiation caused by a single ionization (cp. Chapter III, p. 72), and together with the indications provided by the shape of the survival curve and the independence of inactivation dose on intensity constitutes the experimental evidence for regarding virus inactivation as an example of this type of action.

Relation between virus size and inactivation dose

Having established that the inactivation of viruses is a type of action to which the single ionization target theory should be applicable, Fig. 8A of Chapter III may be used to determine the target size from the experimental inactivation dose. It is of particular interest to compare the target size with the size of the virus. In Table 33 are given the inactivation doses of a number of viruses irradiated with X- or γ-rays, together with the sizes of the viruses.[2]

[1] Luria, S.E. & Exner, F.M. (1941).

[2] We omit from Table 33 and Fig. 18 tobacco mosaic virus and potato virus X, since these are rod-shaped, and, while of definite diameter, have very variable lengths. The length of the minimum infective unit is not known.

Fig. 18 exhibits the same data as a dot-diagram relating inactivation dose to virus diameter, and a correlation in the direction of small virus size being associated with large inactivation dose[1] is unmistakably visible.

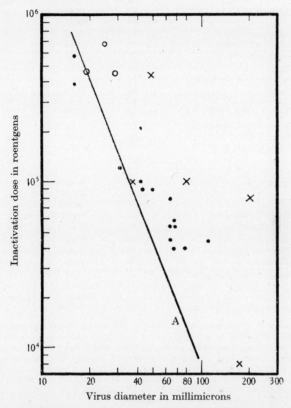

FIG. 18. Relation between virus diameter and inactivation dose (for X- and γ-rays). Curve A is the calculated relation between target diameter and inactivation dose of X-rays of 0·15 A. on the assumption that there is a single spherical target, one or more ionizations in which suffice for the inactivation of the virus. ● Phage; ○ plant virus; × animal virus.

In Fig. 19 is plotted in similar manner the inactivation dose for α-rays against the virus diameter. Only a few viruses have been irradiated with α-rays; the inactivation doses for the six points in Fig. 19, which were measured by Wollman, Holweck and Luria, Lea and Smith, and Lea and Salaman, have been

[1] The existence of a correlation of this sort has been pointed out by Wollman, E. & Lacassagne, A. (1940) for bacteriophages.

TABLE 33. Relation between size of virus and
inactivation dose (of X- or γ-rays)

(All doses in units of 10^6 r. All virus diameters refer to the unhydrated virus.)

Virus	Diameter mμ	Inactivation dose	Reference
Phage S 13	16	0·39	1
,,	16	0·58	2
Tobacco ringspot	19	0·46	3
Tobacco necrosis	25	0·67	3
Tomato bushy stunt	29	0·45	3
Phage C 13	~31	0·12	4
Encephalitis (St Louis)	~37	~0·1	5
Phage C 36	~42	0·10	1
,,	~42	0·21	2
Phage *megatherium*	~43	0·09	1
Shope rabbit papilloma	48	0·44	6
Phage P 28	50	0·09	4
Phage *Staph.* K	64	0·054	1
,, ,,	64	0·045	4
,, ,,	64	0·079	2
Phage C 16	~68	0·054	1
,,	~68	0·040	4
,,	~68	0·039	7
Phage T 105α	68	0·059	1
Fowl plague	~80	~0·1	8
Phage PC	80	0·04	9
Phage *subtilis*	~110	0·044	1
Vaccinia	200	0·08	10
Phage *Streptococcus* B	—	0·20	11
,, ,, C	—	0·11	11
,, ,, D	—	0·06	11
Shope rabbit fibroma	170	0·008	12

1 Wollman, E. & Lacassagne, A. (1940). From the tabulated data given by these authors logarithmic survival curves have been drawn and the 37 % doses read off. The X-ray dose rates given by the authors refer to the surface of a layer of broth 1 cm. deep, and it is necessary to convert them to average dose-rates in the liquid by multiplying by 0·52, which factor we have calculated to take into account diminution of intensity in the liquid by absorption and by increasing distance from the target of the X-ray tube.

2 Lea, D.E. & Salaman, M.H. (unpublished).

3 Lea, D.E. & Smith, K.M. (1942).

4 Luria, S.E. & Exner, F.M. (1941).

5 Moore, H.N. & Kersten, H. (1937). The inactivation dose can only be inferred in order of magnitude from the incomplete data given by the authors.

6 Syverton, J.T., Berry, G.P. & Warren, S.L. (1941).

7 Wollman, E., Holweck, F. & Luria, S.E. (1940).

8 Levin, B.S. & Lominiski, I. (1936). The inactivation dose is inferred in order of magnitude from the fact that a dose of 10^6 to $1·5 \times 10^6$ r. produced about the same effect as a dilution of 5×10^6 times.

9 Luria, S.E. & Anderson, T.F. (1942), quoting Luria, S.E. & Exner, F.M. (unpublished). The inactivation dose is not given explicitly, but is inferred in the statement that the target size is the same as previously determined for phage C 16.

10 Lea, D.E. & Salaman, M.H. (1942).

11 Exner, F.M. & Luria, S.E. (1941).

12 Friedewald, W.F. & Anderson, R.S. (1943).

listed in Table 32. The correlation between inactivation dose and virus size is strongly marked.

In Fig. 8A of Chapter III we gave curves for various radiations of the 37 % dose to be expected for an action of radiation due to

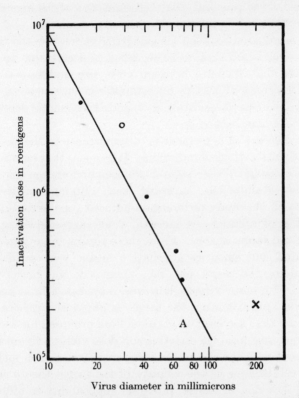

FIG. 19. Relation between virus diameter and inactivation dose (for α-rays). Curve A is the calculated relation between target diameter and inactivation dose of 4eMV. α-rays, on the assumption that there is a single spherical target, one or more ionizations in which suffice to cause the inactivation of the virus. ● Phage; ○ plant virus; × animal virus.

the production of one or more ionizations in a spherical target of given diameter. We have transcribed the appropriate curves from Fig. 8A to Figs. 18 and 19. It is seen that these calculated curves show a relation between inactivation dose and target size very similar in trend to that found between inactivation dose and virus size experimentally.[1] If the inactivation dose were used to

[1] The agreement was even closer in several instances when the published values of virus diameter, based on filtration studies interpreted

calculate the target size, the diameter obtained would be within a factor of two of the virus size for twenty out of twenty-three points in Fig. 18 (fifteen out of eighteen viruses), and within a factor of 1·5 for five out of six viruses shown in Fig. 19. This is the basis of the proposal[1] that determination of the inactivation dose can be used to estimate the size of a virus.

The correlation between the calculated target size and the size of the virus is too close to be regarded as accidental. In many instances the difference between virus size and target size is within the range of possible uncertainties in the size of the virus, calculation of the theoretical curve, and experimental determination of the inactivation dose.

This measure of agreement is of importance in the theory of the biological actions of radiations. It happens that quantitative radiation studies of the viruses have been undertaken only fairly recently. Results had, however, been obtained earlier, particularly in the study of radiation-induced gene mutations and the action of radiations on bacteria, which suggested the application of the target theory. From these results target sizes were calculated, but since no alternative means were available of determining the target size, the explanation was rather hypothetical. For many viruses, however, it appears that at any rate to a first approximation the target is identical with the virus itself, and that the method of calculation proposed for deducing the target size from the inactivation dose yields a figure in fair agreement with the size of the virus by non-radiation methods.

The establishing of the validity of the target theory for the viruses does not, of course, prove its applicability to other biological actions of radiation. It does, however, show that the minute amount of energy represented by a single ionization can produce an observable effect, an assumption which had to many workers seemed unplausible. It shows further that when the target theory is applicable, the size calculated for the target is rather close to the real size of the biological entity concerned, suggesting that the target size calculated for gene mutation, for

using Elford's factor relating virus size to pore diameter of the membrane, were used instead of the somewhat larger sizes now thought more probable.

[1] Gowen, J.W. (1940); Wollman, E., Holweck, F. & Luria, S.E. (1940); Lea, D.E. (1940b); Luria, S.E. & Exner, F.M. (1941).

example, is a reasonably reliable estimate of the actual size of a gene.

Having stressed the importance of the general measure of agreement between the experimental points and theoretical curves of Figs. 18 and 19, we proceed to discuss the detailed discrepancies. It is noticeable that for the larger viruses the discrepancy is in the direction of the target size being less than the

TABLE 34. Inactivation of phage S 13 [1]

Radiation	γ-rays	X-rays (1·5 A.)	α-rays (4 eMV.)
Inactivation dose	0·58	0·99	$3·5 \times 10^6$ r.
Target diameter	15·5	15·9	16·3 mμ

Unhydrated virus diameter = 16 mμ (hydrated 18 mμ).

virus size. This tendency is, however, not noticeable for the smaller viruses. The smallest virus so far studied is the dysentery phage S 13, and it happens that the size of this virus is fairly reliably known.[2] In Table 34 we see that the estimates of the target size by three different radiations, read off for the corresponding inactivation doses from the curves of Fig. 8 A (Chapter III) are mutually consistent although the inactivation

TABLE 35. Inactivation of *Staph.* phage K [1]

Radiation	γ-rays	X-rays (1·5 A.)	α-rays (7 eMV.)	α-rays (4 eMV.)
Inactivation dose	0·079	0·109	0·21	$0·45 \times 10^6$ r.
Target diameter	31	40	58	50 mμ

Unhydrated virus diameter = 64 mμ (hydrated 75 mμ).

doses covered a range of 6 : 1, and that they agree with the size of the virus deduced from filtration, sedimentation and diffusion studies.

As an example of the results obtained with a somewhat larger virus for which the calculated target size is a little smaller than the virus size, we give in Table 35 figures for *Staph.* phage K. It is seen that not only are the calculated target sizes rather

[1] Lea, D.E. & Salaman, M.H. (unpublished).
[2] All the methods of size determination used with phages S 13 and K determine the hydrated sizes. The unhydrated sizes given in Tables 34 and 35 are calculated from the hydrated sizes on the arbitrary assumption that on drying they take the density 1·35 g./cm.^{3}, which is typical for dry plant viruses and dry protein generally.

smaller than the virus size, but that they increase in the order γ-rays, X-rays, α-rays. In Chapter III (pp. 92–98) we discussed possible over-simplications of the simple model on which target sizes are calculated, and in particular the possibility that the probability of an ionization causing inactivation might in certain places have a value intermediate between the extremes nought and unity which alone are contemplated by the simple model. It was shown that over-simplification of this sort could lead to the calculated target size being less for γ-rays than for α-rays, and it is possible that the results with *Staph.* phage K are to be explained in this manner. On the other hand, one might expect the same over-simplification to lead to a similar discrepancy in the case of the dysentery phage, where it is not found.

Alternatively, calculated target sizes less than the virus size could be interpreted to mean that not the whole of the virus was radiosensitive. By radiosensitive we would mean here parts so essential to the infectivity of the virus that chemical change produced in them following ionization would lead to loss of infectivity. By radioinsensitive we would mean parts in which ionization could be produced (and presumably cause chemical change) without loss of infectivity. In higher cells, of course, only a very small proportion of the cell material is so essential that a single ionization can lead to a detectable effect.

Interpreting the results with *Staph.* phage K along the lines suggested on p. 91 one would infer that about 12% of the virus was radiosensitive, the radiosensitive material not being concentrated in a single spherical mass but being either a lamina, or a filament, or being multiple.[1]

Electron micrographs of the larger phages show[2] some internal structure, indicating that the phage particle is not of a uniform composition throughout and thus may well be differentiated into radiosensitive and radioinsensitive parts.

On the other hand, several of the small plant viruses have been crystallized and their particles are believed to be single molecules of nucleoprotein. If this is true also of the even smaller phage S 13, the deduction from the radiation experiments that there is

[1] On the multi-target theory the best fit of the experimental inactivation doses is given by the assumption of 14 targets each of diameter 12·5 mµ.

[2] Luria, S.E. & Anderson, T.F. (1942).

in this phage no differentiation between radioinsensitive and radiosensitive regions is plausible enough.

With viruses of medium size such as *Staph*. phage K which we have just been discussing, while the discrepancy between target size and virus size suggests such a differentiation, one is not prepared uncompromisingly to insist on it, in view of the fact that over-simplification in the model is a possible alternative explanation of a small discrepancy of this sort. However, the position is different in the case of the largest virus so far studied, namely, vaccinia virus, which in Figs. 18 and 19 is represented by the experimental points farthest from the theoretical curves.

TABLE 36. *Non-applicability of the single spherical target theory to a large virus (vaccinia)* [1]

Radiation	γ-rays	X-rays (1·5A.)	α-rays (5eMV.)
Inactivation dose	0·080	0·104	$0·211 \times 10^6$ r.
Target diameter	31	41	70mμ

Mean unhydrated virus diameter = 200mμ (hydrated 235mμ).

The size of this virus is known fairly reliably, and there is no reason to suspect serious error in the radiation data. The discrepancy shown by Table 36 between the target sizes for different radiations and between the target sizes and the virus size, is too great to be ascribed either to experimental error or to over-simplification of the model.

There is therefore in this virus definitely a differentiation between radiosensitive and insensitive material, and the radiosensitive part amounts (in volume) to less than 1 % of the whole virus. In this respect vaccinia resembles bacteria more than the smallest viruses, and we shall defer fuller discussion of it until Chapter IX.

The status of the viruses

In the introduction to the present chapter it was mentioned that in a typical cell the highly specific and genetically important nucleoprotein constituting the chromosomes comprised a relatively small part of the whole cell. In the small crystallizable viruses there is no such differentiation between genetically important chromatin and cytoplasm, the virus being composed of

[1] Lea, D.E. & Salaman, M.H. (1942).

nucleoprotein alone. Regarding the larger viruses, information by direct methods is scanty. It appears that radiation experiments are able to offer some information on the nature of the viruses. The radiosensitive material, a single ionization in which leads to the inactivation of the virus, is to be identified with the genetically important nucleoprotein, since we cannot plausibly imagine a single ionization in the less specific cytoplasm leading to inactivation. Determination of the inactivation dose enables the volume of the sensitive material to be calculated, and comparison with the volume of the virus enables the proportion of radiosensitive material in the virus to be estimated. In this way we are able to show that the smallest viruses are almost completely radiosensitive, in agreement with the chemical finding that they are entirely nucleoprotein. Vaccinia, one of the largest viruses, shows a differentiation between radiosensitive and insensitive material comparable to that in higher cells. Genetically, the smallest viruses may be considered as 'naked genes', while the largest viruses are more akin to single-celled organisms. Radiation experiments afford a means of deciding to which category a given virus belongs.

Inactivation of viruses by ultra-violet light

Not much quantitative work has been done on the inactivation of viruses by ultra-violet light.[1] The survival curves are exponential and the effect of a given dose is independent of the time over which it is spread. These results suggest that the absorption of a single ultra-violet quantum is able to cause inactivation, in the same way that the corresponding results with ionizing radiations suggest that a single ionization or single ionizing particle is able to cause inactivation. With ultra-violet light no additional test is available such as that afforded for ionizing radiations by the increase of inactivation dose with increase of ion-density, and the adequacy of a single ultra-violet quantum to cause inactivation cannot be considered established with the same certainty as is the adequacy of a single ionization. If we accept, however, that inactivation by ultra-violet light is due to

[1] See Hollaender, A. & Duggar, B.M. (1936); Price, W.C. & Gowen, J.W. (1937); Rivers, T.M. & Gates, F.L. (1928); McKinley, E.B., Fisher, R. & Holden, M. (1926); Olitsky, P.K. & Gates, F.L. (1927); Lea, D.E. & Smith, K.M. (1940).

a single quantum, we can calculate the quantum yield of the reaction, i.e. the probability that a single quantum absorbed in the virus particle will cause inactivation. It comes out to be very low, about 0·025 for phage $S13_1$ and 7×10^{-5} for tobacco mosaic virus,[2] each exposed to ultra-violet light. In inorganic gas reactions quantum yields of the order of unity are usual. However, with large organic molecules lower quantum yields are obtained,[3] and a very low value for the much larger molecule of tobacco mosaic virus is not impossible[4] It is evident that it is not possible to calculate virus sizes or target sizes from experiments on the inactivation of viruses by ultra-violet light, since, while the postulate that the production of a single ionization leads inevitably to inactivation seems to be fairly closely verified, the assumption that a single ultra-violet quantum absorbed leads inevitably to inactivation is in patent contradiction to the facts for tobacco mosiac virus.

[1] Latarjet, R. & Wahl, R. (1945, see also Latarjet (1946)), found that the activity of a dilute suspension of phage S 13 in distilled water was reduced to 37% by an incident energy of $1·47 \times 10^4$ ergs/cm.². If we assume that the absorption coefficient of this virus for ultra-violet light of 2536 A. is of the same order ($\sim 10^4$ cm.$^{-1}$) as the measured absorption coefficient of plant-virus protein (cp. Table 1, p. 5), this corresponds to an energy absorption of $1·47 \times 10^8$ ergs/cm.³, or about 40 quanta in the volume $2·2 \times 10^{-18}$ cm.³ of one phage particle (since 1 erg of this radiation corresponds to $1·277 \times 10^{11}$ quanta). Thus if we accept the deduction, from the exponential shape of the survival curve, that inactivation is due to a single quantum, and if we accept the evidence of the experiments with ionizing radiations (Table 34, p. 121) that in phage S 13 there is no differentiation into sensitive and non-sensitive regions, then we must infer that though inactivation, when it occurs, is due to only one quantum, the probability of any given quantal absorption being effective is only about one-fortieth.

There appears to be no cumulative effect of the absorption of individually ineffective quanta, since in this event one would expect, not an exponential, but a sigmoid shape of survival curve.

[2] Uber, F.M. (1941), recalculated using the inactivation data of Lea, D.E. & Smith, K.M. (1940).

[3] E.g. 0·017 for the inactivation of trypsin by 2536A., Uber, F.M. & McLaren, A.D. (1941).

[4] Cp. Jordan, P. (1938b); Crowther, J.A. (1938).

Chapter V

GENETICAL EFFECTS OF RADIATION

The mechanism of heredity[1]

Modern genetics is usually considered to have begun about the year 1900, when the experiments made by Mendel nearly forty years earlier were confirmed and their significance realized, and the basic principles can best be explained in terms of these often-quoted experiments.

Mendel had two pure-breeding strains of garden peas, tall and dwarf. When crossed, the offspring (i.e. F_1 *generation*) were tall. When the F_1 generation were self-fertilized their offspring (i.e. the F_2 generation) were composed of tall and dwarf plants in the ratio of 3 : 1. The dwarf plants were all pure-breeding, and one-third of the tall plants were pure-breeding, but the remaining tall plants were like the F_1 generation, that is, tall in appearance (i.e. *phenotypically* tall) but giving the 3 : 1 ratio on self-fertilization, indicating that *genotypically* they were hybrids. Mendel's explanation, which has not needed modification, is that the *gametes* (germ cells) of the tall plants carry a factor T and the gametes of the dwarf plants a factor t. The F_1 plants thus carry the factors Tt, and since phenotypically these plants are tall the tall factor T is said to be *dominant* to the dwarf factor t which is *recessive*. The gametes of the F_1 plants will carry the factors T or t with equal frequency, and on self-fertilization the combinations TT, Tt, tT, tt, will occur therefore with equal frequency. Thus in the F_2 generation from self-fertilization of the F_1 plants we shall find pure-breeding tall (TT), plants resembling the F_1 plants (Tt, tall but not pure-breeding), and pure-breeding dwarfs (tt), in the ratio 1 : 2 : 1.

---

[1] The condensed introduction to genetics given in the opening sections of this chapter is intended to provide the reader who is not a geneticist with the minimum knowledge of the principles and terminology of the subject to follow the discussion of the genetical effects of radiation given later in the chapter. It may be supplemented by reading the following books: Timoféeff-Ressovsky, N.W. (1937a), *Mutationsforschung*; Darlington, C.D. (1937), *Recent Advances in Cytology*; Waddington, C.H. (1939), *Introduction to Modern Genetics*; Sturtevant, A.H. & Beadle, G.W. (1940), *Introduction to Genetics*; *Cold Spring Harbor Symposia*, vol. 9 (1941) on 'Genes and Chromosomes'.

Other characters of the pea were found to be inherited in a similar fashion. Thus a strain with yellow seeds crossed to a strain with green seeds gave yellow seeds in the F_1, and yellow and green in the ratio of 3 : 1 in the F_2 generation, the factor for green (**y**) evidently being recessive to the factor for yellow (**Y**). Similarly, the factor for wrinkled seeds (**r**) was recessive to the factor for round seeds (**R**). When strains of peas differing in both these characters were crossed the characters were found to be *independently assorted*. That is to say, when pure-breeding yellow round (**YYRR**) and pure-breeding green wrinkled (**yyrr**) plants were crossed, giving F_1 plants of composition **YyRr**, the gametes of the F_1 were not only **YR** and **yr**, but also, with equal frequency, **Yr** and **yR**.

This *independent assortment* is not, however, found with all pairs of characters. For example, in man, haemophilia and colour-blindness are *linked*. In a family tree which contains colour-blind and haemophilic members, if one member is colour-blind and haemophilic it will be rare to find other members who are colour-blind without being haemophilic or vice versa. If, on the other hand, two families, one of which has haemophilic members and the other colour-blind members, intermarry, it will be rare for a descendant to have *both* haemophilia and colour-blindness.

In an organism such as maize, in which many Mendelian characters have been investigated, they can be catalogued into *linkage groups*—ten in the case of maize—such that characters in the same group are usually linked (like haemophilia and colour-blindness in man) and characters in different groups are not linked (like seed colour and seed shape in peas).

The fact that the majority of inherited qualities, e.g. height or mathematical ability in man, do not appear to obey Mendelian laws is explicable on the ground that a great number of independently assorted simple Mendelian characters probably contribute to such qualities. The simplicity of the Mendelian laws in cases such as those we have been describing when single characters can be observed suggests a simple mechanism of heredity.

The *chromosomes* provide this mechanism. Chromosomes are thread-like structures, chemically described as nucleoprotein, found in the nuclei of practically all cells, and readily stained by basic dyes when the cell is in division, though difficult to demonstrate at other times. The germ cells of a plant or animal contain a number n of chromosomes which is widely variable for different species, ranging from one up to several hundred, but constant for a given species, e.g. 24 in the case of man. When at fertilization the sperm and egg cells fuse, a cell is formed with $2n$ chromosomes, e.g. 48 in the case of man. The chromosomes in a germ cell are (in general) all different and can in favourable cases be recognized microscopically by differences in size and shape. Each chromosome contributed by a male germ cell is, with one exception, essentially identical (*homologous*) with a corresponding chromosome contributed by the female germ cell. One pair of chromosomes—the sex chromosomes—in many animals, including man and the fruit-fly, *Drosophila*, is exceptional in that in a fertilized egg (or *zygote*) which is going to develop into a male organism the two chromosomes of the pair are distinguishable, the X chromosome which has been contributed by the egg cell and the Y chromosome which has been contributed by the sperm cell being different. A zygote which is going to develop into a female organism, on the other hand, has two X chromosomes. The sperms are of two kinds, those carrying an X chromosome giving rise to female zygotes and those carrying a Y chromosome to male zygotes. (In birds, moths and butterflies the female is heterozygous (XY). Seed plants and many lower animals do not usually have the XY mechanism of sex determination.)

The fertilized egg divides into two cells, and by repetition of this process the organism is formed. At each cell division an elaborate cycle of operations (*mitosis*) occurs which results in the two daughter cells being identical as regards their chromosome complement. During the first stage of mitosis (*prophase*) the nuclear membrane disappears, and the chromosomes, which in the resting stage between divisions are long thin threads, usually unstainable, contract as a result of spiralization and become loaded with desoxyribose nucleic acid, so that they have the appearance (at *metaphase*) of short thick rods, readily staining with basic dyes by virtue of the nucleic acid content. At this stage the chromosomes are oriented in the equatorial plane of the cell. At an early stage in this *condensation* process (or sometimes in the *interphase* or *resting stage* or even in the preceding mitosis), each chromosome thread has split longitudinally, and when the chromosomes have become fully condensed the two halves of each chromosome separate (*anaphase*) and move to opposite ends of the cell (*telophase*). The forces responsible for the movement of the chromosomes to opposite poles of the cell at anaphase are not understood, but appear to be applied to a particular organ of the chromosome, the *centromere*. If, as sometimes happens after irradiation, a chromosome lacks a centromere, it is liable to be left behind and fail to be included in either daughter nucleus. A nuclear membrane now forms round each group of chromosomes and the cell divides; so that two daughter cells are formed, each having the same number of chromosomes as the mother cell, and being, as regards the chromosomes, a very exact copy of the mother cell. The process of mitotic division thus ensures that every cell in the adult organism contains $2n$ chromosomes, n derived from the n chromosomes originally contributed by its male parent and n derived from those contributed by its female parent. As regards the sex chromosomes, every tissue cell of a female contains two X chromosomes, and every tissue cell of a male contains an X and a Y chromosome.

The adult organism produces germ cells by a process known as *meiosis* in which two successive cell divisions occur with only one chromosome division, so that the chromosome number is halved. During the prophase of meiotic division the chromosomes come together in pairs, the homologues derived from male and female parent lying in close contact along their length (*synapsis*). At (first) anaphase the homologues separate again and move to opposite poles of the cell, and the cell divides. During synapsis each homologue splits longitudinally, and so when the first cell division has occurred each chromosome is already split. The half-chromosomes or *chromatids* later move apart in a second anaphase, and a second cell division follows. The final result is thus four cells, each of n chromosomes, each chromosome equally likely to have derived from the male or female parent. As regards the sex chromosomes, each of the four cells produced at meiosis in a female (only one of which however functions as a germ cell) has an X chromosome, while of four sperms produced at meiosis in a male, two carry an X chromosome (derived from the mother) and two carry a Y chromosome (derived from the father).

The chromosome mechanism affords a satisfactory explanation of the Mendelian laws of inheritance. The physical entity corresponding to a Mendelian character is a *gene*, a small portion of a chromosome. Since each *somatic* cell (i.e. tissue cell in distinction from germ cell) contains $2n$ chromosomes, each gene is normally present in duplicate, one derived from each parent. When strains differ in a Mendelian character, the gene corresponding is in some way different in the two strains; we refer to different *allelomorphs* of a given gene. Thus one of the genes affecting eye colour in the fruit fly *Drosophila melanogaster* exists in several allelomorphs which when *homozygous* (i.e. one allelomorph present in duplicate) give various eye colours described as white, coral, eosin, cherry, buff, ivory, and others, all of which are recessive to the *wild-type* allelomorph which is more common and which gives dark red eyes.

When a cross is made between two strains the offspring will contain two different allelomorphs, i.e. will be *heterozygous* for the particular gene. In general we might expect an intermediate phenotypical effect, and this sometimes occurs, but in practice the phenotype is more often nearly identical with that produced by one of the allelomorphs in homozygous condition, so that the presence of the other allelomorph, which we describe as recessive, is not detectable except by breeding tests.

The behaviour of Mendelian characters in crosses is explained by the behaviour of chromosomes at meiosis and in fertilization. Characters which are not linked correspond to genes in different chromosomes, and characters which are linked are localized in the same chromosome. In *Drosophila melanogaster* and in maize the number of linkage groups (four and ten respectively) genetically determined has been shown to be the same as the *haploid* chromosome number (i.e. n, in distinction from $2n$, the *diploid* number) microscopically observed. Further, the phenomenon of *incomplete linkage* is explained in terms of chromosome behaviour. As mentioned already, haemophilia and colour-blindness in man are linked. Cases have been observed in which linkage fails, indicating that a woman having both recessive genes in one of her X chromosomes and both normal allelomorphs in her other X chromosome has produced a germ cell having an X chromosome with the recessive allelomorph of one gene and the normal allelomorph of the other. The explanation for this incomplete linkage lies in the fact that during the stage of meiosis in which four chromosome strands lie in juxtaposition (*four-strand pachytene*) it is possible for two strands to break at corresponding points and change partners, thus separating genes on opposite sides of the break which were originally in the same chromosome thread and bringing them into separate threads, and vice versa. Thus incomplete linkage is explained by *crossing-over*. The closer together two genes are in a chromosome the smaller is the chance of crossing-over occurring between them. Thus the cross-over frequency between two genes is a measure of their distance apart in the chromosome, and in this way chromosome maps can be constructed in which the genes in the chromosomes are shown in their correct linear order.

Genes localized in the sex chromosomes are described as *sex-linked*. Since the Y chromosome is usually genetically inert and carries very few genes, a sex-linked gene is normally one localized in the X chromosome. If a female is heterozygous for the wild-type allelomorph and a recessive allelomorph of a sex-linked gene, the phenotypical appearance will be wild type. If a male, however, carries the recessive allelomorph of a sex-linked gene it will show up phenotypically since, as there is only one X chromosome in the male, there is no wild-type allelomorph present so that the male is *hemizygous* for the gene in question. It is for this reason that haemophilia, for example, while transmitted by a heterozygous woman, is not shown by her but is shown by those of her sons who receive from her an X chromosome carrying the recessive, instead of a normal, gene.

Mutations

Strains of plants and animals differing in that different allelomorphs of a particular gene are present in the different strains are found naturally. Moreover, a pure-breeding strain occasionally gives rise to an organism having a different allelomorph of a particular gene from the rest of the strain. Such changes are called gene mutations, and with individual genes occur spontaneously with a frequency of the order of 10^{-5} or 10^{-6} per generation. The rate may be somewhat speeded up by rise of temperature in the case of *Drosophila*, in which this procedure is practicable, and the production of mutations by chemical treatment has occasionally been reported. X-rays and other ionizing radiations cause mutations at a rate greatly in excess of the spontaneous rate, and radiation-induced mutations have, since the discovery of this effect by Muller in 1927, been intensively studied, especially in *Drosophila melanogaster*.

Mutation can doubtless occur, spontaneously or by radiation, in any cell, but a mutation in a single cell of an adult organism would be practically impossible to detect. If a mutation occurs, however, in a sperm or egg cell, every cell of the adult organism growing from the germ cell will carry the mutant gene. If a mutation occurs in a cell of the developing organism at some stage intermediate between the zygote and the adult, then in the adult a *mosaic* effect may be noticeable, as a result of the mutant character appearing in the group of cells which have developed from the cell in which the mutation occurred, the remaining cells being unaffected.

A great deal of the work which has been done on the induction of mutations by radiation has been carried out by irradiating *Drosophila* sperm. Male flies are irradiated and mated to untreated females. If a mutation has occurred in one of the sperms of the irradiated males, every cell of the F_1 fly which has arisen from the egg fertilized by this sperm will carry the mutation. If the mutant gene is a dominant, the fly will show the character. If it is a recessive, suitable tests will enable it to be brought to light in a subsequent generation.

Even with X-ray-induced mutations, the rate at which any given mutational step occurs is very low, since the X-ray dosage which can be given without producing sterility is limited to a few thousand roentgens, and the yield of even the most frequently occurring mutations is only about 10^{-5} per thousand roentgens. To accumulate data on the mutation frequency of a particular gene is therefore very laborious, and it has been usual therefore to record the total number of mutations, or the total number occurring in a given chromosome. The study of sex-linked mutations, i.e. mutations localized in the X chromosome, is technically most convenient, since a sex-linked recessive mutation will show up when present in the male, while a recessive mutation in an *autosome* (i.e. a chromosome other than a sex chromosome) will not show up when heterozygous, and to obtain it homozygous requires an additional generation in the breeding tests.

Among the visible mutations in *Drosophila*, recessives are several times more frequent than dominants, and most work has therefore been done on them.

The majority of recessive visible mutations as well as producing, when homozygous, the visible change by which they are recognized, reduce the viability of the organism. This is not surprising, since a mutant form which was advantageous to the organism would by selection probably have become the wild type. There are a great many recessive mutations which, when homozygous, are lethal, so that the adult stage is never attained. The number of these *recessive lethals* is in fact several times greater than the number of visible mutations. By means of a convenient breeding test (the *ClB* method of Muller) it can be arranged that a culture of flies in the F_2 generation from an irradiated grandfather is devoid of males whenever a sex-linked lethal has been produced in a sperm of the grandfather. The detection of lethals can thus be made with greater ease and freedom from subjective error than the detection of visible mutations. For these two reasons a great deal of the work on radiation-induced mutations, particularly quantitative work requiring the recording of a large number of mutations to attain statistical significance, has been made on sex-linked recessive lethals. This is to some extent unfortunate, since there is evidence that lethals are not entirely analogous to the visible gene mutations.

The nature of genes and mutations

Views on the nature of genes are, at present, to a considerable extent speculative. The evidence of Mendelian genetics is that the genes are arranged in linear order in the chromosome. When homologous chromosomes are synapsed, crossing-over takes place without producing effects in the genes, showing that the chromosomes can be broken between consecutive genes without changing the genes. The picture suggested by these facts is of beads on a string, the genes being independent units connected together by genetically inert material which is more readily breakable than the genes themselves. Microscopical examination may be considered consistent with this simple model. Chromosomes in the condensed state at metaphase are too short and tightly coiled for there to be any possibility of resolving individual genes, the number of these in a chromosome of, for example, *Drosophila*, being believed to be of the order of a thousand. The most extended state normally available for observation occurs during meiotic prophase, when a plant chromosome may have a length of 100μ. In some organisms at this stage the chromosomes have a beaded appearance, and during synapsis the beads or *chromomeres*, and not merely the chromosomes as a whole, are accurately paired.

Still more detailed structure can be seen in *salivary gland chromosomes*.[1] In the larvae of *Drosophila* and other two-winged flies the salivary glands grow during the development of the organism from the embryonic stage to the fully grown larva not by the normal process of cell multiplication, but by increase of cell size without cell multiplication. In consequence the cells attain a volume about a thousand times greater than that of ordinary cells. During this process the chromosomes apparently undergo repeated duplications without splitting, so that each salivary gland chromosome is a bundle several microns in diameter, made up of some hundreds of chromosome threads which are not individually resolvable under the microscope. The chromosomes are uncoiled, and homologues are paired as in meiotic prophase.

The salivary gland chromosomes are crossed by disks of more deeply staining material which are presumably made up by the juxtaposition of the chromomeres of the individual chromosome threads. The bands in the salivary chromosomes form a sufficiently characteristic and recognizable pattern for even very small internal losses and rearrangements of the chromosomes to be detected, and for maps to be made in which the individual bands, to the number of some thousands in the nucleus, are numbered.

Ultra-violet light absorption measurements indicate that the disks are made up of nucleoprotein (protein in combination with nucleic acid), and the intervening portions, less deeply staining and less strongly absorbent of ultra-violet light, of protein of globulin and protamin types. The bands are mechanically less extensible than the interband portions.

The number of bands—647 in the X chromosome—is of the order of the probable number of genes. The obvious hypothesis that the bands may mark the positions of genes receives support

[1] Plate III *a, b* (Catcheside, D.G.).

from the fact that there have been established many instances where the absence of a band in the salivary chromosome can be correlated with a genetical effect consistent with a deficiency (i.e. complete absence) of a particular gene, absence of the band invariably being accompanied by the genetic effect. Further, certain parts of the chromosomes known as the *heterochromatic* regions (recognizable cytologically by their retention during the resting stage of the charge of nucleic acid which the ordinary or *euchromatic* regions of the chromosome possess only during cell division) are known to be genetically inert, very few genes having been recognized in these regions. In the salivary chromosomes the heterochromatic regions are, relative to the euchromatic regions, much contracted and do not show a well-defined band structure.

It appears then that in the salivary glands the genes are located in the nucleic acid loaded bands, and it is a reasonable presumption than in ordinary chromosomes the genes are molecules of nucleoprotein located in, or connected by, a protein backbone.

Observationally an upper limit can be set to the size of the gene from the consideration that a number probably of the order of 1000 genes are present in the X chromosome of *Drosophila*, the salivary gland length of which is about 200μ. Thus the length of chromosome thread available to one gene is about $200\,\mathrm{m}\mu$. An estimate of about $100\,\mathrm{m}\mu$ is given by measurements of the length in the salivary chromosome of a deficiency known from its genetic effect to involve at least two genes.[1] The diameter of a single thread cannot be obtained from salivary gland observations, but the volume of a metaphase X chromosome can be measured, and if it be assumed that this is a coiled-up thread of length (when fully extended) 200μ, the thread diameter is calculated by Muller[2] to be not more than $20\,\mathrm{m}\mu$ in diameter. Thus an upper limit for the space taken up by a gene in a chromosome is about $100 \times 20\,\mathrm{m}\mu$, the corresponding molecular weight being about 25 million. The upper limit obtained for the molecular weight of the gene, which may of course greatly exceed the actual value, is comparable with the molecular weight of the crystallizable viruses (e.g. tomato bushy stunt virus, 11 million;

[1] Muller, H.J. & Prokofyeva, A.A. (1935).
[2] Muller, H.J. (1935).

tobacco mosaic virus, 42 million). Chemically viruses are also nucleoprotein, and like the genes they have the property of synthesizing exact copies of themselves from a suitable intracellular substrate. If significance is attached to these analogies, one is led, on the one hand, to regard a crystallizable virus as, genetically, the most primitive form of life, namely, a naked gene, and to regard the genes in higher organisms as autonomous nucleoprotein molecules, regimented in chromosomes to facilitate synchronization in division.

This picture of the nature of the gene may be taken to be one extreme of the range of views currently held regarding the nature of genes. On this view different genes would differ in structure to a marked degree, as do presumably viruses which are serologically unrelated, while the various allelomorphs of a given gene would have smaller differences, such as distinguish serologically related strains of viruses. Mutation of a virus to a related strain has been observed.

Considering different allelomorphs as different stable states of essentially the same molecule, there will presumably be an activation energy for transition from one state to another, and the frequency of transition and its temperature coefficient should be capable of treatment by the methods of chemical kinetics.[1] In particular, the fact that mutation rate is, on the scale of ordinary chemical reactions, extremely slow, leads to the expectation of a temperature coefficient larger than ordinarily found in chemical reactions. This expectation is realized, a temperature rise of 10° C. producing a fivefold increase in spontaneous mutation rate, compared with a two- or threefold increase in ordinary chemical reactions. The activation energy corresponding is about 1·5 eV.

Chromosome structural changes, and the position effect

The picture as we have so far presented it, of genes as autonomous units and of mutation as internal change in a gene, is oversimplified in that it neglects interaction between genes, and mutational effects produced by structural changes in chromosomes, which are readily produced by radiation and which are discussed in greater detail in Chapters VI and VII.

When a chromosome is broken by radiation, the broken ends

[1] Timoféeff-Ressovsky, N.W., Zimmer, K.G. & Delbrück, M. (1935).

are usually left in a joinable condition. They may rejoin restituting the original chromosome, or if a chromosome is broken in two places the four broken ends may rejoin in new ways. Such *illegitimate* union will clearly lead either to a single chromosome in which the portion between the breakage points has been taken out and reinserted in reverse order (*inversion*), or to two chromosome bodies, one a ring formed by the joining of the ends of the segment between the breakage points, and the other a rod formed by the joining of the two remaining segments.[1] One of these two chromosome bodies lacks a centromere, and is likely therefore to be lost at cell division. If the cell is able to survive the loss, we shall then have a cell in which one of the chromosomes is *deficient* of a certain number of genes, the phenomenon being called *deletion* or *deficiency*. If two different chromosomes are broken, illegitimate union between the four broken ends leads to *interchange* (also called *reciprocal translocation*), with the production of two chromosome bodies each having a portion of one of the two original chromosomes. If one new chromosome body has both centromeres (i.e. is *dicentric*) and the other has no centromere (i.e. is *acentric*), as happens as a result of *asymmetrical* interchange,[2] the acentric fragment is likely to be lost at cell division and the dicentric chromosome may be broken, or eventually lost, or cause breakdown of the cell, owing to mechanical difficulties (for example, to the two centromeres sometimes moving to opposite poles at cell division). However, if the two new chromosome bodies each contain a centromere, as happens in *symmetrical* interchange,[3] there is no mechanical disadvantage attaching to the new formation as compared with the original, no portions of chromosome have been lost (unless a minute deficiency occurs at a breakage point, as occasionally happens) and cells containing such chromosomes are likely to survive. Inversions also are likely to survive. *Duplications*, i.e. the presence in a chromosome of an additional piece without there being a corresponding loss of the same piece from its normal place in the same chromosome or in another chromosome, are also known, so that a certain number of gene loci may be present three times in the somatic cells instead of the normal twice.

[1] See Fig. 30 B, C, D, E, p. 194.
[2] Fig. 30 G (p. 194). [3] Fig. 30 F (p. 194).

It is convenient to distinguish between *gross* structural changes and *minute*, the latter being inversions, duplications, deficiencies, or translocations in which the abnormality is for only a small length of chromosome. Duplications and heterozygous deficiencies in *Drosophila* are usually viable, if small, but not if large. A homozygous deficiency, even when minute, is usually lethal, and a proportion of recessive lethals are, in fact, minute deficiencies. A few instances are known when a deficiency produces a visible mutant effect. Thus a deficiency for a portion of the region 1A in the salivary chromosome map behaves as a recessive visible mutation, producing, when homozygous, a fly with a yellow body in place of the wild-type grey body. A deficiency which includes the band $3C7$ behaves as a dominant, producing, when heterozygous, a notching of the wing veins. It is lethal when homozygous.

In addition to the phenotypical effects obtained with duplications and deficiencies and which are due to the absence of some genes or the presence of extra genes, phenotypical effects are sometimes obtained accompanying inversions and eucentric interchanges in which, as far as can be seen in the salivary chromosomes, there has been neither loss nor gain of chromosome material but merely rearrangement.[1] The phenotypical effect produced is an alteration in the characters governed by the genes which are located next to, or close to, the loci of the breaks in the chromosomes, and the possible explanations are either that the ionizing particle which causes the break simultaneously causes mutation in a gene close to the break, or that the effect of a gene is modified by the genes in its immediate neighbourhood, and hence is changed when a structural rearrangement follows a break close to the gene, although no internal change has occurred in any of the genes concerned (*position effect*).

The phenomena are different, depending on whether the two breaks taking part in the structural change are both in euchromatic regions of the chromosomes, or whether one is in heterochromatin. In the latter case, the effect of the structural change is to cause a euchromatic region to be joined to a heterochromatic region. Genes in the euchromatin up to a distance of twenty salivary bands away from the place where it now joins the heterochromatin are liable to be affected. The effect may take

[1] In *Drosophila*; not usually in plant material.

the form of an unstable mutation resulting in a mottled phenotype, owing to some of the cells showing the mutant and others the wild-type character. Thus if the $3C1$ band of the X chromosome is brought into proximity with heterochromatin in this way, the eyes of the male are white with red patches. (The wild-type fly has red eyes; a male carrying the commonest mutant allelomorph of this locus has white eyes.) Conversely, a gene normally located near a heterochromatic region of its chromosome may show a position effect as a result of a chromosome structural change which brings it instead adjacent to a euchromatic region.[1]

There are several lines[2] of evidence which make it tolerably certain that the characteristic effects shown by a gene transferred to or from the neighbourhood of heterochromatin are due to position effect, and not to a simultaneous change induced in the gene by the ionizing particle which broke the chromosome near it. It is possible that the different nucleic acid metabolism in heterochromatin compared with euchromatin interferes with the reproduction of the genes located near a heterochromatic region.

The phenomena are rather different, in the case when both breaks taking part in the structural change are in the euchromatin. In these cases the effect is usually limited to genes immediately adjacent to the break instead of extending up to a distance amounting to several percent of the whole chromosome length. This circumstance makes plausible the explanation, already mentioned, that the ionizing particle which causes the break may simultaneously cause mutation of the gene adjacent to the break. It happens further that the tests referred to,[2] by which a number of changes involving heterochromatin have been shown unmistakably to be position effects, cannot be applied to effects existing only in the immediate proximity of the break.

[1] The example of this is *cubitus interruptus*, located in the fourth chromosome near the heterochromatin of the centromere (Dubinin, N.P., Sokolov, N.N. & Tiniakov, G.G. 1935). An interchange which separates the *c.i.* locus from the centromere and brings it into a euchromatic region of another chromosome may cause the wild-type allelomorph of the *c.i.* locus to lose its dominance over the recessive allelomorph. Interchange which brings the *c.i.* locus into a heterochromatic region of another chromosome has no effect.

[2] Dubinin, N.P. & Sidorov, B.N. (1935); Muller, H.J. (1941).

Thus the explanation of mutational changes accompanying chromosome structural change not involving heterochromatin cannot be considered certainly decided. This is particularly unfortunate since a fair proportion of the sex-linked recessive lethals, on which a great deal of the experimental work on the production of mutations by radiation has been done, are accompanied by chromosome structural changes, and it is necessary for an understanding of the mechanism of action of radiation to know whether these lethal mutations are to be attributed to position effect of a gene adjacent to the break due to its being separated from its normal neighbour and brought into proximity with another gene, or whether they are due to internal changes produced in the gene by the ionizing particle which caused the break.

We shall, from the analysis of the radiation experiments, find support for the latter explanation, and shall interpret lethal mutations accompanying gross chromosome structural change in this way instead of invoking the position effect. (This is not to deny the existence of position effect completely, but merely to conclude that it is not usually the cause of lethal mutations associated with structural change. There are a few well-established cases of position effect not involving heterochromatin, such as Bar eye.[1])

The production of visible mutations by radiation

Mutations have been induced by ionizing radiations in a great many organisms.[2] The method of investigation is naturally different for different organisms. In the case of plant viruses,[3] the method adopted is to inoculate the leaves of a suitable test plant with the irradiated virus solution. The test plant chosen is one which does not give, with the unchanged virus, local necrotic lesions at the points of entry of the virus into the leaves. If some of the virus has been changed by the radiation into a form which does give local necrotic lesions, these can be detected despite the great excess of unchanged virus. The lesions can be cut out and the mutant virus strain isolated and the permanence of the change tested.

[1] Cp. Sutton, E. (1943).
[2] A list of organisms investigated is given by Timoféeff-Ressovsky, N.W. (1937a).
[3] Gowen, J.W. (1941), using tobacco mosaic virus.

VISIBLE MUTATIONS

In the case of bacteria[1] the experiment consists simply in plating out suitable dilutions of the irradiated bacteria on a solid nutrient growth medium, and examining large numbers of colonies. Each colony having grown from a single organism, a mutation in an irradiated bacterium should lead to a colony in which every individual has the mutant character. Mutations in the bacteria which lead to differences in the size, colour, or surface texture of the colony are recognizable by inspection of the colonies.

A major difficulty in the study of bacterial and virus mutation is that contamination of the preparations by bacteria or viruses from outside is difficult to prevent, and such contaminations may be recorded as mutations. The number of different characters which can be recognized is, moreover, small compared with the number available in higher organisms. For these reasons little work has been done on the induction of mutation in bacteria and viruses. It does not appear, however, that the process is essentially different from that in higher organisms.

Much the fullest information on the induction of mutations by radiation is available in the case of the fruit fly *Drosophila melanogaster*. The most usual procedure is to irradiate the male flies, to mate them to untreated females, and to look for mutations in their offspring. The mutations investigated are thus produced in the sperm of the irradiated male. Dominant mutations will be visible in the F_1 generation. The number of dominant mutations is, however, much smaller than the number of recessive mutations, and the latter therefore are usually worked with. A recessive mutation induced in the sperm will not be visible in the F_1 generation unless special means are adopted. If mutations of a particular locus are being investigated, then the irradiated wild-type male can be mated with a female homozygous for a recessive allelomorph of this locus. The great majority of the offspring will be phenotypically wild type, but an occasional one will show the mutant character. This is due either to mutation of the wild type to a recessive allelomorph, or to the complete loss of the locus concerned owing to a minute chromosome deficiency having occurred which includes the locus.[2] Further tests will decide between these possibilities.

[1] Gowen, J.W. (1941), using *Phytomonas stewartii*.
[2] According to Muller, H.J. (1940) about one-third of the mutants found by this method are minute deficiencies. According to Patterson, J.T. (1932) seven-eighths are minute deficiencies.

It is often desired to collect all mutations occurring in a given chromosome, and not only mutations at a selected locus. Mutations in the X chromosome are more easily studied than mutations in autosomes, since the male has only one X chromosome. Normally it inherits this from its mother, but if a special *attached*-X stock of females is used in the matings, then the male receives its X chromosome from its father. Thus, when an irradiated wild-type male is mated to an attached-X female, a sex-linked mutation produced in an irradiated sperm will be revealed by the male offspring which develops from the egg fertilized by that sperm. All that is necessary therefore is to examine the F_1 flies for mutant males.

The proportion of mutations is liable to be underestimated by the attached-X method, not only because it is easy to overlook a single mutant fly in a culture, but also because most mutants are of reduced viability, and the proportion reaching maturity is smaller than in the case of the wild-type flies. Muller's *ClB* method is more accurate, but has the disadvantage of requiring an extra generation. The irradiated males are mated to *ClB* females. The peculiarity of *ClB* females is that one of their X chromosomes carries a minute duplication acting as a dominant mutation, and producing an easily recognized narrowing of the eyes (Bar eye). In addition, there is a long inversion which effectively prevents crossing-over between this chromosome and the other X chromosome during meiosis in the female. Finally, there is a recessive lethal in the chromosome so that a male zygote receiving a *ClB* chromosome will not survive.

The irradiated males are mated with *ClB* females and the half of the female F_1 offspring showing Bar eye is picked out. Each of these females has two X chromosomes, one is a *ClB*, the other has been derived from the X chromosome of the irradiated sperm, and therefore carries a mutation if one was induced in this sperm by the radiation. Each of these females is mated in a separate vial with one of its brothers, and the male F_2 offspring examined. Each of these F_2 males receives its X chromosome from its mother. In view of the lethal factor carried by the *ClB* chromosome, in all the F_2 males which survive the X chromosome is derived from the original irradiated X chromosome of the grandfather. Hence if a mutation occurred as a result of the irradiation, *all* the males of this culture will show it.

Visible mutant types differing only slightly from the wild type are less easily overlooked by this method than by the attached-X method, and reduced viability of mutant as compared with wild type will not lead to a depressing of the apparent mutation rate. A lethal mutation will be shown by the F_2 culture having no males.

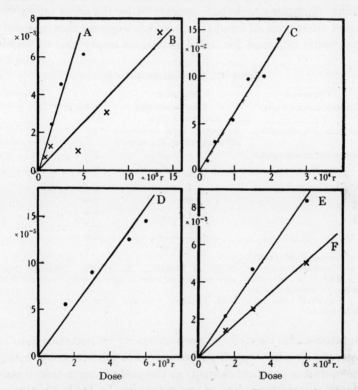

FIG. 20. Proportion of visible mutations induced by X-rays as a function of dose. A–B, tobacco mosaic virus (Gowen): A, type to aucuba; B, aucuba to type. C, *Neurospora* (Demerec *et al.*). D–F, *Drosophila melanogaster* (Timoféeff-Ressovsky & Delbrück): D, a single mutation step (wild type to eosin eye-colour); E, all sex-linked recessive mutations detected by *ClB* method; F, all sex-linked recessive mutations detected by attached-X method.

In Fig. 20 are shown the proportions of visible mutations produced by X-rays in different organisms, plotted against the dose. It is seen that, within the error of the experiments, the yield of mutations is proportional to the dose.

It is believed that the yield of visible mutations produced by a given dose is independent of the wave-length of the radiation over the usable range 0·01–1 A., and of the intensity at which it is delivered. This conclusion has been accurately established for lethal mutations (*vide infra*) in *Drosophila*. Statistically adequate experiments of a similar type using visible mutations would be very laborious owing to the ten to fifteen times smaller rate of induction of visible mutations, but the proportion of visible mutations to lethals appears to be the same under different conditions of irradiation,[1] which suggests that probably the results obtained for lethal mutations apply also to visible

TABLE 37. Mutation constants (α)

Organism	Mutation	α	Reference
Tobacco mosaic virus	Type $\to$ aucuba	$1·6 \times 10^{-8}$	1
	Aucuba $\to$ type	0·5	
Bacteria (*Ph. stewartii*)	Average of three mutations affecting colony appearance	3·7	1
Drosophila melanogaster	$+ \to w^o$	2·6	2
	$w \to w^e$	0·3	
	$w^e \to +$	0·8	
	$+ \to m$	2·4	
	$m \to +$	1·0	
	$+ \to f$	6·6	
	$f \to +$	2·4	
	$+ \to w^x$	15·9	3
	$w \to w^x$	1·2	

1 Gowen, J.W. (1941).
2 Timoféeff-Ressovsky, N.W. & Delbrück, M. (1936). Symbols used: $+$ = wild type; w = white eye colour; w^e = eosin eye colour; m = miniature wings; f = forked bristles.
3 Timoféeff-Ressovsky, N.W. (1933b). w^x = any other allelomorph at the white locus.

mutations. In view of the proportionality of mutation rate to dose, the simplest way of expressing mutation rate is by stating the mutation constant defined[1] as the probability α that a gene should mutate for a dose of one roentgen. In Table 37 a list of mutation constants is given. The general agreement in order of magnitude between the mutation constants of organisms as different as bacteria and *Drosophila* is remarkable.

Recessive lethal mutations in *Drosophila*

The general method employed in the investigation of sex-linked recessive lethals in *Drosophila* is as follows. A batch of male flies are irradiated under the chosen conditions and mated to *ClB* females. The proportion of F_2 cultures which are com-

[1] Timoféeff-Ressovsky, N.W. & Delbrück, M. (1936).

pletely devoid of males is determined; this gives the proportion of (viable) sperm which after irradiation had one or more recessive lethal mutations induced in the X chromosome. A small correction is required for the small proportion of spontaneous lethals, and is determined by carrying out a similar breeding test on unirradiated flies.

Experiments of this sort have been carried out for a number of years by different workers. The principal results are:

(a) The number of lethals obtained increases linearly with increase of dose.

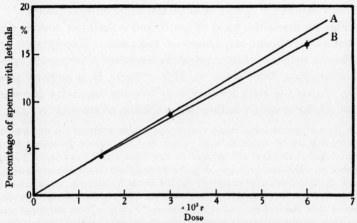

Fig. 21. Percentage of sperm in which sex-linked recessive lethal mutation is induced by X-rays (Timoféeff-Ressovsky).

(b) The effect of a given dose is independent of whether it is concentrated into a short exposure at high intensity or is spread over a prolonged time by fractionation or by the use of low intensity.

(c) Different wave-lengths of X- and γ-rays are equally effective except for a possible slight reduction of efficiency for wave-lengths exceeding 1 A. Neutrons are somewhat less effective than X-rays, for equal ionization in the tissue.

In Fig. 21 the proportion of viable sperm which carry sex-linked recessive lethals is plotted against the dose of X-rays (the spontaneous lethals having been subtracted). These data, accumulated by Timoféeff-Ressovsky[1] over a period of years, are

[1] Timoféeff-Ressovsky, N.W. (1939).

based on the examination of some 60,000 F_2 cultures. A straight line satisfactorily fits these points, the χ^2 *test of goodness of fit*[1] giving $P = 0\cdot16$. However, if the argument of Chapter III (p. 72) is recalled, it will be realized that since the experiment gives the proportion of sperm which carry a lethal, and not the mean number of lethals per sperm, the curve cannot be strictly linear, since at no dose, however high, could the yield exceed 100 %. Instead, we should expect[2] the yield to be proportional to $1 - e^{-mD}$, where D is the dose and m is the initial gradient of the curve of yield against dose. In Fig. 21 B is the curve $1 - e^{-mD}$ and A its initial gradient, which is $2\cdot89$ % per 1000 r. The χ^2 test gives $P = 0\cdot34$ for the fit of B to the experimental points.

Fig. 22 shows the yield of sex-linked lethals per 1000 r. produced by different intensities of radiation.[3] Experiments by different authors taken together extend over a very wide range of intensities, from 0·009 to 2700 r./min. It is evident from Fig. 22 that the yield does not vary by more than a few per cent, if at all, for nearly a millionfold variation of intensity.[4]

[1] The χ^2 test is a statistical test to determine whether the departures between a set of observational values and the values predicted by a hypothesis under test are greater or not than could reasonably be expected on statistical grounds. P is the probability that a departure equal to or greater than that observed should occur by chance. A value of P between 0·9 and 0·1 is interpreted to mean that the data tested are consistent with the hypothesis. A low value of P, e.g. $< 0\cdot05$, indicates that the data are not consistent with the hypothesis being tested, or that some unsuspected source of error is present. A high value of P, e.g. between 0·99 and 1·0, indicates that the statistical variation of the observations is unreasonably low and casts doubt on their reliability.

[2] As pointed out by Oliver, C.P. (1932); Gowen, J.W. & Gay, E.H. (1933); Zimmer, K.G. (1934).

[3] Timoféeff-Ressovsky, N.W., Zimmer, K.G. & Delbrück, M. (1935) using X-rays; Wilhelmy, E., Timoféeff-Ressovsky, N.W. & Zimmer, K.G. (1936) using soft X-rays; Ray-Chaudhuri, S.P. (1944) using γ-rays.

[4] Attention should be directed to the mutual agreement of the yields obtained at the two or three different intensities employed in each experiment, rather than to the slight difference in average yield obtained in the different experiments. Widely different wave-lengths were used by the different authors, and in some experiments lethals only, and in others total sex-linked recessive mutations, were counted.

Hanson, F.B. & Heys, F. (1929, 1932) have also reported that the yield of lethals produced by γ-rays is independent of the intensity. We have not, however, included their results in Fig. 22 on account of some anomalies in their data. When a χ^2 test is made of the goodness of fit of their 1929 and 1932 data respectively to the hypothesis that the yield of

Experiments, in which the effect of a given dose split into a series of fractions with rest periods between has been compared with the effect of the same dose given as a single exposure, have shown no difference in yield produced by fractionation.[1] It may be concluded that the effect of a given dose is independent of the manner in which it is distributed in time.

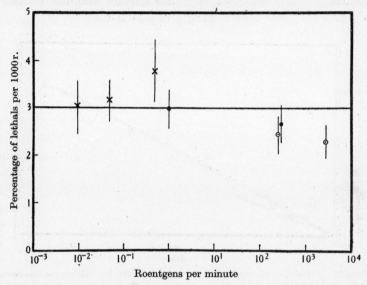

FIG. 22. Yield of sex-linked recessive lethals at different intensities. ● Timoféeff-Ressovsky, Zimmer & Delbrück (X-rays); ⊙ Wilhelmy, Timoféeff-Ressovsky & Zimmer (soft X-rays); × Ray-Chaudhuri (γ-rays).

Experiments using X-rays excited at different kilovoltages down to about 10 kV. (i.e. wave-lengths up to about 1 A.), γ-rays and β-rays have not shown any difference in the yield per roentgen for these radiations, as shown in Fig. 23.[2] Experiments with still softer X-rays of wave-length exceeding 1 A. have given a

mutations is proportional to the dose and independent of the intensity, values of P of 0·993 and 0·9996 are obtained, indicating unreasonably little statistical variation. Further, there appears to be some error in the description of the experimental arrangement employed, since with the amount and disposition of radium described, the doses in the 1932 experiments would have been at least a hundred times smaller than are required to produce the yield of mutations obtained.

[1] See Timoféeff-Ressovsky, N.W. (1937a) for a summary of these experiments.
[2] Timoféeff-Ressovsky, N.W. & Zimmer, K.G. (1939).

somewhat smaller yield. Dosage measurements with these long wave-lengths are liable to error, particularly on account of uncertainty in the correction required for absorption in the tissue overlying the testes of the irradiated flies. Gowen and Gay[1] obtained a yield of 1·2 % per 1000 r. at wave-lengths 1·5 and 2·3 A., which by comparison with other data is probably too low. Wilhelmy, Timoféeff-Ressovsky and Zimmer obtained a yield of $2·23 \pm 0·3$ % per 1000 r. using X-rays of 2–3 A.[2]

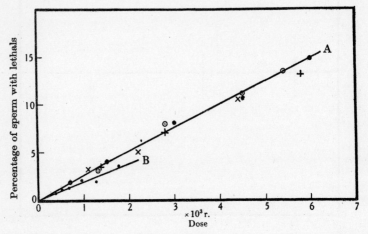

FIG. 23. Induction of sex-linked recessive lethals by different radiations (Timoféeff-Ressovsky & Zimmer). Curve A: ● X-rays; × soft X-rays; ⊙ γ-rays; + β-rays. Curve B: • neutrons.

Experiments using fast neutrons have been made, but, owing to the fact that neutron dosimetry is not yet usually so accurate as X-ray dosimetry, their reliability is on the whole not so high as in the case of X- and γ-rays. For the results to have value it is necessary for the neutron dosages to be expressed in a unit comparable to the unit of X-ray dosage, e.g. in terms of Gray's 'v-unit' (Chapter I, p. 20), which produces in water an ionization equal to that produced by 1 r. of X- or γ-rays. American authors usually measure in terms of 'n-units', and we have converted to v-units in Table 38 on the basis that 1 n-unit = 2·5 v-units.

[1] Gowen, J.W. & Gay, E.H. (1933).
[2] Wilhelmy, E., Timoféeff-Ressovsky, N.W. & Zimmer, K.G. (1936). In view of the difficulties of accurate dose measurements the authors were not convinced that the yield was really lower than the yield of 2·89 % obtained with shorter wave-lengths of X-rays.

The first quantitative measurements, and still the most complete, are those of Timoféeff-Ressovsky and Zimmer,[1] which have been included in Fig. 23, and which show that the yield for equal ionization in the tissue is only about two-thirds as great for neutrons as for X-rays. In Table 38 are collected data by various authors on the number of sex-linked lethals per v-unit of neutrons. These data have been corrected for spontaneous lethals

TABLE 38. *The relative efficiencies of fast neutrons and X-rays in the production of sex-linked lethal mutations in* Drosophila melanogaster

Authors	Radiation	Lethals per 100 X chromosomes per 1000 r. or v-units	Ratio of yields
Timoféeff-Ressovsky & Zimmer [1]	X-rays Neutrons	$2·89 \pm 0·05$ $1·92 \pm 0·15$	$0·66 \pm 0·06$
Giles [2]	X-rays Neutrons	$2·41 \pm 0·7$ $1·76 \pm 0·5$	$0·73 \pm 0·3$
Dempster [3]	X-rays Neutrons	—	$0·75$
Fano [4]	Neutrons	$2·3 \pm 0·3$	$0·80 \pm 0·1$
Demerec, Kaufmann & Sutton [5]	X-rays Neutrons	$0·93 \pm 0·12$ $0·52 \pm 0·07$	$0·56 \pm 0·19$

[1] Timoféeff-Ressovsky, N.W. & Zimmer, K.G. (1938, 1939); Zimmer, K.G. & Timoféeff-Ressovsky (1938).
[2] Giles, N.H. (1943).   [3] Dempster, E.R. (1941a).   [4] Fano, U. (1943b).
[5] Demerec, M. (1938); Demerec, M., Kaufmann, B.P. & Sutton, E. (1942); Kaufmann, B.P. (1941b).

and fitted by the least squares method to the formula $1 - e^{-mD}$, and in the last column of the table the neutron yields per $1000 v$ are compared with the X-ray yields per 1000 r. using the figure 2·89 % per 1000 r. as the X-ray yield in cases where it was not determined as part of the neutron experiment. The experiments of Demerec, Kaufmann and Sutton were made with a stock (Oregon-R) of *Drosophila* having an unusually low mutation rate, but when the neutron yield is compared with the X-ray yield with the same stock the ratio is approximately the same as in the other experiments. The standard deviations attached to the ratios in the last column are based on the number of lethals counted in the various experiments, and take no account of possible systematic errors in dosimetry. There seems to be little doubt that for inducing sex-linked recessive lethals in *Drosophila*

[1] Timoféeff-Ressovsky, N.W. & Zimmer, K.G. (1938); Zimmer, K.G. & Timoféeff-Ressovsky, N.W. (1938).

neutrons are, for equal ionization in the tissue, less effective than X-rays.

The suggestion has been made that, apart from the quantitative difference in the yield of mutations for equal ionization in the tissue, neutrons differ from X-rays in that there is a tendency for a grouping of mutations to occur, i.e. a tendency for several sperm in a given irradiated male to be affected. Experiments do not, however, indicate that the distribution of lethals is other than random. Nagai and Locher[1] obtained a total of 44 lethals in 69 irradiated male flies which were tested, i.e. an average of 0·6377 lethal per male. In the absence of any specific grouping effect we should expect the numbers of males with 0, 1, 2, or $\geqslant 3$ lethals to be given by a Poisson distribution with $m = 0\cdot6377$. The numbers expected on this basis are calculated to be 36·5, 23·3, 7·4, 1·8, and the experimental numbers are 39, 18, 10, 2. The agreement is satisfactory ($\chi^2 = 2\cdot3$, $n = 2$, $P = 0\cdot3$), and Nagai and Locher's experiments therefore show no special grouping effect, but are consistent with the lethals being produced at random. Similar data published by Nishina and Moriwaki[2] are also consistent with the distribution of lethals between different males being random[3] ($\chi^2 = 2\cdot5$, $n = 2$, $P = 0\cdot3$).

Another type of grouping effect has been suggested by Nishina and Moriwaki.[4] This is a tendency for more than one lethal to occur in a given X chromosome. They irradiated male flies with various doses of neutrons which produced lethals in from 5 to 15 % of the sperm, tested some of the lethal-bearing chromosomes, and obtained evidence that in 4 chromosomes out of 16 the chromosome carried more than one lethal. This is five times higher than is to be expected on the basis of a random distribution of lethals in irradiated chromosomes, and suggests a grouping effect ($P = 0\cdot01$). However, Fano[5] failed to obtain any grouping effect of this sort. Out of 998 X chromosomes tested, 60 were found to carry lethals, but only 2 chromosomes carried more than one lethal. This is a much smaller proportion than

---

[1] Nagai, M.A. & Locher, G.L. (1938).

[2] Nishina, Y. & Moriwaki, D. (1939).

[3] Nagai & Locher, and Nishina & Moriwaki, however, incorrectly report their experiments as suggestive of a grouping effect.

[4] Nishina, Y. & Moriwaki, D. (1941). The figures we quote refer to their experiments N I, N IV, N V and N VI.

[5] Fano, U. (1943b).

that reported by Nishina and Moriwaki, and is in close agreement with the number expected on the basis of a random distribution of lethals in irradiated chromosomes. It appears probable therefore that no significant grouping effect exists.

The idea that a grouping effect is to be expected theoretically has its basis in the fact that when a tissue is irradiated by neutrons, the energy is dissipated by a comparatively small number of ionizing particles (protons), each of which produces a large number of ionizations per micron path. The notion is that the passage of a proton through the testes of an irradiated fly will be a rare event, but that when it happens usually several sperm will be affected. Similarly, the passage of a proton through a sperm will be a rare event, but when it happens often more than one mutation will be produced. This idea does not, however, bear further examination. In a tissue irradiated by 1000 v-units of neutrons, about 4 protons cross each square micron of area. Taking the dimensions of the sperm head to be $7 \cdot 4 \times 0 \cdot 37 \mu$, this dose therefore leads to about 11 protons traversing each sperm. The proportion of sperm in which lethals are induced by this dose is 0·02, so that the probability of a lethal being produced in a sperm by a proton which crosses it is less than 0·002. There is thus no theoretical foundation for the idea either that with doses of the order commonly used the passage of a proton through a sperm is a rare event, or that when such a passage occurs the probability of a mutation is high enough for more than one frequently to occur. We conclude that there is no justification either on experimental or on theoretical grounds for a belief that a grouping effect of the sort looked for occurs with neutrons.

There is clearly quite a high probability that a proton which produces a mutation, and which therefore presumably traverses the chromosome thread, should pass through two or more adjacent turns of the spiral chromosome. There is thus the possibility that lethals may occur in compact groups. Several lethals very close together would not, however, be distinguished from a single lethal by the methods used in the experiments cited.

It would be of great interest to determine whether the yield of sex-linked lethals for a given amount of ionization diminishes still further when a still more densely ionizing radiation is employed. Experiments with α-rays are technically difficult on account of the short range (less than 70μ) of the rays in tissue.

Zimmer and Timoféeff-Ressovsky have shown that mutations are in fact produced by α-rays when the flies are made to breathe an atmosphere containing radon,[1] but they were not able to evaluate the dose received by the gonads. The most satisfactory method would be to make use of artificial insemination, which would permit of the sperm being irradiated *in vitro*. In the absence of results by this method, an attempt can be made to deduce the yield of lethals from consideration of an experiment by Ward.[2]

At a certain early stage during the development of a fertilized egg which is destined to give rise to a male fly, the cells which after numerous divisions and meiosis will form the sperm of the adult fly are congregated at one end of the egg sufficiently near the surface to be irradiated by α-rays from a source outside the egg. Ward irradiated the eggs at a stage in which there were about thirty of these *primordial germ cells* in the polar cap of the egg. It is evident that if the radiation produces a recessive lethal in an X chromosome in one of these thirty cells, then a certain proportion of the sperm of the adult fly (namely all those containing an X chromosome derived from the one in which the mutation was induced) will carry the lethal, provided that damage produced by the radiation does not cause breakdown at some stage between irradiation and the formation of the mature sperm. An X chromosome in which a lethal is not induced will, again provided that damage produced by the radiation does not cause breakdown, give rise to a normal sperm. If experimentally it is found that one-quarter of the sperm of an adult fly carry a lethal, we infer that in this fly the radiation has induced a lethal in one X chromosome out of four irradiated.[3] In Ward's experiments, 149 flies were tested and 7 were found to carry lethals, the proportion of sperm carrying lethals varying in different flies

[1] Zimmer, K.G. & Timoféeff-Ressovsky, N.W. (1936).
[2] Ward, F.D. (1935).
[3] Evidently we are assuming that the proportion of lethals is the same in X chromosomes we are able to test as it is in X chromosomes which we are unable to test, on account of breakdown occurring at some stage between irradiation of the primordial germ cell and maturation of the sperm. A similar assumption is implicit in all work on induced mutation obtained by irradiating *Drosophila* sperm, since the doses given are normally sufficient to render non-viable a large proportion of the sperm, and the experiment consists in determining the proportion of mutations induced in those which remain viable.

between 100 and 8 %. In this way one is able to deduce[1] that the radiation induces 7 lethals in about 1000 X chromosomes. The final estimate obtained for the yield of sex-linked recessive mutations by α-rays is $0\cdot 84 \pm 0\cdot 3$ % per 1000 r. This is smaller than the yields obtained with X-rays ($2\cdot 89$ % per 1000 r.) or with neutrons ($1\cdot 9$ % per 1000 r.), and indicates that the yield, for equal ionization in tissue, decreases in the order of increasing ion-density of the radiation. It is, of course, rather unsatisfactory to compare the mutation rate induced in primordial germ cells in the polar cap of the egg by one radiation with the mutation rate induced by another radiation in ripe sperm, and more strictly comparable experiments are urgently required.

However, provisionally accepting the data at their face value, we have drawn up Table 39 showing the yield of sex-linked lethals per 1000 r. by various radiations. The ion-density of the

TABLE 39. Yield of sex-linked lethals per 1000 r. of various radiations

Radiation	β-rays, γ-rays or X-rays	Soft X-rays 2–3 A.	Neutrons	α-rays
Yield of sex-linked lethals per 1000 r. (or v-units)	$2\cdot 89$ %	$2\cdot 23$ %	$1\cdot 90$ %	$0\cdot 84$ %

radiations listed increases from left to right in the table; a regular diminution of yield with increasing ion-density is apparent. The following results have thus been established:

(a) The yield of recessive lethal mutations is proportional to dose.

(b) For a given dose the yield is independent of the time over which the irradiation is extended.

(c) The yield for a given ionization in the tissue diminishes with increase of ion-density of the radiation.

Thus the principal tests by which one recognizes an action of radiation produced by a single ionization (cp. Chapter III, p. 72) are satisfied.

The idea has been current for some years that a mutation can be produced by a single ionization in or in the immediate vicinity of a gene, and in fact result (c) was predicted before the experi-

[1] The method of calculation outlined above differs from that employed by Ward, who arrives, however, at the same qualitative result that α-rays are less effective per ionization than X-rays.

ment had been performed.[1] If mutation is a rather direct result of ionization, one may expect that the yield of X-ray-induced mutations will be independent of the temperature at which the material is held during irradiation. Stadler observed no influence of temperature on induced mutations in barley. The evidence in the case of *Drosophila* is conflicting. (See p. 364.)

TABLE 40. Independence of yield of sex-linked lethals on the temperature (Timoféeff-Ressovsky & Zimmer)[2]

Dose (r.)	Temperature °C.	No. of X chromosomes tested	No. of lethals	% of X chromosomes having lethals
3000	10	802	76	9·48
	35	736	60	8·15
2750	3	1821	154	8·46
	33	1645	134	8·15
1200	7	976	25	2·56
	32	921	29	3·15

Test of hypothesis that yield is independent of temperature gives $\chi^2 = 1·5$, $n = 3$, $P = 0·7$.

Relation between lethals[3] and chromosome structural change[4]

A few years ago it was supposed that a recessive lethal mutation, like a visible gene mutation, was an internal change in the gene, either giving an allelomorph which was lethal, or else destroying the capacity of the gene for reproducing itself, so that the gene was lost. On this basis a consistent picture was built up[5] on which the principal experimental facts concerning both induced and spontaneous mutations were explained. The view that lethals are essentially similar to visible gene mutations is supported by the fact that many lethals have been found to occur at loci of the chromosomes at which visible allelomorphs are known. Thus if a female fly is known to carry a recessive lethal in one of its X chromosomes, which is otherwise wild type, and the white-eye allelomorph in the other X chromosome, then if the fly is phenotypically white one infers that the lethal involves the white locus. Many other lethals are not connected with

[1] Timoféeff-Ressovsky, N.W. (1937).
[2] Timoféeff-Ressovsky, N.W. & Zimmer, K.G. (1939).
[3] In this section by *lethal* is to be understood *sex-linked recessive lethal*.
[4] The section follows the treatment given by Lea, D.E. & Catcheside, D.G. (1945a).
[5] Cp. Timoféeff-Ressovsky, N.W., Zimmer, K.G. & Delbrück, M. (1935).

known loci, and here the inference is that either there are no viable allelomorphs of these genes other than the wild type, or else that they do not produce phenotypes recognizably different by inspection.

More recently the situation has been complicated by the realization that the recessive lethals do not form a homogeneous class, but may be subdivided into the following three types:

(A) The chromosome containing the lethal may be structurally unchanged as far as can be determined by an examination of the salivary gland chromosomes. This type may reasonably be regarded as mutation to a lethal allelomorph, or loss by the gene of the power to reproduce itself.

(B) There may be a minute deficiency, revealed by the absence of one or more bands in the salivary chromosome, the deficiency including the locus at which the lethal occurs. The smallest deficiencies, e.g. where a single band is deleted, may perhaps be caused by the destruction of the capacity for reproduction of a single gene. Somewhat larger deficiencies involving several bands are probably due to the simultaneous breakage of the chromosome at two places due to the passage of a single ionizing particle. Still larger deficiencies, up to 50 bands or nearly one-tenth of the whole chromosome, are probably due to independent breakage of the chromosome in two places by separate ionizing particles.[1] Deficiencies exceeding about 50 bands are not observed, doubtless because they behave as dominant lethals, i.e. are not viable even in the presence of a non-deficient homologous chromosome.

(C) There may be a gross structural change (inversion or interchange) not usually involving any cytologically detectable deficiency, one of the breaks coinciding with or being very close to the locus of the lethal. As already mentioned, there are two possible explanations of this. One is that the lethal mutation is a position effect due to the separation of the gene concerned from its usual neighbour or to its being brought into contact with a different gene. The alternative explanation is that the lethal is an internal change in the gene (mutation to a lethal allelomorph or loss of the power to reproduce) which was caused by the ionizing particle which broke the chromosome.

[1] This distinction between deficiencies caused by one or by two ionizing particles will be justified later.

A decision between these two possible explanations of a lethal associated with a gross structural change should be possible on the basis of the experimental data of the variation of yield with dose. Lethals of class (A), not associated with structural change, should increase linearly with dose, being produced by single ionizing particles. Those lethals of class (B) which are associated with very small deficiencies produced by a single ionizing particle should also increase linearly with dose. A minority of the class (B) lethals are associated with larger deficiencies caused by two ionizing particles, and these will increase with a higher power of the dose.

If the association of a lethal with a gross structural change is an example of the position effect, then the lethals in class (C) form a separate type the number of which is proportional to the number of gross structural changes produced. Now this number increases more rapidly than the first power of the dose, since gross structural changes involve two or more breaks which (with X-rays) are produced by separate ionizing particles. Cytological observation shows that in the dose range 1000–4000 r. the number of gross structural changes is proportional to $(dose)^{\frac{3}{2}}$. On the position effect explanation therefore the observed lethals should be the sum of two types, one type not associated with gross structural change, the number of which is proportional to dose, and the other type, associated with gross structural change, the number of which is proportional to $(dose)^{\frac{3}{2}}$. The total number should therefore increase more rapidly than the first power of the dose.

On the other hand, if the lethals associated with gross structural change are not essentially different from other lethals, and the association with the structural change is due to the circumstance that the ionizing particle which caused the lethal also caused a break which happened to take part in a chromosome rearrangement, then the *total* number of lethals of types A and C should be proportional to the dose. The number associated with gross structural change will increase as $(dose)^{\frac{3}{2}}$; the residual number not associated with gross structural change will increase less rapidly than the first power of the dose.

Oliver[1] made some experiments in which he determined the number of sex-linked lethals as a function of the dose, and at

[1] Oliver, C.P. (1932).

each dose determined the proportion of the lethals which were associated with gross structural change. His results are reproduced in Fig. 24. It is evident that the number of lethals associated with gross structural change increases more rapidly than dose (curve I is the curve (dose)$^{\frac{3}{2}}$), but that it is the *total* number of mutations (curve III) not the number unassociated with gross structural change (curve II) which increases linearly with dose.

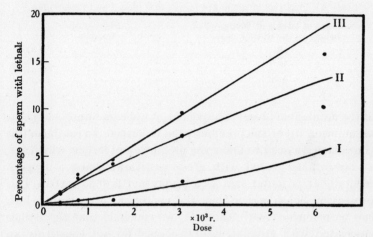

FIG. 24. Analysis of sex-linked recessive lethals (Oliver). I, lethals associated with gross structural change; II, lethals not associated with gross structural change; III, total lethals.

To make the test objective, we have assumed in turn (i) that the mean number of lethals per chromosome is proportional to dose, (ii) that it is the sum of two terms, proportional to (dose)^1 and (dose)$^{\frac{3}{2}}$ respectively, and have determined what is the maximum proportion of the (dose)$^{\frac{3}{2}}$ lethals which can be admitted without disagreement with experiment. The χ^2 test was used as a measure of agreement with experiment; spontaneous lethals were allowed for, as also was the distinction (discussed on p. 146) between the mean number of lethals per chromosome and the proportion of chromosomes carrying lethals. The experimental data used were those shown in Fig. 21, obtained by Timoféeff-Ressovsky[1] and based on some 60,000 cultures. The results of the χ^2 tests are given in Table 41, in which the proportion of (dose)$^{\frac{3}{2}}$ lethals postulated is indicated by stating in the first

[1] Timoféeff-Ressovsky, N.W. (1939).

column the proportion of the total number of lethals which at 3000 r. are of this class.

It is clear that the data provide no evidence for any (dose)$^{\frac{3}{2}}$ component, that a proportion as high as 17·5 % is improbable, and that a proportion as high as 22 % can practically be ruled out. Remembering that such of the deficiencies of type B as involve two independent ionizing particles will account for some

TABLE 41. Analysis of sex-linked lethals into (dose)^1 and (dose)$^{\frac{3}{2}}$ classes

Percentage of lethals at 3000r. which belong to the (dose)$^{\frac{3}{2}}$ class	χ^2	Degrees of freedom	P
0	2·3	2	0·32
12·5	3·6	2	0·16
17·5	6·4	2	0·04
22·0	9·6	2	0·008

of the admissible (dose)$^{\frac{3}{2}}$ component, it appears improbable that assumption (ii) of the previous page is correct. To establish this conclusion we need to know the proportion of lethals which are, at 3000 r., associated with gross structural change. Experimentally it is found that this proportion is $0·35 \pm 0·04$.[1] This proportion being higher than the maximum proportion which can be reconciled with Table 41, we conclude that the lethals associated with gross structural change do not constitute an additional class of lethals caused by position effect and requiring two ionizing particles for the expression of the lethal, but merely represent those cases where the ionizing particle which caused the lethal caused also a break which took part in structural change. We shall develop the subsequent discussion on this basis.

There is good reason to believe[2] that not all the chromosome breaks primarily produced take part in structural change, but that many of the broken ends rejoin and the restituted chromosome is cytologically indistinguishable from an unbroken chromosome. Whether restitution or structural change occurs appears not to depend on a difference in the breakage process, but mainly on whether other breaks are available with which interchange can occur. It seems necessary to accept therefore

[1] Based on data by Oliver, C.P. (1932); Demerec, M. (1937); Demerec, M. & Fano, U. (1941); reviewed by Lea, D.E. & Catcheside, D.G. (1945a).
[2] See Chapter VII.

THREE TYPES OF RECESSIVE LETHALS

that since some of the breaks which take part in structural change are lethals, so also some of the breaks which restitute are lethals. Such lethals will be recorded as type A lethals (p. 155), i.e. lethals without any cytologically detectable chromosome change. We have no reason to suppose *a priori* that a type A lethal cannot be produced without the chromosome at the same time being broken. However, admitting the necessity, on other grounds, for a considerable number of restitutional breaks, a large part of the type A lethals must be restitutional breaks, and we shall see how far a consistent picture can be obtained on the basis that *all* the type A lethals are restitutional breaks.

As a beginning we need to know the numbers of the three types of lethal produced by a given dose. The experimental result is that 3000 r. produce in 1000 X chromosomes 87 lethals, of which 39 are type A, 18 are type B, and 30 are type C.[1] Now the number of breaks in the euchromatin of the X chromosome which take part in gross structural change when a dose of 3000 r. is given to the sperm is 80 per 1000 X chromosomes;[2] 30 of these carry lethals (i.e. the 30 type C lethals). Evidently the probability that a chromosome break shall cause a lethal is $30/80 = 0.38$.

There are 18 minute deficiencies (i.e. the type B lethals), which are lethal because one or more loci are deleted. There is reason for believing that when a chromosome is broken in two places, the probabilities are approximately equal that the segment between the breaks shall be *deleted* and that it shall be *inverted*.[3] We presume therefore that there are also 18 minute inversions. An inversion will not, on our view, behave as a lethal *per se*, but since it involves two breaks each of which has a probability 0·38 of being a lethal, the probability is $1 - (1 - 0.38)^2 = 0.62$ that at least one of the breaks will be a lethal. Thus of the 18 minute inversions $18 \times 0.62 = 11$ will behave as lethals, and will therefore be included in the type A lethals, since a minute inversion will rarely be recognized cytologically.

This leaves 28 type A lethals which are restituted breaks.

[1] See Lea, D.E. & Catcheside, D.G. (1945a) for details.
[2] Deduction by Fano, U. (1941) from salivary gland observations of Bauer, H. (1939a).
[3] Demerec, M., Kaufmann, B.P., Sutton, E. & Fano, U. (1941); Demerec, M. & Fano, U. (1941).

Since only 38% of breaks are lethals, the total number of restituted breaks must be $28/0.38 = 74$. The total number of breaks of all sorts produced by 3000 r. in the euchromatin of 1000 (viable) X chromosomes is therefore 226, made up of 36 in minute deletions (2 breaks per deletion), 36 in minute inversions, 80 in gross structural changes, and 74 which restitute. Thus of the 226 breaks at 3000 r. a proportion $80/226 = 35\%$ take part in structural change. At greater doses the proportion will be higher, at smaller doses it will be lower. It is to be noted that 11 out of 39 or about 30% of the type A lethals are expected to be minute inversions. Slizynski,[1] from observation of salivary chromosomes, suspected that some of the non-deficiency lethals were minute inversions.

In this way it is possible to build up a consistent picture[2] of the production of recessive lethals. To summarize: Breaks are caused in the chromosomes by the passage through them of ionizing particles. Sometimes an ionizing particle causes two breaks close together, in which case deletion or inversion may occur of the segment of chromosome between the breakage points. The remaining breaks are available for taking part in gross structural change, the probability of a break doing so being dependent on the availability of other breaks, and thus increasing with increase of dose. Those which do not take part in gross structural change rejoin in the original formation.[3]

Any break irrespective of its subsequent history may result in a lethal change in the gene at or adjacent to the breakage point, this change probably often being destruction of the power of reproduction. The probability of a break causing a lethal in this way is 0·38. Thus any type of chromosome structural change—minute deletion, minute inversion, or gross structural change—is liable to have a lethal change in the gene adjacent to a breakage point. Apart from this, a deletion will usually have a recessive lethal effect on account of the absent loci.

[1] Slizynski, B.M. (1938).
[2] Some additional evidence supporting the picture outlined is given by Lea, D.E. & Catcheside, D.G. (1945a).
[3] We are limiting our discussion at present to viable changes. A third alternative behaviour of a broken chromosome leads to a dominant lethal (*vide infra*).

Dominant lethals in *Drosophila*

A recessive lethal mutation in a diploid organism is a gene change (or deletion) which in the homozygous or hemizygous condition results in the organism being non-viable. When the cells of the organism are heterozygous for the lethal, the

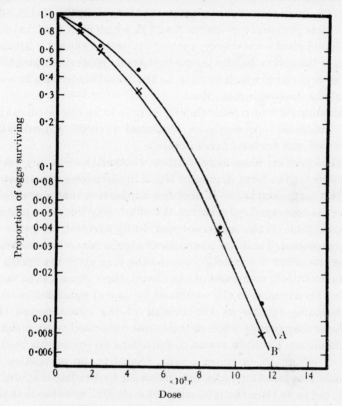

FIG. 25. Proportion of eggs fertilized by irradiated sperm which attain the stage of: A, larvae; B, adult flies (Catcheside & Lea).

organism is viable. Since there are among visible gene mutations both dominant and recessive mutations, we may by analogy expect also dominant lethals to exist. An egg fertilized by a sperm carrying a dominant lethal will by definition not give rise to an adult organism. In consequence a dominant lethal cannot be studied in successive generations and cannot be obtained in a salivary chromosome, and we can only infer the occurrence of a dominant lethal in irradiated *Drosophila* sperm, for example, by observing that a proportion of the eggs fertilized by such sperm fail to develop into adult flies.

A number of authors have made experiments to determine how the proportion of eggs which hatch, or which reach the pupal or adult stages, diminishes with increase of the dose of radiation received by the sperm which fertilized the eggs. Fig. 25 shows the results[1] obtained in an experiment of this sort. The close proximity of curves A and B, which refer to larval and adult survival respectively, shows that death caused by irradiation of the sperm usually occurs in the embryonic state, and that those organisms which survive to the larval stage usually successfully develop into adults.

Examination of the developing embryo in an egg fertilized by an irradiated sperm shows abnormal division figures with clumped and broken chromosomes.[2]

The question immediately arises whether there is any justification for the term dominant lethal in discussing experiments of this sort, carrying as it does the implication that a genetical effect is concerned, or whether the effect may not be a *physiological effect*[3] on the sperm not specifically affecting the genes or chromosomes. There are a number of arguments[4] which indicate that the effect is a genetic one. In the first place the sperm is almost entirely composed of chromatin, the volume of the head being about equal to the combined volume of the chromosomes it contains (taking as the volume of the chromosomes the volume they occupy when in their most condensed state, meiotic metaphase). Thus the action of radiations on sperm can hardly be on cytoplasm or nuclear sap. Secondly, the proportion of female flies hatching from eggs fertilized by irradiated sperm is reduced more than the proportion of male flies, showing that the X-bearing sperm are more sensitive than the Y-bearing sperm. If the action is a genetic one on the chromosomes, then this result is to be expected since fewer breaks are produced in the Y chromosome than in the X, and it is genetically practically

[1] From an experiment of Catcheside, D.G. & Lea, D.E. (1945a). A summary of other work on this subject is given in this paper. In Fig. 25 correction has been made for deaths in the controls.

[2] Sonnenblick, B.P. (1940).

[3] This vague term is conveniently employed to denote an effect other than a direct effect on a single structure such as a gene or a chromosome.

[4] Muller, H.J. (1940).

inert. Finally, it will be found that a quantitative explanation of the curve relating the yield of dominant lethals to the dose can be given on the basis that the cause of dominant lethals is the breakage of chromosomes by the radiation.

It appears that dominant lethals are for the most part chromosome changes, rather than gene mutations proper. Many deletions, involving a few bands of the salivary chromosome and one or a few known loci, have been studied, and have mostly been found to behave as recessive lethals. It appears therefore that deficiency for a single gene usually has a recessive lethal and not a dominant lethal effect. It seems improbable that a gene mutation should have a more strongly lethal effect than a complete deficiency for the gene concerned, so that dominant lethal mutations in the strict sense of intra-gene changes are probably rare. Larger deficiencies than about 50 salivary bands are, however, not found even in the heterozygous condition, nor are dicentric chromosomes ever found in salivary chromosomes. There is reason to believe,[1] however, that deletions are produced about as frequently as inversions of the same size, and that asymmetrical interchanges, producing dicentric and acentric chromosomes, are about as frequent as symmetrical interchanges producing monocentric chromosomes. Larger deficiencies must therefore be presumed to behave as dominant lethals. An acentric chromosome, since it lacks a centromere which is responsible for the commencement of the migration of the chromosome to the pole of the dividing cell, is liable to be left out of both daughter nuclei. The two centromeres of a dicentric chromosome may attempt to migrate to opposite poles, or if they migrate towards the same pole the sister chromatids may interlock, in either event leading eventually to loss of the chromosome or breakdown of the dividing cell.[2] There is little doubt therefore that the larger deletions, and asymmetrical interchanges, behave as dominant lethals. However, calculations based on the frequencies found for the observable types of aberration show that these aberrations are not frequent enough to account for the whole of the dominant lethals.[3] Moreover, the shape of the survival curve (Fig. 25) is not consistent with the explanation that dicentric interchanges and larger deletions are entirely re-

[1] Demerec, M., Kaufmann, B.P., Sutton, E. & Fano, U. (1941).
[2] Pontecorvo, G. (1942).   [3] Fano, U. (1941).

sponsible for dominant lethals. The survival curves, which in Fig. 25 are plotted on a logarithmic scale, are initially linear. As explained in Chapter III, a survival curve which is linear on a logarithmic scale suggests that a single ionizing particle causes the effect studied. Now an interchange or a deletion (other than a minute deletion) involves the production of two chromosome breaks by two separate ionizing particles. We conclude that in addition to the production of asymmetrical interchanges and deletions, there is some other mechanism, involving a single ionizing particle, which has a dominant lethal effect. At small doses this mechanism predominates, giving the approximately linear curve. At higher doses the two-break aberrations, the number of which increases more rapidly than the first power of the dose, will become increasingly important, thus explaining the increased gradient shown at higher doses in Fig. 25.

Chromosome loss, caused by a single chromosome break, has been plausibly suggested[1] as the mechanism in question. When a chromosome is broken, the two broken ends usually join, either with one another restituting the original chromosomes, or with other broken ends which may be available in the cell giving one of the various familiar types of chromosome aberration (interchanges, inversions or deletions). It appears, however, that in a small proportion of cases the two broken ends remain unjoined. If still unjoined when the chromosome splits into two chromatids, it appears likely[2] that sister-union of the chromatids will occur at the breakage points. At anaphase therefore we have one acentric fragment and one dicentric chromosome. As already explained, such chromosomes are likely to be lost at cell division.

Now loss of a chromosome is certainly lethal to *Drosophila* in the case of the large autosomes II and III, though a fly can exist in the absence of one of its very small IVth chromosomes. Moreover, X/O males are viable, and such individuals may be considered as normal males which have lost a Y or as normal females which have lost an X chromosome. Thus losses of the IVth chromosome, or of the sex chromosome as a result of irradiating sperm, may be viable. However, since about 80% of all the breaks produced occur in the chromosomes II and III,

[1] Muller, H.J. (1940); Pontecorvo, G. (1941, 1942).
[2] From analogy with known types of aberration in plant material. See Chapter VI.

unjoined single breaks, if they do occur, will usually behave as dominant lethals.

That chromosome losses induced by a single ionizing particle do in fact occur has been shown in experiments[1] specifically designed to study viable loss of the sex chromosomes, and in which the zygote which had sustained the loss could be distinguished from one which had not. The percentage of losses induced was found to be proportional to the dose, indicating that loss occurred by a single break process.

The number of viable losses obtained in these experiments was small, much smaller than the number of unjoined breaks expected in the sex chromosomes on the basis of the number of unjoined breaks in the autosomes needed to explain dominant lethals. It appears that though the complete absence of one sex chromosome from the zygote would have been viable, the process of loss often caused death. The explanation is possibly to be sought in the mechanical upset of mitotic division caused by the chromosome bridge which is formed when a dicentric chromosome is present.

We now proceed to obtain a mathematical expression for the yield of dominant lethals as a function of the dose of radiation given to the sperm on the basis of the mechanisms we have been discussing, namely, unjoined single breaks and asymmetrical interchanges.[2] To make the mathematical analysis manageable it is necessary to simplify the problem to some extent. The principal simplifying assumptions are that the possibility of more than one break occurring in a single chromosome arm is neglected, and joining between the various broken ends in the cell is supposed to be at random. The first simplifying assump-

---

[1] Muller, H.J. (1940); Pontecorvo, G. (1941). The method in Pontecorvo's experiment was as follows. The irradiated males had a special Y chromosome containing the wild-type allelomorph (y^+) of the X chromosome gene y (yellow body). They were mated to attached-X females homozygous for y. The female offspring are normally of the constitution $X.X/Y$, deriving the attached-X chromosome $X.X$ from the mother and the Y chromosome from the father, and are grey-bodied since the y^+ in the Y masks the recessive y genes in the $X.X$ chromosome. If, as a result of irradiation of the sperm, the sex chromosome (either X or Y) is lost from the zygote, a female of constitution $X.X/O$ is formed which will have a yellow body since the y genes in the $X.X$ are no longer masked.

[2] The treatment given is due to Lea, D.E. & Catcheside, D.G. (1945a).

tion will not be a serious source of error.[1] The second assumption is plausible for *Drosophila* sperm though not for some other materials (see Chapter VII).

We take p to be the probability that a given break shall neither restitute nor take part in interchange, but shall lead to chromosome loss and therefore to a dominant lethal effect. $q = 1-p$ is the probability that it shall either restitute or interchange. If there are r breaks in the cell, q^r is taken to be the probability that all either restitute or interchange.

Suppose that with dose D the mean number of breaks per sperm is $m = \alpha D$. The proportion of sperm having r breaks per sperm is given by the Poisson distribution, and is $e^{-m}.m^r/r!$. The probability that a sperm shall have no breaks is e^{-m}. The probability that it shall have one break is $m.e^{-m}$. Sperm with one break will contribute $(1-q)m.e^{-m}$ to the number of dominant lethals, and $qm.e^{-m}$ to the number of viable nuclei without aberration.

Of the $\frac{1}{2}m^2.e^{-m}$ sperm with two breaks per sperm, $(1-q^2)\frac{1}{2}m^2.e^{-m}$ will be dominant lethals owing to failure of one or both breaks either to restitute or to interchange. In $\frac{1}{2}m^2.q^2.e^{-m}$ sperm the four broken ends will all join. Under the assumption of random joining, in one-third of these sperm there will be restitution, giving viable sperm without aberration, in one-third there will be symmetrical interchange giving viable sperm with chromosome aberration, and in one-third there will be asymmetrical interchange adding a further quota to the dominant lethals. Thus of the sperm with two breaks, $\frac{1}{6}m^2.q^2.e^{-m}$ will be viable without aberration, $\frac{1}{6}m^2.q^2.e^{-m}$ will be viable with aberration, and the remaining $\frac{1}{2}m^2.e^{-m}(1-\frac{2}{3}q^2)$ will carry dominant lethals.

In general there will be $e^{-m}.m^r/r!$ sperm having r breaks. In $e^{-m}.m^r.q^r/r!$ sperm no breaks will remain unjoined. In a sperm of this class the r breaks can join in

$$1.3.5.....(2r-1) = (2r)!/(r!\,2^r)$$

ways, of which one way is viable without aberration, $r!-1$ ways are viable with aberration, and the remainder are inviable.[2]

[1] It can be avoided at the cost of some complication of the calculations. Haldane, J.B.S. & Lea, D.E. (unpublished).
[2] Catcheside, D.G. (1938a); cp. also Fano, U. (1943a). These results follow from the postulate of random joining of broken ends.

Collecting the contributions from sperm with various numbers of breaks, and replacing m by its value αD, we have:

Proportion of cells which are viable and without aberration is

$$X = e^{-\alpha D} S_1,$$

where $\quad S_1 = 1 + \alpha q D + \tfrac{1}{6}(\alpha q D)^2 + \ldots + \dfrac{(2\alpha q D)^r}{(2r)!} + \ldots . \qquad$ (V-1)

Proportion of cells which are viable (with and without aberration) is

$$Y = e^{-\alpha D} S_2,$$

where $\quad S_2 = 1 + \alpha q D + \tfrac{1}{3}(\alpha q D)^2 + \ldots + \dfrac{(2\alpha q D)^r . r!}{(2r)!} + \ldots . \qquad$ (V-2)

Total number of primary breaks formed in viable sperm, per total sperm is

$$Z = e^{-\alpha D} S_3,$$

where $\quad S_3 = \alpha q D + \tfrac{2}{3}(\alpha q D)^2 + \ldots + \dfrac{(2\alpha q D)^r . r . r!}{(2r)!} + \ldots . \qquad$ (V-3)

The sums S_1, S_2 and S_3 of the infinite series in equations (1), (2) and (3) can be shown to be given by the following algebraic expressions:

$$S_1 = \cosh \sqrt{(2\alpha q D)}, \qquad \text{(V-4)}$$

$$S_2 = 1 + \sqrt{(\tfrac{1}{2}\pi \alpha q D)}\, e^{\frac{1}{2}\alpha q D}\, \mathrm{erf}\sqrt{(\tfrac{1}{2}\alpha q D)}, \qquad \text{(V-5)}$$

$$S_3 = \tfrac{1}{2}\alpha q D \left\{ 1 + \frac{1 + \alpha q D}{\sqrt{(\tfrac{1}{2}\alpha q D)}} \frac{\sqrt{\pi}}{2}\, e^{\frac{1}{2}\alpha q D}\, \mathrm{erf}\sqrt{(\tfrac{1}{2}\alpha q D)} \right\}, \qquad \text{(V-6)}$$

where $\cosh x = \tfrac{1}{2}(e^x + e^{-x})$ is the hyperbolic cosine, and $\mathrm{erf}\, x = \dfrac{2}{\sqrt{\pi}} \displaystyle\int_0^x e^{-x^2} dx$ is the error-function. In Table 42 values of S_1, S_2 and S_3 are tabulated for a suitable range of values of $\alpha q D$.

One observable quantity is the proportion of viable sperm which have chromosome structural changes. The theoretical expression for this proportion is evidently $(1 - X/Y) = (1 - S_1/S_2)$, and is listed in Table 42 as a function of $\alpha q D$. In Fig. 26 we show experimental data of the proportion of viable sperm having chromosome structural changes as a function of the dose,[1] together with the theoretical curve $(1 - S_1/S_2)$, which has been fitted to the data by taking $\alpha q = 0.57$ per 1000r.

[1] Catcheside, D.G. (1938a); Bauer, H., Demerec, M. & Kaufmann, B.P. (1938).

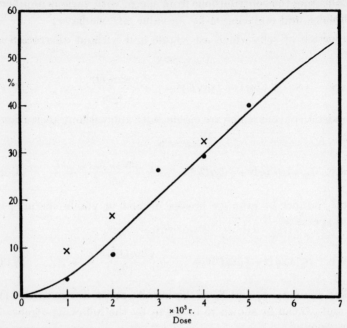

Fig. 26. Percentage of viable sperm having gross chromosome structural change. Curve theoretical, points experiments of: ● Bauer, Demerec & Kaufmann; × Catcheside.

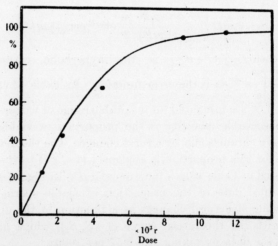

Fig. 27. Percentage of dominant lethals as a function of dose. Curve theoretical, points experiments of Catcheside & Lea.

A second observable quantity is $(1-Y)$, the proportion of total sperm which are non-viable. In Fig. 27 the experimental observations[1] on the proportion of eggs fertilized by irradiated sperm which fail to attain the adult stage are plotted, together

TABLE 42. Functions involved in the dominant lethal theory

αqD	S_1	S_2	S_3	$(1-S_1/S_2)$	S_3/S_2
0·32	1·337	1·356	0·3952	0·0140	0·2914
0·50	1·543	1·592	0·6942	0·0309	0·4360
0·72	1·811	1·920	1·152	0·0572	0·5996
0·98	2·151	2·373	1·849	0·0935	0·7793
1·28	2·577	2·996	2·915	0·1396	0·9731
1·62	3·107	3·858	4·553	0·1945	1·180
2·00	3·762	5·060	7·090	0·2565	1·401
2·88	5·557	9·172	17·29	0·3941	1·885
3·92	8·253	17·78	43·23	0·5357	2·432
5·12	12·29	36·82	112·2	0·6663	3·046
6·48	18·31	81·57	304·6	0·7755	3·734
8·00	27·31	193·6	870·8	0·8590	4·497

with the theoretical curve. In computing the formula for Y (see equation (2)) we already know that $\alpha q = 0.57$, which enables S_2 to be calculated for each dose with the aid of Table 42; α still remains arbitrary. The value $\alpha = 0.75$ was found to give the best fit of the theoretical curve to the experimental points.

It follows that $q = 0.57/0.75 = 0.76$, so that we have the figures:

$\alpha = 0.75$ is the number of primary breaks produced per sperm per 1000 r.

$q = 0.76$ is the probability that a break shall join, either in restitution or in interchange.

$p = 1-q = 0.24$ is the probability that a break shall remain unjoined, and shall behave as a dominant lethal.

It is with these values of α, p and q that the theoretical curves in Figs. 26 and 27 have been computed.

It is of interest to calculate the mean number of breaks primarily formed per *viable* sperm (which will be a little less than αD). Referring back to equations (2) and (3), this is seen to be S_3/S_2. S_3/S_2 is tabulated against αqD in Table 42. Using the value $\alpha q = 0.57$ just found and interpolating in Table 42, we obtain the estimates given in Table 43 of the mean number of primary breaks per viable sperm. We can calculate from these figures the number of primary breaks in the euchromatin of the

[1] Catcheside, D.G. & Lea, D.E. (1945a).

X chromosome, by making use of the result that a fraction 0·162 of all observed breaks occur there.[1] Thus at 3000 r.

$$0.162 \times 1.23 = 0.199$$

primary breaks occur in the euchromatin of the X chromosome.

TABLE 43. Mean number of primary breaks per viable sperm

Dose (r.)	1000	1500	2000	3000	4000	6000
Mean number of primary breaks per viable sperm	0·49	0·69	0·89	1·23	1·54	2·15

The analysis we have just given of the dominant lethals and chromosome aberrations leads to an estimate 0·199 of the number of primary breaks produced by 3000 r. in the euchromatin of the X chromosome in sperm which remain viable. Our analysis of recessive lethals on the basis of the hypothesis that recessive lethals are rejoined breaks led independently to an estimate of this same quantity, the value obtained (p. 160) being 226 per 1000 sperm, i.e. 0·226 per sperm. The agreement between 0·199 and 0·226 is satisfactory, and shows that the two theories are compatible.

We have mentioned that in the progeny of irradiated males there are more males than females, indicating that fewer dominant lethals are produced in a Y-bearing sperm than in an X-bearing sperm by a given dose. This can be ascribed to the known fact[2] that fewer breaks are produced by a given dose in a Y chromosome than in an X chromosome, in the ratio of 79:100. The calculation we have just given can readily be extended to predict the ratio of females to males in the progeny of males which have received a given dose of X-rays, and the predicted curve,[3] together with the experimental points,[4] is given in Fig. 28A.

It has been found[5] that if the irradiated males are of a special

[1] Deduction by Fano, U. (1941) from the observations of Bauer, H. (1939a).

[2] Deducible from the observations of Bauer, H., Demerec, M. & Kaufmann, B.P. (1938).

[3] Details of the calculation are given in Lea, D.E. & Catcheside, D.G. (1945a).

[4] Hanson, F.B. (1928); Muller, H.J. (1928); Gowen, J.W. & Gay, E.H. (1933); Bauer, H. (1939b); Catcheside, D.G. & Lea, D.E. (1945a). The experimental data are reviewed in the last of the references cited.

[5] Bauer, H. (1939b).

stock in which the X chromosome, instead of being in the form of a rod, is ring-shaped, then the sex-ratio distortion is much greater. The reason for this is that not only does a break which fails to join have a dominant lethal effect, but also a proportion

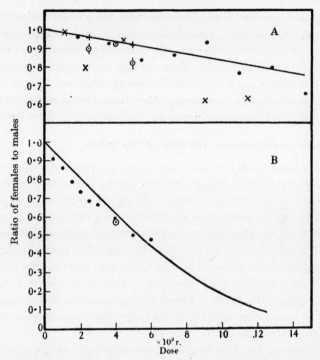

FIG. 28. Depression of the sex-ratio in the progeny of irradiated males. Curves theoretical, points experiments of: A, rod-X stock: + Hanson; ⌀ Muller; ● Gowen & Gay; ⊙ Bauer; × Catcheside & Lea. B, ring-X stock: ● Bauer; ⊙ Catcheside.

of the breaks which restitute. For, if the broken ends rotate through half a revolution relative to one another before rejoining, then when the chromosome splits into two chromatids, instead of these forming two separate rings, they will form a single ring of twice the size. Such a chromosome, being dicentric, will be lost and thus behave as a dominant lethal. It is known[1] that in approximately half of the cases restitution is inviable. Thus the probability that a single break shall have a dominant

[1] From study of the relative frequency of inversions in ring-X and rod-X chromosomes; Catcheside, D.G. & Lea, D.E. (1945b).

lethal effect is $p+\tfrac{1}{2}q$ in the case of a ring chromosome, against p in a rod chromosome.

Taking account of this phenomenon the sex-ratio depression in ring-X stock can be predicted, and the agreement of the experimental points and the calculated curve [1] is shown in Fig. 28B.

Further, we can, from the experimentally observed proportions of the sex-ratio distortion in rod-X and ring-X stocks, deduce the ratio of $p+\tfrac{1}{2}q$ to p, which leads to values for p and q, namely, $p=0.26$, $q=0.74$.[2] These values are in good agreement with the values $p=0.24$, $q=0.76$ already obtained (p. 169).

We conclude that dominant lethal production is satisfactorily explained by the mechanisms and calculations we have given.

Deductions concerning the size of the gene [3]

At the time when the induction of gene mutations by radiation was discovered, the 'target theory' of the biological action of radiations was already current. When it appeared that the evidence on the production of mutations by radiation was consistent with the view that a single ionization could cause mutation of a gene, it was natural that the radiation data should be used to calculate the target size for mutation, and that the target should be identified with the gene. More recent writers have usually adopted a more cautious attitude, and some have gone so far as to say that the size of the target has nothing to do with the size of the gene.[4]

The reasons for doubting that the size of the target is a fair approximation to the size of the gene are principally:

(a) It is argued that ionization outside the gene may perhaps cause mutation.

(b) Ionization inside the gene may not always cause mutation, but may have a probability of doing so considerably less than unity.

We have already discussed in Chapter III the modifications made in target theory calculations by introducing these con-

[1] See Lea, D.E. & Catcheside, D.G. (1945a) for details of the calculation.
[2] Catcheside, D.G. & Lea, D.E. (1945b).
[3] This section follows the treatment of Lea, D.E. & Catcheside, D.G. (1945b).
[4] E.g. Timoféeff-Ressovsky, N.W. & Delbrück, M. (1936); Muller, H.J. (1940).

siderations. For (a) to be of serious importance it is necessary for sufficient spread of the effect of an ionization to occur for there to be appreciable probability of mutation in a gene when an ionization is produced at a distance of the order of a gene diameter or more outside it. The possibility of some spread of the effect of an ionization has been discussed in Chapter II. A spread over a range of the order of $1\,m\mu$ can be understood. Spread over a distance equal to the separation of two sister chromatids at prophase ($\sim 100\,m\mu$) has been shown experimentally not to occur.[1] The notion that the effect of an ionization can spread appreciable distances has been adopted by some geneticists[2] because the yield of minute chromosome aberrations has been found to be proportional to the dose of radiation (Chapter VI), and this has been taken to mean that a single ionization can produce two chromosome breaks an appreciable distance apart. However, the proportionality to dose does not necessarily mean that a single *ionization* causes the two breaks, but that a single *ionizing particle* causes them. As explained in Chapter VII we should in any event expect that two breaks primarily produced at a separation of less than about $100\,m\mu$ would be produced by a single ionizing particle and not by two separate ionizing particles. Thus the finding that the yield of minute aberrations is proportional to dose confirms expectation, but has no bearing on the question of whether two breaks can be produced by a single ionization or not.

It is probably safe to neglect the possibility of gene mutation resulting from ionization outside the gene by comparison with the probably much higher chance that ionization inside will be effective, though the point is one which requires more investigation. However, that considerations (a) and (b) together do not in practice prevent target-theory calculations giving essentially correct results is made clear by the argument we now proceed to develop.

The obvious method of testing the correctness of the method by which it is proposed to calculate the size of the gene from

[1] Long wave-length X-rays, which dissipate their energy in tissue by means of photoelectrons of range less than the separation of sister chromatids, are unable to break both chromatids simultaneously (Catcheside, D.G. & Lea, D.E. 1943; Lea, D.E. & Catcheside, D.G. 1945b).

[2] Muller, H.J. (1940).

radiation experiments is to apply the same method in the case of some object as nearly as possible resembling a gene but of known size. Enzymes and viruses appear to satisfy the required conditions. As shown in Chapter II, a single ionization in an enzyme molecule leads to its inactivation, and as shown in Chapter IV, there is, in the case of the macromolecular viruses, a close relation between the 'target' size for virus inactivation and the size of the virus particle itself. It is true that the target diameter, although often closer, sometimes differs by a factor of 2 from the virus diameter, but in view of the extreme paucity of our knowledge of the size of genes by other methods,[1] an estimate which was probably correct to a factor of 2 would constitute much the most precise estimate at present available.

The fact that the probability of producing a gene mutation is proportional to the dose of radiation and independent of the intensity makes it likely that a gene mutation is produced by the passage of a single ionizing particle through the gene. In the case of γ-rays the ionizing particle will be a fast electron, successive clusters of ionizations (i.e. successive *primary* ionizations) being produced at intervals of the order of 1μ along its path. We can be quite sure that the diameter of the gene is considerably less than 1μ, so that with γ-rays there is little chance of more than one ion-cluster falling in the gene. From the fact that γ-rays are, per ionization, at least as effective as more densely ionizing radiations, we infer that a single ion-cluster can cause a gene mutation. The majority of ion-clusters are single ionizations,[2] and only a small proportion of the total number of ionizations are located in clusters larger than about three ionizations. Thus we can fairly safely deduce that a single ionization, or at most a cluster of two or three, suffices to cause gene mutation, though we have, of course, not yet given any proof that the probability of a single ionization in a gene causing mutation approaches unity.

What changes may be expected to follow the ionization of an atom in a gene? The energy given to an atom in the process of ionization exceeds the energy necessary to break the chemical

[1] The various methods which have been proposed, e.g. by Muller, H.J. (1935), only give upper limits to the size of the gene.
[2] 'One ionization' involves the production of a positive and a negative ion (see Chapter I).

bonds uniting the atom to its neighbours in the molecule. We should expect therefore ionization of an atom to result in chemical change in the molecule. The review of the chemical effects of radiation in Chapter II was consistent with this expectation, an ionic yield of the order of unity being the typical result obtained in radiochemical studies. There are some cases in which a chain reaction is instituted so that many more molecules react than there are ionizations produced. There are other cases where the ionic yield is low, probably owing to recombination of the products of decomposition of the molecule. However, the more complicated the molecule the less probable seems exact restitution, and we can probably safely assume that the gene will suffer chemical change as a result of the ionization of an atom in it.

In attempting to picture the probable results of chemical change in a gene we shall make use of the now widely adopted view that genes have a good deal in common with the macromolecular viruses. It is known that a certain amount of chemical change can be tolerated by a virus without producing any permanent inherited change in it.[1] Thus the chemical change produced in a gene by an ionization may sometimes not produce any genetically important effect.

Sometimes the change produced by the ionization will be a permanent stable change, so that the gene reproduces itself in the changed form. This clearly happens in the case of gene mutation, when the chemical change in the gene is recognized by a detectable change in its behaviour.[2] It must be realized that we cannot be sure of detecting every permanent change in the gene by a change in gene behaviour. For example, in the white-eye allelomorphs in *Drosophila* we have a large number of different states of the gene which are distinguishable because eye colour is a character in which many quantitatively slight differences can be recognized. The different allelomorphs also affect the colour of the Malpighian tube in the larva, but if we had to rely

[1] Miller, G.L. & Stanley, W.M. (1941) showed that 70% of the amino groups of tobacco mosaic virus protein could be acetylated by chemical treatment without reducing the infectivity of the virus. After multiplication of the treated virus in the host plant, the virus had returned to its original chemical composition.

[2] We are not accepting the extreme view that all mutations are position effects, and are supposing that in a typical point mutation there is chemical change in the gene.

only on the colour of the Malpighian tube we should recognize fewer allelomorphs and hence in radiation experiments often fail to recognize the occurrence of a mutation when mutation had in fact occurred. Even using the more sensitive eye colour as a means of detecting mutation, changes in the gene can occur without being detected, since it is known that allelomorphs exist having the same eye colour but differing in other properties (e.g. viability or fertility[1]).

It is evident therefore that the experimentally determined mutation frequency of a given locus will usually underestimate the frequency with which permanent and viable changes are produced in the gene.[2] This consideration probably applies especially forcibly to the experiments described earlier in the chapter (p. 140) on the induction of mutations in a plant virus. Here mutation was recognized by the production of a local lesion on a leaf inoculated with virus, in place of the usual mottle. It would clearly be unreasonable to assume conversely that, if after irradiation the virus still produced a mottle on the leaf, there was therefore no permanent change induced in the virus molecule by the radiation. The fact that an ionization in the virus had rather a small chance (of the order of 10^{-3}) of producing the particular effect studied does not necessarily mean that an ionization in the virus has only this small probability of producing a viable inherited change.

The complete deficiency of a gene, when homozygous, is usually lethal in *Drosophila*. If as a result of ionization a gene suffers a change which causes the loss of the power of reproduction or alternatively the loss of all its characteristic activity, a recessive lethal will usually be recorded. Lethal mutation at a given locus (excluding deficiency caused by deletion of a portion of chromosome) appears to be not much more frequent than visible mutation,[3] which is perhaps surprising. The probability that an ionization produced in a gene shall cause a lethal mutation is a quantity of which we can make some estimate by appeal to experiments on enzyme and virus inactivation, since, if we take

[1] Timoféeff-Ressovsky, N.W. (1933b).
[2] This point has been emphasized by Timoféeff-Ressovsky, N.W. (1937a).
[3] Data bearing on this question have been given by Patterson, J.T. (1932), Demerec, M. (1937), Muller, H.J. (1940).

SIZE OF THE GENE 177

enzyme and virus inactivation to be analogous to lethal mutation in these substances, the ratio of the size of the target to the size of the molecule will be equal to the probability that an ionization in the molecule shall cause lethal mutation. In Table 44 we list the target size and the particle size for the inactivation of some enzymes and viruses. If we are correct in arguing by analogy from the inactivation of viruses and enzymes to lethal mutations in genes, the figures in Table 44 show that the target

TABLE 44. *Enzyme and virus inactivation*

Enzyme or virus	Molecular weight	Target weight	Ratio	Reference
Ribonuclease	1.5×10^4	3.2×10^4	2.1	1
Phage S 13	1.7×10^6	1.5×10^6	0.9	2
Tobacco ringspot virus	3.4×10^6	2.27×10^6	0.67	3
Tobacco necrosis virus	7.2×10^6	1.57×10^6	0.22	3
Bushy stunt virus	10.6×10^6	2.32×10^6	0.22	3

1 Lea, D.E., Smith, K.M., Holmes, B. & Markham, R. (1944).
2 Lea, D.E. & Salaman, M.H. (unpublished).
3 Lea, D.E. & Smith, K.M. (1942).

size agrees with the gene size within a factor of 5 in weight (which is a factor of 1·7 in diameter). Paying principal attention to the virus data, the difference between target size and virus size is in the direction of target size being less than virus size, which is to be expected in view of the fact, already discussed, that some changes in the virus produced by an ionization leave it still infective.

To determine the size of a gene from radiation data we need to derive from the observations an estimate of the probability that, with a given dose, ionization shall be produced in the gene. From the discussion which we have given, this means the sum of the probabilities that there shall be visible mutation to any other allelomorph, or lethal mutation (excluding chromosome deletion), plus the probability that there shall be change in the gene which is either not permanent or which gives no detected change in its properties.

Studying any given locus we are usually only in a position to determine the probability that a recognizable viable mutation shall occur, which will be an underestimate of the total probability, and will thus lead to an underestimate of the size of the gene. The mutation frequencies in Table 37 (p. 144) are of the order 1–10×10^{-8} per gene per roentgen, i.e. a dose of the order 10^7–10^8 r. is required for an average of one mutation per gene. Consulting Figs. 8 and 9 we deduce that the molecular weight of the gene is 10,000–100,000, or its diameter (if spherical) is 2–6mμ. As explained, these estimates are liable to be underestimates for the sizes of the genes to which the data of Table 37 refer, though they may not necessarily be underestimates for the average size of a gene in *Drosophila*, since the genes of which the mutation rates are known include those most frequently suffering mutation under irradiation.

Turning now to lethal mutations at random loci, we have the advantage of knowing the relative efficiency of different radiations (Table 39). As discussed already, lethals do not constitute a single type. However, according to the view put forward earlier in the chapter, types A and C lethals do not differ in origin. Type B lethals, namely, those involving cytologically detected deficiencies, are, however, of different origin. At any rate, deletions for more than one band probably involve two breaks close together and the deletion of the part of the chromosome between them, and therefore do not necessarily involve the ionization of the gene responsible for the lethal effect. We should therefore estimate the mutation yield after excluding the chromosome deletions, and calculate the relative efficiency of different radiations also after their exclusion. It is at present not clear whether deficiencies for single salivary chromosome bands should be excluded, or whether such deficiencies are due to ionization in a particular gene causing loss of the power of reproduction, in which case they should be retained. The proportion of X-ray random lethals which are deficiencies for one or more bands is about 30%;[1] the proportion which are deficiencies for more than one band is lower; there is as yet no information available regarding the proportion of random lethals produced by other radiations which are cytologically detectable deficiencies. Lacking at the present time the information necessary to correct Table 39 for the proportion of lethals due to cytologically detectable deficiencies, we shall, for the purposes of the present calculation, have to neglect this correction.

We shall first make the calculation on the assumption that one or more ionizations anywhere in the gene leads to lethal muta-

[1] Slizynski, B.M. (1938, 1942).

TARGET DIAMETER FOR MUTATION

tion, and then consider the modification introduced by allowing for the probability of lethal mutation being less than unity. As explained in Chapter III, knowledge of the relative efficiencies of soft X-rays, neutrons and α-rays compared to γ-rays or hard X-rays enables an estimate to be made of the target diameter. In Table 45 we give the inferred target diameters (assuming density 1·35 g./cm.³) deduced from the relative efficiencies listed in Table 39 with the aid of Fig. 10. The estimate of the target

TABLE 45. Target diameter for mutation

Radiation	Hard X-rays or γ-rays	Soft X-rays 2–3A.	Neutrons (Li+D)	α-rays ~3eMV.
Relative dose for equal yield of mutations	1·00	1·30	1·45	3·44
$2r\rho$ (in mμ)	—	6	12	9
Target diameter (in mμ)	—	4·4	9	6·6

diameter obtained from the relative efficiency of different radiations is very sensitive to error in the determination of this relative efficiency, and may also be appreciably affected by oversimplification of the calculation.[1] It is satisfying, however, that the order of target diameter given by Table 45 (4–9 mμ) is consistent with that already obtained on p. 178 (2–6 mμ) by an entirely different method.

We may make an estimate of the number of genes in the X chromosome (strictly the number of genes capable of showing a recessive lethal mutation) as follows. The yield of sex-linked lethals by X- or γ-rays is 2·89 % per 1000 r., so that

$$1000/0\cdot 0289 = 3\cdot 46 \times 10^4 \text{ r.}$$

is the dose required for an average of one lethal mutation per X chromosome. For a given size of gene we can read off from Fig. 8A the dose required for an average of one mutation per gene. Taking the gene diameter to be 4 mμ, the smallest value which seems consistent with Table 45, we find that the dose of X-rays (0·15 A.) required for an average of one mutation per gene is $2\cdot 9 \times 10^7$ r. The number of genes in the X chromosome is therefore deduced to be $2\cdot 9 \times 10^7 / 3\cdot 46 \times 10^4 = 838$. Larger sizes for the

[1] E.g. a target diameter of about 4 mμ was deduced from essentially the same data when this calculation was first given (Lea, D.E. 1940a), owing to the neglect of δ-rays and some other simplifications.

gene are permitted by Table 45 and would lead to smaller estimates of the number of genes in the X chromosome. Thus if the gene diameter is $6\,m\mu$ we obtain 281 as our estimate of the number of genes in the X chromosome. These estimates are of the right order, a figure of about 500 or 1000 having been suggested by several authors.[1]

The calculation as we have so far made it is based on the assumption that the probability of a lethal mutation occurring when an ionization is produced in the gene is unity. A *priori* it may be less than unity, though from the evidence of Table 44 we do not expect it to be an order of magnitude less. However, in Chapter III (p. 95) we went into the question of the effect upon calculations of this sort of the probability, p, being less than unity, and concluded that the effect was to cause the estimate of the size of the gene to be too small in the ratio of $p:1$, and the estimate of the number of genes to be too great in the ratio of $1:p^2$. However, it is clear that there is not room for our estimate of the number of genes to be too great by any large factor, certainly not by a factor of 10. Hence p^2 cannot be much less than unity, certainly not lower than $0\cdot1$. Hence p cannot be as low as $0\cdot3$. We deduce that our estimate of 4–$9\,m\mu$ cannot be greatly in error on this account. It is probable that the average gene diameter does not exceed $10\,m\mu$, and we shall take 4–$8\,m\mu$, as suggested by Table 45, as the most probable value.

Cosmic rays and mutations

Following Muller's discovery in 1927 that ionizing radiations are able to cause mutations, the idea occurred to several workers that spontaneous mutations may be due to natural radiation, i.e. cosmic rays from outside the earth, γ-rays due to the radioactive content of the floor and walls of the room, or α- and β-rays due to the content of radioactive matter in the organism itself. This attractive possibility, however, has been conclusively disproved in the case of *Drosophila*. In the first place, the intensity of the natural radiation is insufficient.[2] With X-rays and γ-rays the rate of production of sex-linked lethals is roughly 3 % per 1000 r., and is proportional to dose. Experiments demon-

[1] E.g. Muller, H.J. (1929); Gowen, J.W. & Gay, E.H. (1933); Demerec, M. (1934).

[2] Muller, H.J. (1930).

strating the proportionality of yield to dose do not, of course, extend to an intensity as low as that of natural radiation. They do, however, cover a very wide range (Fig. 22), and if the theoretical interpretation that a single ionizing particle is responsible for a mutation is correct, we should expect the yield to be proportional to dose however low the intensity. A generous estimate of the dose rate due to natural radiation is 50 ionizations per cm.³ of air per second, or $2 \cdot 5 \times 10^{-8}$ r./sec., which corresponds to about 0·05 r. in the lifetime of a fly. Thus the rate of production of sex-linked lethals per generation by natural radiation should amount to 0·00015%, which is about a thousand times smaller than the observed rate of spontaneous sex-linked lethals.

This calculation, based on an estimate of the intensity of the natural radiation derived from measurements of the natural leak of ionization chambers, is not entirely convincing since living organisms sometimes concentrate radioactive matter in their tissues. Mott-Smith and Muller,[1] however, made measurements of the radium content of flies and concluded that it was insufficient to account for the spontaneous mutation rate.

A further argument comes from the fact that the rate of induction of mutations by radiation is independent of temperature (Table 40), while the rate of spontaneous mutation is increased markedly by rise of temperature.

It thus appears certain that spontaneous mutations in *Drosophila* are not due to natural radiation.[2] There is some evidence with less well-investigated organisms pointing in the opposite direction,[3] but further experiments are required here.

Genetical effects of ultra-violet light

The induction of mutations by ultra-violet light has been studied in *Drosophila*, in maize, in *Antirrhinum*, and in some lower plants.

The small penetrating power of the radiation complicates the experiments with *Drosophila*. Some experiments have been made by irradiating fertilized eggs in the polar cap stage,[4] as in

[1] Muller, H.J. (1930).
[2] See Rajewsky, B.N. & Timoféeff-Ressovsky, N.W. (1939) for further discussion and experiments strengthening this conclusion.
[3] Rajewsky, B.N., Krebs, A. & Zickler, H. (1936).
[4] Altenburg, E. (1934, 1936).

Ward's α-ray experiments already described (p. 152). More usually adult males have been irradiated. The flies are held with their abdomens compressed between quartz plates and their undersides irradiated.[1] By this technique the testes are brought nearer to the surface, and a sufficient intensity of ultra-violet light reaches the sperm to give a few per cent of mutations with the largest doses which the fly is able to sustain. Absorption by the intervening tissue is, however, large and variable, and it is not practicable to determine, for example, whether the yield is accurately proportional to the dose, or to compare the doses to the sperm of radiations of different wave-lengths required to produce mutation in equal percentages of the sperm. In practice the greatest percentage yields are obtainable with a rather long wave-length, 3130 A., since the flies will tolerate greater intensities of this wave-length than of shorter wave-lengths, and a greater fraction of the incident energy penetrates to the testes. Expressed in terms of mutation frequency per unit energy absorbed in the sperm, shorter wave-lengths are, however, probably more efficient.

In X-ray experiments a practical limit is set to the percentage of irradiated sperm in which mutation can be induced by the fact that with large doses the proportion of sperm remaining viable is greatly reduced by the induction of dominant lethals. In ultra-violet light experiments the practical limit seems to be set by the damage to the flies by irradiation of their tissues rather than by the induction of dominant lethals in the sperm, and in occasional flies the percentage of viable sperm having sex-linked recessive lethals is as high as 50 %.[2]

Of considerable interest is the question whether recessive lethals induced by ultra-violet light are entirely gene mutations (type A) or whether, as with X-rays, some are minute deletions (type B) and some are associated with gross structural changes (type C). Gross structural changes induced by ultra-violet light are very rare by comparison with recessive lethals, if they occur at all,[3] so that the proportion of type C lethals is negligible.

[1] Mackenzie, K. & Muller, H.J. (1940). Note that ultra-violet light intensities stated in this paper are, by error, 1000 times too great. Also Demerec, M., Hollaender, A., Houlahan, M.B. & Bishop, M. (1942).

[2] Demerec, M., Hollaender, A., Houlahan, M.B. & Bishop, M. (1942).

[3] Isolated instances of chromosome interchanges in ultra-violet irradiated material have been reported by Muller, H.J. (1941a) and by Demerec, M., Hollaender, A., Houlahan, M.B. & Bishop, M. (1942).

According to Slizynski, some ultra-violet-induced lethals involve minute deficiencies detectable in the salivary chromosomes,[1] but other reports[2] suggest that ultra-violet light does not cause minute structural changes. This important question is therefore not yet settled.

The small penetrating power of ultra-violet light is of less importance when small objects such as fungal spores are irradiated. Experiments in which the yield of mutations as a function of dose has been studied have shown that the yield is proportional to dose at low doses, but that at sufficiently high doses the proportion of the viable spores which carry mutants fails to rise further and may even diminish.[3] This unexpected result appears at doses at which the fraction of irradiated spores which survive is less than 1%. It is probably due to the mutant spores having a lower viability than those which have not suffered mutation.

As the result of extensive work by Stadler and his co-workers, a considerable amount of information is available about the genetical effects caused by irradiating maize pollen with ultra-violet light, and about the differences between the effects of X-rays and ultra-violet light.[4] The maize pollen is spread in a single layer and irradiated from above. After irradiation it is used to pollinate a maize plant. The maize seeds developing are examined, and, if desired, sown to obtain F_1 plants. At the time of irradiation the pollen grain contains two haploid *sperm nuclei*. When they enter the embryo-sac (which prior to fertilization contains eight haploid nuclei as the result of three consecutive divisions of one of the products of meiosis), one of the sperm nuclei fuses with the *egg nucleus* to form a diploid zygote, which by multiplication forms the *germ* or *embryo* of the seed and

[1] Slizynski, B.M. (1942). In 21 lethal-bearing X chromosomes he found 5 deficiencies, viz. 1 case of 1 band deficient, 3 cases of 2 bands deficient, and 1 case of 14 bands deficient.

[2] Mackenzie, K. & Muller, H.J. (1940); Muller, H.J. (1941a) reporting the preliminary results of an experiment by Bridges, P.N. & Muller, H.J.

[3] Emmons, C.W. & Hollaender, A. (1939 a, b); Hollaender, A. & Emmons, C.W. (1941), using *Trichophyton mentagrophytes*; Demerec, M., Kaufmann, B.P., Fano, U., Sutton, E. & Sansome, E.R. (1942) using *Neurospora crassa*.

[4] Experiments with ultra-violet light are described by Stadler, L.J. (1939, 1941); Stadler, L.J. & Sprague, G.F. (1936 a, b, c); Stadler, L.J. & Uber, F.M. (1938, 1942). Comparable experiments with X-rays are described by Stadler, L.J. (1928a, 1930a,b, 1931); Stadler, L.J. & Sprague, G.F. (1937).

eventually the F_1 plant. The other sperm nucleus fuses with two of the other haploid nuclei (the *polar fusion nuclei*) to produce a triploid nucleus, which by multiplication develops into the *endosperm*, the starchy tissue which constitutes the bulk of the seed but does not persist in the F_1 plant.

The radiation may produce genetical effects in either of the two sperm nuclei. A dominant lethal in the sperm which fuses with the polar fusion nuclei will result in a miniature seed having a normal embryo but defective endosperm. A dominant lethal in the sperm which fuses with the egg nucleus will lead to a germless seed, a condition which is detectable by examination of the seed. Visible mutations induced in this sperm may be detected by sowing the seed, when any dominant visible mutations will be revealed by inspection of the F_1 plants and the (more frequent) recessive visible mutations will be revealed by self-fertilizing the F_1 plants and looking for mutants in the F_2 generation.

A large part of Stadler's work was concerned with the production of endosperm deficiencies. The pollen irradiated came from a stock having the dominant allelomorphs of a number of genes affecting the endosperm (e.g. modifying its colour or surface texture). The pollen was used to pollinate a plant homozygous for the recessive allelomorphs of these genes. The seeds developing after fertilization normally showed the dominant phenotype. As a result of pollen irradiation, however, some of the seeds showed the recessive phenotype. This loss of effect of the dominant gene could be explained as being due either to mutation to the recessive allelomorph, or to removal of a portion of the chromosome containing the locus concerned. In practice the latter explanation (*deficiency*) is believed to be in almost all cases the correct one.[1]

A similar genetical technique employing marker genes affecting plant characters instead of (or as well as) endosperm characters

---

[1] With a given dose of radiation to a chromosome, deficiencies are probably not more frequent than are (random) gene mutations. Since, however, one deficiency involves many genes, a *specified* locus is more likely to be affected by a deficiency than it is to suffer gene mutation. There is also some experimental evidence for the endosperm effects being due to deficiencies rather than to gene mutations, namely, the observation that when a chromosome contains two marked loci, these often lose their effects together, suggesting the deletion of a portion of the chromosome which includes both genes.

enables deficiencies at selected loci to be detected which are produced in the sperm nucleus which fertilizes the egg nucleus. In this case it is necessary to sow the seeds and examine the F_1 plants.

Deficiencies can also be detected without the use of marked loci by the fact that, while maize is fairly tolerant of deficiencies when they are heterozygous in diploid or triploid tissue,[1] after meiosis has occurred and haploid cells are produced, all except the smallest deficiencies are usually lethal in the male *gametophyte*. Consequently half of the pollen produced by an F_1 plant which is heterozygous for a deficiency will be defective, and can be recognized as such by inspection of the pollen.

If the F_1 plant is heterozygous for a chromosome interchange, then about half of the pollen it produces will be defective, having a deficiency of one chromosome segment and a duplication of another. It is possible to decide whether a plant producing defective pollen is heterozygous for a deficiency or for an interchange by examining meiotic figures in the anthers.[2]

The principal interest of the maize experiments lies in a number of differences between the effects of ultra-violet light and X-rays. Since dosages of ultra-violet light and X-rays are not measured in comparable units,[3] absolute yields per unit dose of the two radiations cannot be compared, and the comparison rests therefore upon qualitative differences and upon differences in the relative frequencies of different types of genetic effect. These differences we now proceed to enumerate.

[1] Deficiencies of a large part of a chromosome arm (Singleton, W.R. 1939; Singleton, W.R. & Clark, F.J. 1940) or even of a whole chromosome (Stadler, L.J. 1931) are sometimes found in heterozygous condition in the F_1 plants following pollen irradiation. Such plants are visibly defective in growth and may not flower.

[2] In meiosis homologous chromosomes are paired. If an interchange between two non-homologous chromosomes has occurred, one group of four chromosomes and eight groups of two will be seen at diakinesis instead of ten groups of two (Stadler, L.J. 1931). The larger deficiencies can also be seen in meiosis at the pachytene stage (Singleton, W.R. 1939; Singleton, W.R. & Clark, F.J. 1940).

[3] It would, of course, be possible in principle to express both ultra-violet light and X-ray dosages in terms of ergs per cubic micron dissipated in the chromosomes, but the experimental data at present do not permit this. It is likely, in view of results with viruses (Chapter IV), that in terms of these units ultra-violet light would be found to be much less effective than X-rays in producing all types of genetic effects.

(a) The ratio of the number of visible gene mutations to chromosome deficiencies in the F_1 plants is higher with ultra-violet light than with X-rays.[1]

(b) In the majority of cases an endosperm deficiency produced by ultra-violet irradiation of the pollen affects only about half of the endosperm. This is taken to mean that the chromosomes in the pollen grain are already split into two chromatids at the time of irradiation, and the ultra-violet light breaks only one of them.[2] In the majority of cases endosperm deficiencies caused by irradiating pollen with X-rays affect the whole chromosome, indicating that both chromatids are usually broken. This difference is readily understandable if it is remembered that ionizations (but not ultra-violet quantum absorptions) are localized on the paths of ionizing particles. The chromatids being closely juxtaposed, an ionizing particle which passes through one will usually traverse both.

(c) As already mentioned, chromosome deficiencies involving selected loci may be detected either in the endosperm or in the F_1 plants, and with X-rays they occur with approximately equal frequency as is to be expected, since there is no reason to anticipate a difference in sensitivity between the two apparently identical sperm nuclei, one of which fuses with the egg nucleus and the other with the polar fusion nuclei. With ultra-violet

[1] This is shown by the ratio of total visible mutations appearing in the F_2 generation compared with total deficiencies inferred from the segregation of defective pollen by the F_1 plants. It is also shown in the case of a selected gene (A) by the fact that among 200 cases where X-ray treatment caused loss of the dominant effect, all involved some loss of viability and were probably deficiencies, none was simply due to mutation to the recessive allelomorph, while among a smaller number of cases where ultra-violet light treatment caused loss of the A effect, four were shown to be true mutations to a recessive allelomorph (Stadler, L.J. 1941).

[2] In experiments in which endosperm deficiencies for the characters A, Pr, Su were recorded, 79·8 % of all deficiencies were *fractional* (figure obtained by grouping all data given by Stadler, L.J. & Uber, F.M. 1942). It is possible that the 20 % of *entire* deficiencies represent cases where, by chance, both sister chromatids have been independently broken. That such chance coincidences will not be infrequent is suggested by the fact that in the same experiments 13 % of the seeds were observed to be simultaneously deficient for two or more of the characters A, Pr, Su, although some coincident deficiencies are not observable, being phenotypically indistinguishable from single deficiencies, and others probably behave as dominant lethals.

light, however, there is a great difference, deficiencies being much more frequent in the endosperm than in the F_1 plants. The difference is presumably due to restitution of an ultra-violet-induced break being less probable in the endosperm than in the embryo.[1]

(d) Following X-irradiation of pollen, deficiencies and interchanges are found with comparable frequency in the F_1 plants. But after ultra-violet irradiation of pollen, interchanges are much less frequent than deficiencies, and such as have been found are incomplete.[2] If it is borne in mind that the ultra-violet deficiencies observed in the embryo constitute a small fraction only of those primarily produced (according to the evidence in (c) above), it is clear that ultra-violet light is able to produce breaks but that these have much less chance of combining to form interchanges than when the breaks are produced by X-rays. The reason is not clear; the phenomenon appears to be general, being exhibited by *Drosophila* and *Tradescantia* as well as maize. It is not to be explained on the single ionizing particle argument used in (b), since it is known (in the cases of *Drosophila* and *Tradescantia*) that the two breaks taking part in an X-ray-induced interchange are usually produced by separate ionizing particles.

(e) Ultra-violet-induced deficiencies appear to be entirely *terminal*,[3] while X-ray-induced deficiencies are mainly or entirely *interstitial*. An interstitial deficiency involves exchange between two breaks, so that we have here an additional instance of the smaller probability of exchange occurring between two

[1] McClintock, B. (1939) has shown that if a (mechanically) broken chromosome is present in endosperm or gametophyte, sister-union will occur at the breakage point when the chromosome splits. In the embryo, however, sister-union at a breakage point does not occur. Once sister-union has occurred, restitution is clearly impossible, and thus the lower probability of restitution in endosperm than in embryo suggested by Stadler's ultra-violet light experiments may be due to sister-union occurring in endosperm and not in embryo. The difference will not be shown by X-ray deficiencies, since these usually involve both chromatids of the sperm chromosome, and sister-union will therefore presumably occur before fertilization. (But in any case X-ray deficiencies are not usually terminal.)

[2] I.e. of the four breakage ends, only two unite. As a result an acentric fragment is left unattached and is lost. Such incomplete interchanges are also produced by X-rays, but less frequently than complete interchanges.

[3] Singleton, W.R. (1939); Singleton, W.R. & Clark, F.J. (1940).

breaks produced by ultra-violet light than between two breaks produced by X-rays.

The relative efficiency of different wave-lengths of ultra-violet light in inducing genetical change have been compared with

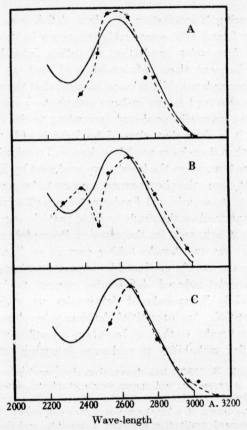

FIG. 29. The relative efficiencies of different wave-lengths of ultra-violet light in inducing mutations. A, maize pollen (Stadler & Uber); B, fungal spores (Hollaender & Emmons); C, liverwort spores (Knapp & Schreiber). The solid curve shows the relative absorption coefficients of thymonucleic acid at different wave-lengths.

several experimental materials, and the results are given in Fig. 29. It is interesting to note the similarity between the various curves of genetic effect and the absorption spectrum of nucleic acid. The inference is that it is absorption of ultra-violet light by the nucleoprotein of the chromosome which leads to the genetical effect.

Chapter VI

THE PRODUCTION OF CHROMOSOME STRUCTURAL CHANGES BY RADIATION[1]

Experimental materials

The production of structural changes in the chromosomes by the irradiation of *Drosophila* sperm and of maize pollen has been briefly described in Chapter v in connexion with the genetical effects associated with these changes. In the present chapter we continue this discussion, and describe also chromosome structural changes in other organisms.

In *Drosophila melanogaster*, which has been very extensively studied genetically, chromosome structural changes can be recognized by genetical means, based on the changes in the linkage relations of genes which follow their rearrangement in the chromosome set and on numerous special methods. In recent years, however, studies of chromosome changes in *Drosophila* have increasingly been made cytologically by examination of the salivary chromosomes, and with other organisms which have been less extensively studied genetically the cytological method[2] is the normal one. For the detailed study of structural changes it is necessary to use nuclei in which the chromosomes are large and few in number, and in relatively few species are these conditions satisfied. In view of the almost complete universality of the chromosome mechanism, it may reasonably be anticipated that the main conclusions derived from the study of chromosome changes in the relatively small number of favourable experimental materials will be of wider application. Some caution is, however, needed in proceeding in this fashion, as is evidenced by certain detailed differences in the mechanisms of production of structural changes shown by the different organisms which have been investigated.

Chromosomes are, in general, only observable during cell division, and observations of structural change are made by

---

[1] This chapter is devoted principally to a statement of the experimental results. Detailed interpretation is deferred to Chapter VII.
[2] See Darlington, C.D. & La Cour, L.F. (1942), *The Handling of Chromosomes*, for the experimental technique of preparing material for cytological examination.

examination of metaphase or anaphase figures in stained preparations. Irradiation during these phases does not cause immediate chromosome breakage, perhaps because of the existence of a *matrix* of protein and nucleic acid which binds together the coiled chromosome, so that even though the chromosome thread may be broken, the chromosome does not separate into two pieces. For the production of structural changes it is necessary to irradiate during the resting stage, or early prophase.[1] Thus, the general procedure is to irradiate, to fix the material some hours or days subsequently, and to examine metaphase and anaphase figures in the fixed preparations.[2] Aberrations seen, for example, in metaphase are then due to radiation which was given to the cell at a stage prior to metaphase corresponding to the time interval which elapsed between irradiation and fixation. If the time scale of the life cycle of the material is known, the stage at which the radiation was given is known. Some uncertainty, however, is introduced by the fact that irradiation is liable to cause a lengthening of the time scale.[3]

This method has been adopted in studying chromosome aberrations in rapidly dividing root tips of seedlings of onion, bean, tomato and other species.[4] Root tips have the merits of small size and ready availability, and also the merit that in the growing regions of the tip the proportion of cells in mitosis is high.

Another method of studying chromosome aberrations in plant material which has been found convenient is to irradiate flower buds and to study division figures in the developing pollen. Either the meiotic division of the *pollen mother cell* may be used,[5]

[1] Irradiation during metaphase or anaphase may lead to aberrations being discoverable at a *subsequent* division. Thus if *Sciara* oocytes are irradiated during meiotic first metaphase or anaphase, aberrations can be detected in the salivary gland chromosomes of the F_1 larvae (Metz, C.W. & Bozeman, M.L. 1940; Reynolds, J.P. 1941; Bozeman, M.L. 1943).

[2] In favourable materials it may be possible to observe breaks in prophase chromosomes and in this event fixation may be carried out immediately after irradiation. Thus Bishop, D.W. (1942) was able to demonstrate the existence of breaks in chromatids at the pachytene or diakinesis stages in *Orthoptera* 3 min. after irradiation. In general, however, observations are made at metaphase or anaphase.

[3] See Chapter VIII.
[4] E.g. Marshak, A. (1937); Marquardt, H. (1938); Sax, K. (1941a).
[5] E.g. Marshak, A. (1935).

or the first[1] or the second[2] haploid mitosis. In Swanson's method of studying the second haploid mitosis, which in *Tradescantia* occurs after germination of the pollen, the pollen is germinated on a smear of artificial medium on a glass slide. An hour or two after germination the generative nucleus enters the pollen-tube which has a diameter[3] of only 6μ. Consequently the chromosomes are accessible to irradiation even by weakly penetrating radiations such as ultra-violet light[4] or soft X-rays.[3] The pollen-tubes are fixed and stained about 24 hr. after germination and the metaphase chromosomes are examined. It is convenient[4] to add acenaphthene or colchicine to the artificial culture medium. These chemicals prevent spindle formation in the dividing cell so that the chromosomes are seen lying in tandem formation in the pollen-tube in a manner favourable to the observation of breaks.

When chromosome breaks in *Drosophila* are being studied cytologically the method is to irradiate male flies, to mate them to untreated females, and to collect the larvae from the culture. The salivary glands of the mature larvae are dissected out, stained, smeared and mounted, and the giant salivary chromosomes examined.[5] Since many divisions occur between irradiation of the sperm and fixation of the salivary glands, it is evident that inviable chromosome changes will not be observable by this method. This is a disadvantage of the use of salivary chromosomes which is, however, for many purposes outweighed by the fact that chromosome changes can be studied in a detail not possible in other materials.

The chromosomes in the mature sperm are in an inactive state. This results in a further difference between experiments on *Drosophila* sperm and experiments on cells which are either in interphase or in prophase but in either event are developing. Chromosome breaks which are induced in *Drosophila* sperm by irradiation do not take part in structural rearrangement until the sperm enters the egg. On the other hand, in developing cells rearrangement takes place during and immediately following the irradiation.

[1] E.g. Sax, K. (1940).   [2] E.g. Swanson, C.P. (1940).
[3] Catcheside, D.G. & Lea, D.E. (1943).
[4] Swanson, C.P. (1940, 1942, 1943).
[5] E.g. Catcheside, D.G. (1938a). Cp. Plate III a,b.

An animal with large chromosomes, which has been found convenient for the study of chromosome structural changes at the division following irradiation, is the grasshopper.[1] Either mitosis and meiosis of the germ cells in the testis are observed, or mitosis of the neuroblasts of the embryo.

Structural changes and physiological changes in chromosomes [2]

The chromosome changes with which we are concerned in this chapter are breakages, and structural rearrangements resulting from the joining in various fashions of the several breakage ends present in a nucleus in which two or more chromosome breaks have occurred. This action of radiation appears to be direct, in the sense that breakage is caused by an ionizing particle passing through or in the immediate vicinity of the chromosome at the point at which the breakage occurs. Breakages of this sort and the resulting rearrangements are, however, not the only type of change produced in chromosomes by the irradiation of cells. Another change consists in the alteration of the surface properties of the chromosomes so that they tend to stick together. This results in chromosomes at metaphase adhering where they happen to touch, and to sister chromatids failing to separate completely at anaphase, giving bridges. In severe cases the chromosomes may remain clumped at metaphase so that no further division stages ensue, or the bridges at anaphase may fail to break so that separate daughter nuclei cannot be formed. The changes of this type do not appear to be due to localized damage to the chromosomes at individual points, such as could be explained by the passage of ionizing particles through these points, but to a general change in surface properties covering the whole surface of the chromosomes. We can conveniently denote this change as a *physiological* effect of radiation on the chromosomes, in contrast to the term *structural* change which we reserve for breakages, and rearrangements by union of breakage

[1] White, M.J.D. (1937); Carlson, J.G. (1938a,b, 1941a,b); Bishop, D.W. (1942).

[2] See Plate III for micrographs of division figures showing chromosome structural changes (Catcheside, D.G.), and Plate IV for micrographs of division figures showing chromosome physiological changes (Lasnitzki, I.; Marshak, A.; Carlson, J.G.).

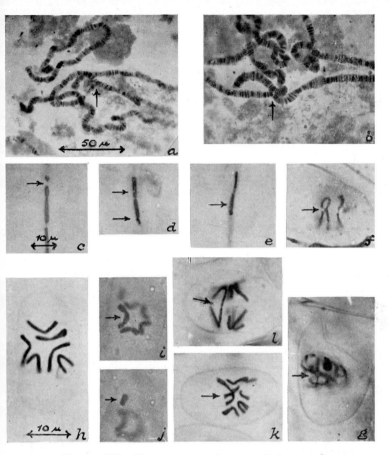

PLATE III. Chromosome structural changes

a, b. Drosophila melanogaster salivary chromosomes: *a*, inversion; *b*, interchange. *c–e. Tradescantia* pollen-tube chromosomes (acenaphthene metaphase): *c*, isochromatid break; *d*, chromatid break (upper arrow), isochromatid break (lower arrow); *e*, chromatid break. *f–l. Tradescantia* pollen-grain mitosis: *f*, chromatid inter-arm intrachange (prophase); *g*, chromatid interchange (prophase); *h*, normal metaphase; *i, j*, asymmetrical chromosome interchange (same cell at two foci: *i*, dicentric chromosome; *j*, acentric fragment); *k*, chromatid interchange (metaphase); *l*, anaphase bridge (due to isochromatid break).

(Photomicrographs by D. G. Catcheside)

ends, which can be attributed to a localized action of the radiation on the chromosome thread.[1]

The most definite hypothesis which has been suggested for the surface stickiness of the chromosomes characteristic of the physiological effect is that the matrix of nucleic acid deposited on the chromosome when it is in the condensed and spiralized form assumed in division is, after irradiation, in a fluid unpolymerized state instead of in the polymerized non-sticky form.[2] The surface stickiness can be produced by means other than irradiation, and appears to be a much less specific effect than the structural changes.

In favourable cytological material in which each cell contains a small number of large, and in some cases individually recognizable chromosomes, it is possible to distinguish with certainty between the two types of chromosome abnormality.[3] Thus when two chromosomes are attached at metaphase, it is possible to decide whether the cause is a chromatid interchange (F 3, Fig. 31), which is a structural change, or is due to the chromosome surfaces sticking together along a portion of their length as a result of the physiological effect. When a chromosome which is normally V-shaped assumes the form of a ring it is possible to decide whether a ring-and-fragment structural change (D 4, Fig. 30) is concerned, or whether the ends of the chromosome have simply stuck together as an expression of the physiological effect of radiation.

By experiments on such favourable materials it has been established that the physiological effect of radiation, resulting in a stickiness of the matrix, is exhibited in cells which are already

[1] The terms employed in the literature are *primary effect* of radiation for what we describe as physiological effect, and *secondary* effect for what we describe as structural change (Alberti, W. & Politzer, G. 1923, 1924; Pekarek, J. 1927; Marquardt, H. 1938). We have preferred not to use these terms, since it is rather confusing in a treatment which deals mainly with the mechanism of action of radiation to describe as a 'secondary effect' what is probably a rather direct action of radiation on the chromosome thread, and as a 'primary effect' what is possibly a less direct action of radiation.

[2] Darlington, C.D. (1942).

[3] Cp. Marquardt, H. (1938), pollen-grain mitosis in *Bellevalia romana*; Carlson, J.G. (1941 b), neuroblast mitosis in *Chortophaga*; Sax, K. (1941 a), mitosis in *Allium* root tips; Sax, K. & Swanson, C.P. (1941), Koller, P.C. (1943), pollen-grain mitosis and root-tip mitosis in *Tradescantia*.

in division at the time of irradiation. Such cells continue division, perhaps with some lengthening of the duration of the various stages. These cells, particularly those in prophase at the

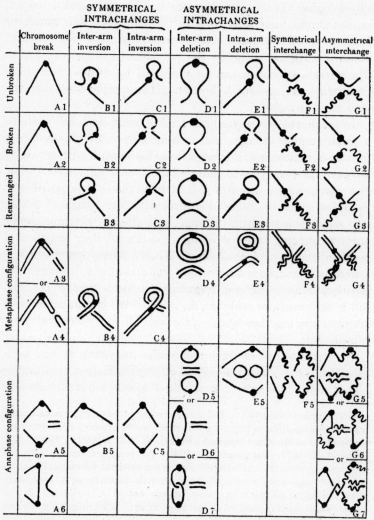

FIG. 30. Structural changes induced in unsplit chromosomes.

time of irradiation, are liable to show the stickiness of the chromosome surface typical of the physiological effect, leading to clumped metaphases and bridges at anaphase. With large

TYPES OF STRUCTURAL CHANGE

doses excessive clumping may prevent mitosis being completed.[1]

Cells not already in division at the time of irradiation, but

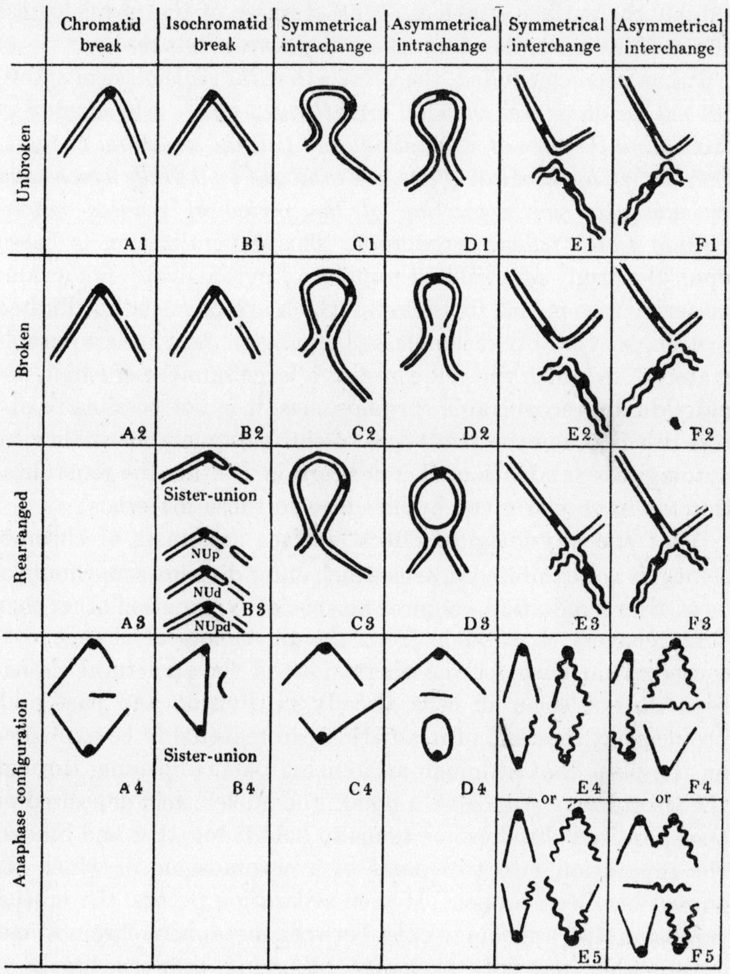

FIG. 31. Structural changes induced in split chromosomes.

nearing the end of interphase,[2] experience a delay, and enter mitosis some hours or days later, the duration of the delay

[1] See Chapter IX for further details.
[2] Or in the initial stages of prophase, Carlson, J.G. (1941 a).

depending on the material and the dose.[1] By the time these cells enter division the cell has recovered from the physiological effect,[2] which appears to be a reversible process, for the chromosomes in the metaphase and anaphase of these divisions do not exhibit the surface stickiness characteristic of the physiological effect. They may, however, show structural changes.

It has been concluded, therefore,[3] that *the physiological effects, but not the structural changes, are exhibited by the cells already in division at the time of irradiation, and that the structural changes, but not the physiological effects, are exhibited by the cells which enter division after the expiration of the period of reduced mitotic activity which follows irradiation.* This generalization is based upon the study of a limited number of cytologically favourable materials (see p. 193, footnote 3), which are however not limited to a single type of tissue or class of organism. In a less favourable material, in which the cell contains a large number of small, not individually recognizable chromosomes, it is not possible to distinguish between the two types of chromosome abnormality by cytological examination. It is not certain whether the generalization we have just given applies also to these materials.

That the physiological effect (surface stickiness of chromosomes) is not exhibited in cells which enter division some hours or days after irradiation, requires no special explanation other than that recovery is possible from the physiological change concerned. That chromosome aberrations of the structural change type do not occur in cells already in division and past mid-prophase at the time of irradiation can reasonably be explained on the basis that although an ionizing particle passing through the thread may still cause a break, the nucleic acid deposited on the spiralized chromosome tends to hold it together and hinders the separation into two parts of a chromosome of which the thread has been broken. If such separation occurs, the mutual repulsion that appears to exist between metaphase chromosomes presumably prevents exchange occurring between breaks in different chromosomes.[4] In *Drosophila* sperm the chromosomes

[1] See Chapter VIII.

[2] At any rate in the materials mentioned in footnote 3, p. 193. It is possible that recovery is not complete in the case of some animal tissues (see Chapter IX). [3] Marquardt, H. (1938).

[4] Irradiation during meiotic metaphase and anaphase does lead to structural changes in *Sciara* (Bozeman, M.L. 1943). Interchanges

are believed to be in a highly condensed state similar to their state during mitosis, and there is clear evidence that the chromosome thread can be broken by irradiation of the sperm but that breakage ends cannot join until the chromosomes assume a less condensed state.[1]

Fragmentation is sometimes found in cells examined at anaphase in experiments in which, according to the generalization given above, only the physiological effect of radiation is expected. Most authors[2] suppose these fragments to be due to mechanical breakage of chromosomes proceeding to opposite poles at anaphase, but stuck together as a result of the physiological effect. It is not impossible, however, that some breaks of the structural type are induced in chromosomes irradiated sufficiently late in prophase to escape the inhibition of division experienced by cells not so far advanced.[3] Experiments such as those of Marshak[4] in which root tips of various plants are fixed soon after irradiation, and the proportion of anaphases which show fragments or bridges recorded, probably deal with a mixture of changes of the physiological and structural types, and are consequently difficult to interpret

TYPES OF STRUCTURAL CHANGE

The principal types of chromosome structural change produced by ionizing radiations are illustrated diagrammatically in Figs. 30 and 31, Fig. 30 applying to cells in which the chromosomes are unsplit at the time of irradiation, and Fig. 31 to cells in which the chromosomes are split. At metaphase or anaphase, when the aberrations are observed, the two sister *chromatids* formed by the longitudinal splitting of each chromosome are separately visible. The exact time prior to metaphase at which splitting occurs has

between chromosomes are rare compared to structural rearrangements (inversions, deficiencies, duplications) *within* chromosomes.

[1] Muller, H.J. (1940).
[2] Marshak, A. (1937); White, M.J.D. (1937); Marquardt, H. (1938); Sax, K. (1941a); Sax, K. & Swanson, C.P. (1941).
[3] Thus Bishop, D.W. (1942) was able to detect breaks at diakinesis in *Chortophaga* 3 min. after irradiation. The chromosomes were not stuck together, and these breaks cannot be explained as mechanical breakage of chromosomes attempting to separate but stuck together as a result of the physiological effect.
[4] Marshak, A. (1937, 1939b); Marshak, A. & Hudson, J.C. (1937).

been a matter for dispute, opinions ranging from early prophase of the same division to as far back as prophase of the preceding division. X-ray experiments themselves furnish some information on the time of split.[1] If, at metaphase, all aberrations are seen to involve both chromatids in the same way, this is evidence that at the time when the structural rearrangement took place the chromosome was single. If, on the other hand, it is observed

TABLE 46. Transition from chromosome to chromatid types of aberration as metaphase is approached

Tradescantia pollen grain mitosis[1]

Hours before metaphase		34	33	32	31	30	29
Interchanges per 100 cells:	Chromosome	12·0	9·9	9·9	11·1	8·1	5·1
	Chromatid	0·0	0·6	0·0	3·9	4·2	3·3
Hours before metaphase		28	27	26	25	23	
Interchanges per 100 cells:	Chromosome	9·9	3·9	0·0	0·0	0·0	
	Chromatid	2·1	5·1	6·6	10·8	17·7	

Chortophaga neuroblast mitosis[2]

Hours before metaphase		72	48	36	24	12
Interchanges per 100 cells:	Chromosome	25	26	15	8	0
	Chromatid	0	1	3	12	12

[1] Sax, K. (1941b), 160 r. Numbers given by Sax have been multiplied by 3 to convert from breaks per 100 chromosomes to aberrations per 100 cells.
[2] Carlson, J.G. (1941b), 125 r.

that one chromatid takes part in a structural rearrangement in which the other does not take part, it is evidence that the chromosome was already split into chromatids at the time when the structural rearrangement occurred. The evidence of radiation experiments is not completely conclusive,[2] but we shall for the purpose of classification of types of aberration be content to describe as unsplit a chromosome which behaves as such in radiation experiments.

In Table 46 data are given illustrating the manner in which the numbers of chromatid and chromosome aberrations per 100 metaphase figures vary with the time elapsing between irradiation and fixation.[3] It is seen that if the radiation is given 26 hr.

[1] Mather, K. & Stone, L.H.A. (1933); Riley, H.P. (1936); Sax, K. & Mather, K. (1939).
[2] See p. 204.
[3] Additional data for plant material are given by Newcombe, H.B. (1942a).

or less before fixation, only chromatid aberrations are seen in *Tradescantia* pollen-grain metaphases. If the radiation is given 32 hr. or more before fixation, practically only chromosome aberrations are seen. The indication is that splitting in these experiments occurred between 26 and 32 hr. prior to metaphase.

Chromosome breaks

The simplest type of aberration produced in an unsplit chromosome is a simple *chromosome break* or *terminal deletion*, as illustrated in column A of Fig. 30. Terminal deletions can be observed in experiments in which the cell is fixed at metaphase or anaphase of the division next following irradiation. At anaphase the chromatid fragments formed by the splitting of the deleted

TABLE 47. Number of fragments per cell at various times after irradiation [1]

Days elapsing between irradiation and fixation	1·3	2·1	3	5	6	8	9	14
Mean number of fragments per cell	2·48	1·94	1·86	1·30	0·93	1·00	0·55	0·05

fragment lag behind the centric chromosomes which are migrating to the poles. Lagging is due to the absence of the centromere, and usually leads to failure of the two fragments to be incorporated in the daughter nuclei.[2] Consequently, if many divisions elapse between irradiation and fixation, acentric fragments will not be observed. Table 47 shows how the number of fragments observed per cell in spermatogonial divisions of *Locusta migratoria* diminishes with lapse of time between irradiation and fixation.[3] The whole time covered by the table is enough for about four successive divisions.

In view of these results it is not surprising to find that in *Drosophila* experiments, in which sperm are irradiated and

[1] White, M.J.D. (1935).
[2] An observation of Carlson, J.G. (1938b) on *Chortophaga* suggests that in this material one chromatid fragment is included in each daughter cell in the majority of divisions. But in *Chortophaga* (and in other materials, e.g. onion root tips, Sax, K. 1941a) the chromatid fragments usually fail to be included in the cell nuclei and form micronuclei which eventually degenerate.
[3] White, M.J.D. (1935). The disappearance of fragments in successive divisions has also been observed in other materials, e.g. in onion root tips (Sax, K. 1941a).

salivary gland chromosomes are examined, no acentric fragments are found.

The two new ends formed by the breakage of an unsplit chromosome usually join either with each other, restituting the original chromosome, or with other breakage ends in the nucleus giving various types of structural rearrangement. The terminal deletions observed at metaphase are thus only a small fraction of those initially produced. The magnitude of this fraction can be estimated, though only by rather indirect methods. It is about one-twentieth for breaks produced in *Tradescantia* microspores.[1] The proportion of breakage ends produced by irradiation of *Drosophila* sperm which are still unjoined at the time of chromosome split is about one-quarter.[2] In *Tradescantia* the sister chromatids formed by the splitting of a broken chromosome do not fuse at the breakage point but remain separate,[3] as indicated in diagrams A3 and A5 of Fig. 30. In some other materials however sister chromatids fuse at the breakage points giving a dicentric chromatid and a single acentric fragment (diagrams A4 and A6, Fig. 30). The dicentric chromatid forms a bridge at anaphase, which usually eventually breaks. The breakage appears often to be at the point at which fusion occurred,[4] but this is not invariable, so that duplications and deficiencies may result from the anaphase bridge. If the original break occurred near the centromere, so that the two centromeres in the dicentric chromatid are close together, the dicentric chromatid may remain on the metaphase plate and so be completely lost.[5]

When an anaphase bridge breaks, each daughter cell is left with a chromosome with a broken end. It is possible that in such cases sister-union again occurs when this broken chromosome splits, leading to further bridges,[6] until loss of the chromosome or death of the cell eventually occurs. The manner in which anaphase bridges are gradually eliminated in the successive cell divisions, probably by death of the cells carrying the aberrant

[1] Lea, D.E. & Catcheside, D.G. (1942).
[2] Chapter v, p. 169.
[3] Sax, K. & Mather, K. (1939).
[4] Carlson, J.G. (1938a).
[5] Cp. discussion by Pontecorvo, G. (1942).
[6] As demonstrated in maize gametophyte and endosperm divisions by McClintock, B. (1938).

chromosomes, is illustrated in Table 48,[1] which shows how the proportion of abnormal anaphases diminishes during the development of an onion root growing from a bulb irradiated in the dormant state and subsequently germinated.

TABLE 48. Gradual disappearance of abnormal anaphases

(Onion root tips, Sax, 1941 a.)

Root length (cm.)	1	2–3	6–8	10–13	20–25
Abnormal anaphases (%)	46	33	21	11	0

The gradual disappearance of abnormal mitotic figures after irradiation has been demonstrated in chick fibroblast cells growing in culture.[2]

Chromatid breaks

Irradiation at a stage in which the chromosomes are already split may produce a break in one only of the sister chromatids, giving a *chromatid break*, as illustrated in column A of Fig. 31, and Plate III *d, e*. The broken fragment remains close to the unbroken sister chromatid fragment in metaphase, possibly being held to it by a matrix which envelopes both chromatids.[3] The chromatid breaks vary in distinctness, sometimes appearing only as achromatic lesions in the chromatid, and are not easy to score with accuracy. In some materials[4] constrictions are seen in the chromatids at anaphase, and these may be half-chromatid breaks, i.e. breaks in one only of the two threads of the chromatid, which must (on this interpretation) be inferred to be already split into two threads in anticipation of the next cell division.

As with chromosome breaks, it is believed that the chromatid breaks observed at metaphase constitute only a minority of those initially produced, the majority restituting or uniting with other breakage ends to give the more complex structural changes.

[1] Sax, K. (1941 a). The dicentrics recorded in this experiment are due not only to the sister-union of broken chromosomes, but also to asymmetrical interchange between non-homologous chromosomes.

[2] Lasnitzki, I. (1943 a).

[3] Swanson, C.P. (1942), observing the pollen-tube mitosis in *Tradescantia*.

[4] E.g. *Chortophaga* neuroblast mitosis, Carlson, J.G. (1938 a); *Tradescantia* pollen-tube mitosis, Swanson, C.P. (1943); *Tradescantia* pollen-mother-cell meiosis, Nebel, B.R. (1937).

Isochromatid breaks

In addition to cases where a single chromatid is broken, there are instances where both chromatids are broken at the same level. This type of aberration is termed an *isochromatid break*, and is illustrated in column B of Fig. 31 and Plate III c, d. It is believed to be produced by a single ionizing particle which passes through, and breaks, both of the chromatids of the already split chromosome. It is clear that the metaphase configurations (B 3, Fig. 31) are similar to the metaphase configurations possible in the case of a chromosome which has suffered a terminal deletion in the unsplit stage (A 3, 4, Fig. 30), and has subsequently split. The question obviously arises whether we are justified in regarding isochromatid breaks as twin breaks produced at the stage when the chromosome is already split, or whether they are not, in fact, produced by the breakage and *subsequent* splitting of a chromosome which is unsplit at the time of irradiation.

The existence of the isochromatid break as a distinct type of aberration is conclusively established in the case of *Tradescantia* microspores; the argument runs as follows. It is known that interchange between breaks in different chromosomes occurs within a few minutes of the break being formed. This follows from the manner in which the yield of interchanges produced by a given dose depends on the intensity (see p. 262). It is evident therefore that if it is observed that the interchanges produced when a microspore is irradiated are chromatid interchanges involving one but not both sister chromatids (E, F, Fig. 31), this shows that at this stage the chromosomes are split. But isochromatid breaks are obtained when microspores are irradiated at this stage, and occasionally even in the same cells and the same chromosomes in which chromatid interchanges are also found.

The conclusion that isochromatid breaks are caused by a single ionizing particle traversing and breaking both of the sister chromatids of an already split chromosome is strengthened by the fact that ultra-violet light does not produce isochromatid breaks, though it does produce chromatid breaks.[1] Quantum absorp-

[1] Demonstrated in *Tradescantia* pollen-tubes, by Swanson, C.P. (1940, 1942), confirmed by Catcheside, D.G. & Lea, D.E. (unpublished). Inferred also in maize pollen (see Chapter v, p. 186).

tions in an ultra-violet light experiment are not localized in lines in the manner in which ionizations are localized along the paths of ionizing particles in an X-ray experiment, so we should not expect with ultra-violet light to obtain breaks in both sister chromatids at the same level (except by chance coincidence of independently produced breaks).

The ratio of the number of isochromatid breaks to the number of chromatid breaks obtained with a given dose of X-rays varies with the stage of prophase at which the chromosomes are irradiated. The ratio is higher in early prophase than in late prophase, as is shown in Table 49. At early prophase the chromatids are

TABLE 49. The ratio of the numbers of isochromatid and chromatid breaks induced at different stages of prophase

Tradescantia, prophase of pollen-grain division [1]

Hours before metaphase	24	12	6			
Ratio isochromatid : chromatid	2·43	1·47	0·79			

Tradescantia, prophase of pollen-tube division [2]

Hours after germination of the pollen grain	0	1	2	3	4	5	6–15
Ratio isochromatid : chromatid	2·50	0·68	0·41	0·24	0·21	0·18	0·14

[1] Sax, K. (1941*b*).   [2] Swanson, C.P. (1943).

closely associated. An ionizing particle which passes through one has therefore a high probability of also passing through the other. As prophase advances the chromatids spiralize in separate spirals, and their distance apart increases. The probability of an ionizing particle which passes through one chromatid thread also passing through the other thus diminishes. The chromosomes were visibly split into two chromatids 2 hr. after germination in the *Tradescantia* pollen-tube experiments[1] quoted in Table 49.

In view of the smooth rather than discontinuous change from the stage at which most breaks affect both chromatids to the stage at which isochromatid breaks are much less frequent than chromatid breaks, it is evident that the X-ray method cannot give conclusive evidence of a chromosome being *unsplit*. An alternative view would always be that the chromosome is split, but the sister chromatids are still so closely juxtaposed that an ionizing particle which breaks one usually breaks both. We have, however, already pointed out that the X-ray method can give

[1] Swanson, C.P. (1943).

conclusive evidence of the chromosome being *split*. We conclude therefore that *if the X-ray experiments indicate that the chromosomes are split, this can be relied upon, but X-ray evidence for a chromosome being unsplit is not final.*

The typical appearance of an isochromatid break is that illustrated in the uppermost of the four alternative forms shown in diagram B 3 of Fig. 31. After breakage, sister chromatids

TABLE 50. Frequency of non-union of sister chromatids at isochromatid breaks in *Tradescantia*

(Irradiated at room temperature, except where otherwise stated.)

Division examined	Radiation		Types of isochromatid break					% of ends not showing union	Reference
			SU	NUp	NUd	$NUpd$	Total		
Pollen grain	X-rays		—	—	—	—	—	6	1
	X-rays		240	28	23	17	308	13·8	2
	X-rays	1°	1801	80	110	14	2005	5·4	3
		20°	1027	98	68	33	1226	9·5	
		30°	302	33	21	6	362	9·1	
	α-rays		71	32	27	286	416	75·8	4
Pollen tube	X-rays		45	1	2	2	50	7·0	5
	α-rays		18	2	3	7	30	31·7	5

1 Catcheside, D.G. & Lea, D.E. (1943) in distal fragments; proximal non-unions not scored.
2 Kotval, J.P. (unpublished).
3 Catcheside, D.G., Lea, D.E. & Thoday, J.M. (1946a).
4 Kotval, J.P. & Gray, L.H. (unpublished). I am grateful to these workers for access to their provisional results prior to publication.
5 Catcheside, D.G. & Lea, D.E. (1943), with additional unpublished data.

undergo union (SU) at the breakage point. Less frequently sister-union fails to occur, either on the side of the break nearer the centromere (i.e. no sister-union *proximally*, NUp), or on the side further from the break (i.e. no sister-union *distally*, NUd), or on both sides ($NUpd$). The frequency of non-union seems to depend on the type of radiation employed, being more common with α-rays than with X-rays, as is clearly apparent in Table 50.[1] The obvious explanation[2] is that the more densely ionizing radiation (α-rays) does more damage to a chromosome that it

[1] Thoday, J.M. (1942) did not keep a record of the frequency of non-union isochromatid breaks in his X-ray and neutron experiments, but states that they are rare. It is to be presumed therefore that the frequency of non-union obtained with neutrons approximates more closely to the X-ray than to the α-ray frequencies listed in Table 50.
[2] Catcheside, D.G. & Lea, D.E. (1943).

breaks, and leaves less readily joinable breakage ends than does the less densely ionizing radiation, X-rays.[1]

Chromosome intrachanges

When two breaks are formed in the same chromosome, the four breakage ends may (apart from restitution) join in either of two ways, one way (*symmetrical intrachange*) leading to the *inversion* of the part of the chromosome between the breaks (columns B and C of Fig. 30), and one way (*asymmetrical intrachange*) leading to the removal of the portion of the chromosome between the breakage points (i.e. *intercalary deletion*), and thus to a ring and a rod, one of which will be without a centromere[2] (columns D and E of Fig. 30). The behaviour of the various aberrations is shown in Fig. 30. When the ring chromosome formed in this way splits, the chromatids may freely separate (D 5 of Fig. 30), or interlock (D 7), or form a single double-size dicentric ring[3] (D 6) depending on the amount of twist of the chromosome which occurred between breakage (D 2) and joining of the breakage ends (D 3). A dicentric ring, or a pair of interlocked rings, will give rise to bridges at anaphase. After breakage of these bridges, fusion of the two breakage ends in each daughter nucleus will probably occur reforming a ring chromosome in each nucleus.[4] Acentric rings or fragments will usually

[1] Since it is known that radiation given in early prophase retards to some extent the course of mitosis, and since different doses and types of radiation may retard to different extents, it is arguable that in the case of observations made at the pollen-grain mitosis, the cells were not necessarily all in the same stage of prophase when irradiated, though in all cases irradiated 24 hr. before metaphase. This might explain the difference between X-rays and α-rays if the joinability of breakage ends were greatly affected by the exact stage at which the chromatids were broken. Considerations of this sort should not, however, apply in the pollen-tube experiments, in which the stage at the time of irradiation is determined by the time elapsing between germination of the pollen and irradiation.

[2] It is possible (McClintock, B. 1938) for both the ring and the rod to have a centromere, when one of the X-ray-induced breaks divides the centromere itself into two functioning portions. This is probably a very rare event.

[3] Sax, K. (1940, 1941a).

[4] Arguing by analogy with the behaviour of interlocked and dicentric rings at somatic divisions in a ring-chromosome stock in maize (McClintock, B. 1938, 1941b).

be lost as a result of failure to be included in either daughter nucleus, the eventual result being often death of the cell.[1]

By inspecting Fig. 30 and comparing column B with column D and C with E, it will be realized that inversion and deletion are alternative aberrations resulting from two breaks in a chromosome. It is of some interest to know whether the alternatives are equally probable, but it is not very easy to obtain information on this point. In plant material, irradiated before chromosome split and observed at metaphase or anaphase following irradiation, deletions can be observed but inversions are not usually distinguishable from unchanged chromosomes. In *Drosophila* experiments in which salivary chromosomes are observed, inversions can be recognized, but deletions, unless very small, are not found since they are inviable. However, there is some information indicating that deletions and inversions of the same length are equally frequent in the X chromosome of *Drosophila*.[2]

In both *Drosophila* and *Tradescantia*, minute interstitial deletions, i.e. asymmetrical intrachanges between breaks produced close together in the chromosome, occur with high frequency. In *Drosophila* they are usually identified by genetical means[3] and their extent then determined by examination of salivary chromosomes. In *Tradescantia* they are observed[4] at the pollen-

---

[1] Different organisms vary in their tolerance of heterozygous deficiencies.

[2] Fano, U. (1941), quoting experiments of Bishop, M. Normal males were irradiated and mated to females having attached-X chromosomes. Zygotes receiving an X chromosome from their father and an attached-X from their mother are normally inviable. However, if a large part of the X chromosome of the father has been deleted by the action of the radiation, a viable female results, which may be distinguished from $X.X/Y$ females by recessive characters for which the $X.X$ chromosome is homozygous being suppressed by normal allelomorphs of these loci carried in the remaining part of the deleted X chromosome. In this way the frequency with which deletions of a given range of sizes occur can be calculated, and compared with the frequency with which inversions of the same range of size are found in salivary gland chromosomes.

[3] Thus Demerec, M. & Fano, U. (1941) mated irradiated male flies and selected F_1 females showing the character Notch (notched wings). Many of these flies had small deficiencies involving the locus concerned (band 3*C*7), which were examined cytologically in female larvae of the Notch stocks.

[4] Rick, C.M. (1940); Newcombe, H.B. (1942*a,b*); Giles, N.H. (1943).

grain mitosis as small bodies (*isodiametric fragments*), presumed to be rings, and most commonly of about 1μ diameter.

Chromatid intrachanges

The corresponding types of aberration occurring in a chromosome after it is split, i.e. inversions and deletions involving a single chromatid only, are shown in columns C and D of Fig. 31 and Plate III *f*. To avoid excessive multiplication of the number of diagrams, only *inter-arm intrachanges* have been shown, and *intra-arm intrachanges* (in which both breaks occur in the same chromosome arm) have not been illustrated, although they occur.

TABLE 51. Relative frequency of symmetrical and asymmetrical types of chromatid intrachange at the pollen-grain metaphase in *Tradescantia*[1]

Temp. °C.	No. of intrachanges			Proportion of intrachanges which are asymmetrical
	Symmetrical	Asymmetrical	Total	
1	321	206	527	0.39 ± 0.02
20	141	129	270	0.48 ± 0.03
30	32	24	56	0.43 ± 0.07

Further, intrachanges between two breaks, one of which is in one chromatid and the other of which is in the sister chromatid, have not been shown. It is not usually possible to decide by examination of the metaphase chromosome whether the intrachange has occurred between breaks in the same chromatid or in sister chromatids. Probably these alternatives are about equally frequent.

The interarm intrachanges can be classified observationally into symmetrical and asymmetrical types, though subject to some error. In the event of the intrachange being between breaks in the same chromatid, asymmetrical intrachange means intercalary deletion (column D, Fig. 31) and symmetrical intrachange means inversion (column C, Fig. 31). Some experimental data of the relative frequencies of the symmetrical and asymmetrical inter-arm intrachanges are given in Table 51.

Interchanges

If each of two (unsplit) chromosomes is broken, joining of the four breakage ends can give rise either to a pair of centric

[1] Catcheside, D.G., Lea, D.E. & Thoday, J.M. (1946*b*). A constant dose of 150r. of X-rays was delivered at various dose-rates.

chromosomes (*symmetrical interchange*, also called *eucentric interchange*, column F of Fig. 30), or to a dicentric chromosome and an acentric fragment (*asymmetrical interchange*, also called *dyscentric* or *aneucentric interchange*, column G of Fig. 30 and Plate III i, j). The behaviour of these aberrations at anaphase is shown in the diagrams. Symmetrical interchange does not lead to any mechanical difficulty, nor are the daughter cells deficient for any portion of the chromosome. Symmetrical interchanges should therefore be viable, and they are frequently observed in salivary chromosomes. They are not readily distinguished in mitosis from unchanged chromosomes. Asymmetrical interchanges result in the formation of an acentric fragment, which usually fails to be included in either daughter nucleus. At anaphase the dicentric chromatids formed when the dicentric chromosome splits may either go to separate poles (G 5 of Fig. 30), or form a criss-cross bridge (G 6), or interlock (G 7). The relative frequencies with which these types are seen in onion root-tip[1] anaphases are 2 : 3 : 1; in grasshopper neuroblast[2] anaphases 9 : 5 : 1. These figures suggest that types G 5 and G 6 are approximately equally frequent and that type G 7 is rarer.

Symmetrical and asymmetrical interchanges involving a single chromatid only are shown in columns E and F of Fig. 31, and Plate III g, k.

Asymmetrical interchanges, whether chromosome or chromatid, since they involve the formation of acentric fragments, are likely to be non-viable. Apart from the deficiency caused by the loss of the fragment, the existence of dicentric chromosomes which sometimes form bridges also causes mechanical difficulties at division. In consequence the proportion of cells carrying acentric fragments or dicentric chromosomes gradually diminishes in successive divisions following irradiation. The data we have already given in Tables 47 and 48 illustrate this.

The death of cells as a result of asymmetrical interchange, or of sister-union following a chromosome or isochromatid break, is due not only to genic unbalance resulting from losses of the chromosomes affected (though in many organisms this would be a sufficient cause for death) but sometimes also to other causes, perhaps the mechanical difficulties experienced by dividing cells in which bridges are formed at anaphase. For in *Drosophila*

[1] Sax, K. (1941*a*).   [2] Carlson, J.G. (1941*b*).

chromosome loss following either type of aberration is lethal in a large majority of cases, even in experiments where by use of special stocks it is contrived that the organism could be viable without one or two of its chromosomes.[1]

It is of interest to compare the frequency with which symmetrical and asymmetrical interchange occurs. In unsplit chromosomes this is difficult, since symmetrical interchanges often

TABLE 52. Relative frequencies of symmetrical and asymmetrical chromatid interchanges

(X-rays at room temperature, except where otherwise stated.)

		Nos. of interchanges		
Material irradiated	Division examined	Symmetrical	Asymmetrical	Reference
Tradescantia microspores:	pollen-grain mitosis	26	66	1
		182	402	2
	1°	956	798	3
	20°	508	472	
	30°	149	154	
Tradescantia pollen grains:	pollen-tube mitosis:			
Ungerminated		49	46	4
		15	70	5
Germinated		53	60	5
		12	9	6
Chortophaga embryos:	neuroblast mitosis	5	9	7

1 Sax, K. & Mather, K. (1939).
2 Sax, K. (1940).
3 Catcheside, D.G., Lea, D.E. & Thoday, J.M. (1946a).
4 Newcombe, H.B. (1942a).
5 Swanson, C.P. (1942).
6 Catcheside, D.G. & Lea, D.E. (1943).
7 Carlson, J.G. (1941b).

cannot be distinguished from unchanged chromosomes at metaphase, so that only asymmetrical interchanges can be scored with accuracy, while in salivary chromosomes, in which symmetrical interchange can be scored with certainty, asymmetrical interchanges are absent, being non-viable. In chromosomes which are

[1] Pontecorvo, G. & Muller, H.J. (1941). Irradiated wild-type male *Drosophila* were mated to triploid females homozygous for *brown* in chromosome II and *ebony* in chromosome III. Loss of chromosomes II and III as the result of asymmetrical interchange between them could be viable if the sperm concerned fertilized an egg containing two each of these chromosomes, and the offspring formed would be recognized by the brown and ebony characters otherwise suppressed by the paternal chromosomes. Nevertheless, such *brown-ebony* offspring were extremely rare, showing that the *process* of loss was almost invariably lethal.

split when the interchange occurs, so that only one chromatid of each chromosome is involved, symmetrical and asymmetrical interchanges can be separately scored. The data collected in Table 52 suggest that symmetrical and asymmetrical interchanges are approximately equally frequent. Though in some of the data statistically significant departures from equality occur, it is doubtful if the reliability of the distinction between symmetrical and asymmetrical interchange is sufficient to make such a difference certain.

The indications afforded by the data we have reviewed are that in both intrachanges and interchanges symmetrical and asymmetrical types are equally probable. These results are clearly related; they indicate that there is no evidence in the union of radiation-induced breaks of any polarization in the chromosome such as would prevent random joining of breakage ends of a bar magnet broken into pieces. Muller[1] has pointed out that this fact makes it improbable that breakage and union occur in linkages in a polypeptide chain, since such linkages are polarized.

In *Tradescantia* it is found that some interchanges are incomplete, only two of the four breakage ends joining. The proportions of incomplete interchanges which have been found in various experiments are listed in Table 53. The proportion of interchanges which are incomplete is greater for α-rays than for X-rays. The explanation is presumably the same as that offered in the case of non-union isochromatid breaks, namely, that the densely ionizing α-particle does more damage in traversing the chromatid than does a less densely ionizing electron or proton, so that the breakage end is less likely to be joinable. There is a general similarity between the two sets of data as is, indeed, only to be expected since sister-union at an isochromatid break is a type of interchange.

The ion-density of the protons which traverse tissue irradiated by neutrons is much higher than the average ion-density of the electrons which traverse tissue irradiated by X-rays. It might have been expected therefore that the proportion of incomplete interchanges and non-union isochromatid breaks would with neutrons have been intermediate between the values found with X-rays and α-rays, instead of approximating to the X-ray value as appears to be the case. However, the ion-density of an elec-

[1] Muller, H.J. (1941).

tron varies considerably along its path, and there is evidence[1] that it is mainly the more densely ionizing parts, where the ion-density is not much less than that of a proton, which are effective

TABLE 53. Proportion of chromatid interchanges which are incomplete in *Tradescantia*

Division examined	Radiation	Proportion of interchanges which are incomplete	Reference
Pollen-grain metaphase	γ-rays	22 in 341 = 6·5 %	1
	γ-rays	58 in 444 = 13·1 %	2
	X-rays	20 in 350 = 5·7 %	1
	X-rays	12 in 206 = 5·8 %	3
	X-rays: 1°	219 in 1754 = 12·5 %	4
	20°	132 in 980 = 13·5 %	
	30°	50 in 303 = 16·5 %	
	Neutrons	51 in 777 = 6·5 %	5
	α-rays	39 in 165 = 23·6 %	6
Pollen-tube metaphase	X-rays	2 in 27 = 7·4 %	7

1 Kotval, J.P. (unpublished).
2 Catcheside, D.G., Lea, D.E. & Thoday, J.M. (1946a).
3 Catcheside, D.G., quoted by Thoday, J.M. (1942).
4 Catcheside, D.G., Lea, D.E. & Thoday, J.M. (1946b). A constant dose of 150r. was given at dose-rates ranging from 5 to 300 r./min. There are no significant differences in the proportions of interchanges which are incomplete at different dose-rates, nor are the proportions significantly different at the three different temperatures.
5 Thoday, J.M. (1942).
6 Kotval, J.P. & Gray, L.H. (1947). The proportion is independent of the dose.
7 Catcheside, D.G. & Lea, D.E. (1943).

in breaking *Tradescantia* chromosomes. The fact that the proportion of incomplete interchanges and non-union isochromatid breaks is approximately the same for X-rays and neutrons may be regarded as supporting this view.

Relative frequency of interchanges and intrachanges

It is obvious that exchange between two breaks can only occur if the breakage ends have opportunity to come into contact. It is of interest in this connexion to compare the frequency with which exchange occurs when the two breaks are

(A) in the same arm of a chromosome, or are

(B) in opposite arms of the same chromosome, or are

(C) in different chromosomes.

The experimental frequencies can be compared with frequencies calculated on the assumption that union between breaks is random, i.e. that the chromosomes are sufficiently intimately

1 Lea, D.E. & Catcheside, D.G. (1942), discussed in Chapter VII.

mingled during the time union is occurring for the probability of exchange to be independent of whether the breaks are in the same or different chromosomes.

If there are in the cell m chromosomes having a centrally placed centromere (i.e. each chromosome having two arms) and t chromosomes having a terminally placed centromere (i.e. each chromosome having one arm), then if breaks are produced equally frequently in all the arms, the expected relative frequency of the three types A : B : C is readily shown to be[1]

$$\frac{1}{(2m+t)} : \frac{2m}{(2m+t)^2} : 1 - \frac{4m+t}{(2m+t)^2}. \qquad \text{(VI-1)}$$

TABLE 54. *Proportion of intrachanges in which both breaks are in the same chromosome arm*

(X-rays at room temperature unless otherwise stated.)

Material irradiated	Examined at	Chromosome or chromatid	No. of major chromosomes		Proportion of intrachanges in which both breaks are in same arm		Reference
			m	t	Experimental	Expected	
Drosophila sperm	Salivary glands	Chromosome	2	1	27 in 41 = 0·66	0·55	1
Tulip microspores	Pollen-grain division	Chromosome	12	0	18 in 33 = 0·55	0·50	2
Tradescantia pollen grains (ungerminated)	Pollen-tube division	Chromosome	6	0	74 in 121 = 0·61	0·50	2
Tradescantia pollen grains (germinated)	Pollen-tube division	Chromatid	6	0	2 in 6 = 0·33	0·50	3
Tradescantia microspores	Pollen-grain division	Chromatid	6	0			4
			1°		204 in 527 = 0·39	0·50	
			20°		101 in 270 = 0·37	0·50	
			30°		26 in 56 = 0·46	0·50	

1 Catcheside, D.G. (1938 a) using X-bearing sperm.
2 Newcombe, H.B. (1942 a).
3 Catcheside, D.G. & Lea, D.E. (1943).
4 Catcheside, D.G., Lea, D.E. & Thoday, J.M. (1946 a).

In Table 54 are collected experimentally determined proportions of intrachanges having both breaks in the same arm, together with the theoretical values.[2]

1 Newcombe, H.B. (1942 a).
2 When intrachanges are produced in chromosomes not yet split at the time of irradiation, it is not possible to observe both the symmetrical (inversion) type and the asymmetrical (deletion) type in the same organisms, and the figures in the case of *Drosophila* are based on the observation of inversions only, and in the case of plant material on the

There are some instances in Table 54 where departures from expectation are recorded which are statistically significant, but taking the table as a whole it appears that there is not a very marked preference for intrachanges between breaks in the same, as compared with breaks in opposite arms, or vice versa. However, taking into account the fact that intra-arm intrachanges are more likely to be overlooked than inter-arm intrachanges, a certain amount of preference for intra-arm intrachanges cannot be considered excluded. Further, *minute interstitial deletions* have not been included in the intra-arm intrachanges listed in Table 54. If these are included, the proportion of (asymmetrical) chromosome intrachanges which have both breaks in the same arm (in *Tradescantia* microspores) is 0·84, according to Rick.[1]

In Table 55 are collected data on the proportion of the total number of exchanges in which both breaks are in the same chromosome.[2]

The majority of the experiments cited (viz. 10 out of 12) indicate that the proportion of exchanges in which both breaks are in one chromosome is somewhat higher than is to be expected on the basis of random union between breakage ends.

When unfertilized eggs of *Sciara*[3] or of *Drosophila*[4] are irradiated, the preference for exchange between breaks in the same chromosome as compared to breaks in different chromosomes is extreme, the number of interchanges being very small compared to the number of intrachanges. This exceptional behaviour is presumed to be due to the chromosomes in the unfertilized eggs being in meiosis when irradiated.

observation of deletions only. In the case of chromatid intrachanges (produced in split chromosomes), where both types are observable, the figures refer to total intrachanges, symmetrical and asymmetrical.

[1] Rick, C.M. (1940).

[2] In *Drosophila* the observations are limited to symmetrical exchanges. In plant material in which unsplit chromosomes are irradiated, the observations are limited to asymmetrical exchanges, and in some experiments (noted in the footnotes to Table 55) intra-arm intrachanges are not recorded.

In plant material in which the chromosomes are split at the time of irradiation, all types of chromatid interchange and intrachange are intended to be included, but the possibility of some intra-arm intrachanges being overlooked cannot be excluded.

[3] Bozeman, M.L. (1943).   [4] Glass, H.B. (1940).

TABLE 55. Proportion of exchanges having both breaks in the same chromosome

(X-rays at room temperature except where otherwise stated.)

Material irradiated	Examined at	Chromosome or chromatid	No. of major chromosomes		Proportion of exchanges in which both breaks are in the same chromosome			Reference
			m	t	Experimental		Expected	
Drosophila melanogaster sperm	Salivary glands	Chromosome	2	1	41 in	81 = 0·506	0·36	1
Drosophila pseudoobscura sperm	Salivary glands	Chromosome	1	3	143 in	361 = 0·396	0·20	2
Onion root tip	Root-tip mitosis	Chromosome	16	0		0·1	0·032	3
Tulip microspores	Pollen-grain division	Chromosome	12	0	33 in	191 = 0·173	0·083	4
Tradescantia microspores	Pollen-grain division	Chromosome	6	0	2408 in	10,344 = 0·233	0·091	5
			3°–12°		249 in	908 = 0·274	0·091	6
			28°–36°		117 in	477 = 0·245	0·091	
			Neutrons		106 in	337 = 0·315	0·091	7
Tradescantia microspores	Pollen-grain division	Chromatid	6	0				8
			1°		527 in	2,281 = 0·23	0·167	
			20°		270 in	1,250 = 0·22	0·167	
			30°		56 in	359 = 0·16	0·167	
			Neutrons		68 in	774 = 0·088	0·167	9
			α-rays		16 in	181 = 0·088	0·167	10
Tradescantia pollen grains (ungerminated)	Pollen-tube division	Chromosome	6	0	121 in	336 = 0·360	0·167	4
Tradescantia pollen grains (germinated)	Pollen-tube division	Chromatid	6	0	6 in	27 = 0·222	0·167	11

1 Catcheside, D.G. (1938a). Salivary glands of female larvae were examined.
2 Koller, P.C. & Ahmed, I.A. (1942), in agreement with Helfer, R.G. (1940). Salivary glands female larvae were used. Inter-arm intrachanges were classed as interchanges, but this will not le to any appreciable distortion of the ratio as only one of the four major chromosomes has a medi centromere.
3 Sax, K. (1941a). Centric ring chromosomes and dicentric chromosomes were scored.
4 Newcombe, H.B. (1942a).
5 Sax, K. (1940). Centric ring chromosomes and dicentric chromosomes were scored. The prop tion of intrachanges was independent of the dose.
6 Sax, K. & Enzmann, E.V. (1939), Tables 2 and 6. Centric ring chromosomes and dicent chromosomes were scored.
7 Thoday, J.M. (1942). Centric ring chromosomes and dicentric chromosomes were scored. T proportion of intrachanges was independent of the dose.
8 Catcheside, D.G., Lea, D.E. & Thoday, J.M. (1946a). The proportion of intrachanges was independent of the intensity.
9 Thoday, J.M. (1942).
10 Kotval, J.P. & Gray, L.H. (1947).
11 Catcheside, D.G. & Lea, D.E. (1943).

Location of breaks in chromosomes

It is of interest to determine whether breaks occur at random in the chromosomes, or whether certain points are specially liable to be broken. Breaks can be located with high precision in *Drosophila melanogaster* by observation of the salivary gland chromosomes; the observations are then limited to breaks which have taken part in viable types of structural change, namely inversion and symmetrical interchange. The first point to be considered is the distribution of breaks between euchromatin and heterochromatin. The parts of the chromosomes nearest the centromeres (*proximal heterochromatic regions*) differ from the bulk of the chromosomes in being genetically *inert*, i.e. in containing few genes (or at any rate few genes detectable by their producing sharply alternative effects in different allelomorphs). They are also distinguishable cytologically by differences in their staining properties at mitosis and meiosis, believed to be due to a difference in the amplitude or timing of the nucleic acid cycle. These heterochromatic regions occupy an appreciable fraction of the whole length of the chromosomes at mitosis or meiosis (one-third in the case of the X chromosome), but only a minute fraction of the length of the salivary chromosomes. It has been found that the relative frequency with which breaks occur in the euchromatic and heterochromatic regions approximates to the relative lengths of these regions in the mitotic chromosomes, not to their relative lengths in the salivary chromosomes. Thus Kaufmann finds that about 30% of all breaks in the X chromosome occur in the proximal heterochromatin,[1] which occupies one-third of the length of the chromosome as seen in mitosis.

The conclusion that the break frequency is proportional to the mitotic length rather than to the salivary length is confirmed by the fact that the break frequency in the Y chromosome, which is mainly heterochromatic, and very short in the salivary glands, is comparable with that in the X chromosome and in the four arms of the autosomes.[2]

The salivary X chromosome map is divided into 120 lettered segments of approximately equal length. Part of the last six represents the proximal heterochromatin. Kaufmann[3]

[1] Kaufmann, B.P. (1946a). The observed proportion of breaks in the heterochromatin was 25%, and the figure 30% makes allowance for exchanges entirely within the chromocentre, which escape detection.

[2] Kaufmann, B.P. & Demerec, M. (1937); Bauer, H., Demerec, M. & Kaufmann, B.P. (1938). The X and Y chromosomes, and the four autosome arms $2L$, $2R$, $3L$, $3R$, are of comparable lengths in mitosis.

[3] Kaufmann, B.P. (1946); see also Prokofyeva, A.A. and Khvostova, V.V. (1939). Similar but less extensive data are given for the autosomes by Bauer, H., Demerec, M. & Kaufmann, B.P. (1938); Bauer, H. (1939); and for *Drosophila pseudoobscura* by Helfer, R.G. (1940).

determined the distribution of 1048 breaks among 114 lettered segments. An apparently high frequency of breaks in certain intercalary heterochromatic regions may be explained as due to the known compression of heterochromatin relative to euchromatin in salivary gland chromosomes. We conclude that in *Drosophila* breaks are fairly uniformly distributed along the length of the chromosome, and that, providing that mitotic length and not salivary length is considered, euchromatic and heterochromatic regions are approximately equally breakable.

In other organisms the location of breaks cannot be determined to such fine limits. In *Tradescantia* microspores the relative numbers of breaks induced by X-rays in successive fifths of a chromosome arm from the proximal fifth to the distal fifth have been determined.[1] The data show a tendency for more breaks to occur per unit length near the centromere (which is in the centre of the chromosome) than near the free end. When *Tradescantia* pollen tubes are irradiated, however, either by X-rays or by ultra-violet light, breaks are more frequent near the free ends than near the centromere.[2] The differences in break frequency between the proximal fifth and the distal fifth found in either microspores or pollen tubes, though statistically significant, do not exceed a factor of 2.

An experiment on *Tradescantia* microspores in which the frequency of production of breaks in ordinary centric chromosomes was compared with the frequency of production in acentric fragments showed that the frequency of breaks observed per unit length was only one-ninth as great in acentric fragments as in centric chromosomes.[3] The explanation suggested is that, since the centromere is largely responsible for movement of the chromosome during division, the strains would be less in an acentric fragment than in a centric chromosome, thus making restitution more probable.

[1] Sax, K. & Mather, K. (1939); cp. also Sax, K. (1938), who shows that the distribution is the same for single breaks and for breaks taking part in exchange.

[2] Swanson, C.P. (1942).

[3] Sax, K. (1942). Fragments were produced by a prior irradiation with X-rays before the chromosomes were split. The aberrations produced then were *chromosome* structural changes and could be distinguished from the *chromatid* structural changes induced by the second dose given three days later after the chromosomes had split.

Frequency relations

As well as the distribution of breaks in chromosomes, the distribution of numbers and types of aberrations in cells has been investigated.

If a single chromosome break is produced by the direct action of a single ionizing particle on a chromosome and is unaffected by the presence or absence of other breaks in the cell, we may expect the relative frequency with which cells are found containing different numbers of such aberrations to be distributed according to the Poisson formula. That is to say, if m is the mean number of aberrations per cell, $e^{-m}.m^r/r!$ is the proportion of cells expected to contain r aberrations. This expectation has been confirmed for chromatid breaks produced by neutrons in *Tradescantia* microspores, and for breaks produced in grasshopper neuroblast chromosomes. The experimental figures, together with the expected figures, are set out in Table 56. The χ^2 test shows that the agreement between the experimental and expected frequencies is satisfactory. Results included in the table also show that in *Tradescantia* the frequencies of occurrence of different numbers of isochromatid breaks, of interstitial deletions, and of exchanges are approximately in accord with the Poisson distribution.

In experiments in which *Drosophila* sperm are irradiated and the salivary chromosomes examined, the relative frequency with which different numbers of exchanges, and also exchanges of different degrees of complexity, are found is affected by their viability, since only viable configurations survive to be classified. In attempting to interpret experimental data in *Drosophila* the usual procedure is to assume (on the basis of the experiments cited in Table 56) that the numbers of breaks primarily produced in different sperm are distributed in a Poisson distribution, and then to calculate, on the basis of certain assumptions regarding the conditions affecting the combination of primary breaks, the relative frequency of different types of viable aberration. Comparison with experiment then affords evidence on the extent to which the assumptions are correct.

All viable structural changes in *Drosophila* are symmetrical exchanges of varying degrees of complication. The simplest is an exchange between two breaks, which we term type 2.

The only way in which a nucleus in which three breaks persist can be viable is for cyclic exchange to occur (type 3). With four breaks persisting one may either have a cyclic exchange involving all four breaks (type 4), or two separate two-break exchanges (designated $2+2$). Similarly, with six breaks

TABLE 56. *Numbers of cells containing* 0, 1, 2, 3, *etc., aberrations*

Material	Aberration	Aberrations per cell					χ^2 test	Reference
		0	1	2	3	$\geqslant 4$		
Chortophaga neuroblasts 62·5 r. X-rays	Chromatid + isochromatid breaks	174 178·4	112 104·2	28 30·5	5 5·9	1 1·0	$\chi^2 = 1·0$ $n = 3$ $P = 0·8$	1
Tradescantia microspores 46·6 v neutrons	Chromatid breaks	408 425·7	263 233·1	56 63·8	8 11·6	1 1·8	$\chi^2 = 6·9$ $n = 3$ $P = 0·08$	2
	Isochromatid breaks	483 480·4	204 205·0	38 43·7	10 6·2	1 0·7	$\chi^2 = 3·4$ $n = 3$ $P = 0·3$	2
	Chromatid exchanges	478 475·8	202 207·5	49 45·3	7 6·6	0 0·8	$\chi^2 = 1·4$ $n = 3$ $P = 0·7$	2
		0	1	2	$\geqslant 3$			
Tradescantia microspores 400 r. X-rays	Chromosome interstitial deletions	614 619·6	219 209·6	33 35·5	3 4·3		$\chi^2 = 1·0$ $n = 2$ $P = 0·6$	3
	Chromosome asymmetrical exchanges	665 677·8	192 168·5	12 20·9	0 1·8		$\chi^2 = 6·2$ $n = 2$ $P = 0·05$	3
Tradescantia microspores 150 r. X-rays	Chromatid interchanges	2278 2280·2	273 269·2	15 15·9	0 0·7		$\chi^2 = 0·8$ $n = 2$ $P = 0·7$	4

The upper figure is the experimental number of cells with the stated number of aberrations; the lower figure is the number expected on the Poisson distribution.

1 Carlson, J.G. (1941 a).
2 I am indebted to Mr Thoday for these unpublished data, which were obtained in the course of the experiments he describes in his 1942 paper (Thoday, J.M. 1942). 3 Rick, C.M. (1940).
4 Catcheside, D.G., Lea, D.E. & Thoday, J.M. (1946 a); irradiation at 30°.

persisting there are four alternatives $(6; 4+2; 3+3; 2+2+2)$. Experimental data exist of the relative frequencies of these various types in aberrant nuclei, and are reproduced in Table 57.[1]

The 'expected' frequencies listed in Table 57 for comparison with the experimental frequencies are calculated by an extension of the theory[2] described in Chapter V in connexion with dominant lethals. The proportion of sperm irradiated by dose D in which n breaks are primarily produced which unite in viable combinations was there found to be $e^{-\alpha D}(2\alpha q D)^n \dfrac{n!}{(2n)!}$. It can

[1] The experimental figures in Table 57 are taken from the analysis by Fano, U. (1941) of data of Bauer, H., Demerec, M. & Kaufmann, B.P. (1938), and Bauer, H. (1939 a). Similar data are available for *Drosophila pseudoobscura* (Koller, P.C. & Ahmed, I.A. 1942).

[2] Lea, D.E. & Catcheside, D.G. (1945 a).

be demonstrated that of such sperm a fraction $1/2(n-2)!$ will show a type 2 exchange, a fraction $1/3(n-3)!$ will show a type 3 exchange, a fraction $1/4(n-4)!$ will show a type 4 exchange, a fraction $1/8(n-4)!$ will show type $2+2$ and so on.[1] These formulae are based on the same assumptions as were used in

TABLE 57. Percentages of cells containing different types of chromosome exchanges

(*Drosophila* sperm irradiated by X-rays. Upper figures experimental, lower figures calculated assuming random union of breakage ends.)

Dose roentgens	Type of exchange %					Total no. of cells
	2	3	4	$2+2$	More complicated	
1000	80	13·3	6·7	—	—	15
	93·5	6·0	0·3	0·2	—	
2000	73·4	12·5	—	12·5	1·6	64
	87·3	10·8	1·1	0·6	0·2	
4000	59·6	11·8	4·2	14·6	9·8	144
	75·1	18·1	3·6	1·8	1·4	

Chapter V, namely, random joining between breakage ends, and each break supposed in a separate chromosome arm. The expected frequencies calculated on this basis are seen to agree tolerably with the experimental frequencies for 2-, 3-, or 4-break cyclic exchanges. The serious discrepancies[2] lie in the fact that a greater number of cells showing more complicated types of aberration is found experimentally at the higher doses, and in the fact that, experimentally, aberrations comprising four breaks are usually type $2+2$, while, theoretically, type 4 aberrations are expected to be twice as frequent as type $2+2$. The cause of these discrepancies is not yet understood.

Among viable structural changes induced by irradiation of maize pollen, and detected at diakinesis in the pollen mother cells of the F_1 plants, 11 configurations of type $2+2$ and 3 of

[1] In general, the fraction of viable sperm of n primary breaks which will show an aberration totalling r breaks, made up of α 2-break exchanges, β 3-break exchanges, γ 4-break exchanges, etc., is

$$\frac{1}{(n-r)!\,\alpha!\,\beta!\,\gamma!\ldots 2^\alpha.3^\beta.4^\gamma\ldots}.$$

[2] As pointed out by Fano, U. (1941). See also Kaufmann, B.P. (1941*b*, 1943); Fano, U. (1943*b*).

type 4 were observed, showing a similar departure from the expected ratio of 1 : 2 to that found in the *Drosophila* experiments.[1]

Modifying factors

The probability of restitution of a break primarily induced by radiation can be modified by other factors besides the position of the break in the chromosome and the presence or absence of a centromere in the chromosome. If *Tradescantia* buds are centrifuged during irradiation, the yield of structural changes is increased, as shown in Table 58.[2] The explanation of this effect is presumably that the increased stresses in the chromosome thread caused by centrifugation tend to separate the breakage ends at a newly formed break and thus reduce the probability of restitution.

TABLE 58. Increased yield of chromosome structural changes owing to centrifugation during irradiation

(*Tradescantia* microspores; Sax, 1943.)

Intrachanges and interchanges produced by ~120 r. in unsplit chromosomes
    Without centrifugation   6·6 %
    With centrifugation   13·2 %
Isochromatid breaks produced by ~150 r. in split chromosomes
    Without centrifugation   27·6 %
    With centrifugation   45·6 %
Intrachanges and interchanges produced by ~150 r. in split chromosomes
    Without centrifugation   26·7 %
    With centrifugation   36·3 %

What may be regarded as the complement of this experiment was performed on onion root tips.[3] Two batches were exposed to the same dose of X-rays, 300 r. One batch had been treated with colchicine, the other had not. The yield of chromatid aberrations in the colchicine-treated series was only one-third as great as in the untreated series, which was believed to be owing

[1] Experimental data of Stadler, L.J. & Sprague, G.F. (1937), and of Catcheside, D.G. (1938b) combined. See Table 2 of Catcheside's paper.

[2] Sax, K. (1943). Centrifugation alone did not produce aberrations. Sax's figures have been converted from breaks per chromosome to aberrations per cell by multiplying by six in the case of single-break aberrations, and by three in the case of two-break aberrations.

[3] Brumfield, R.T. (1943).

to the colchicine reducing the movement of the prophase chromosomes and so favouring restitution compared with rearrangement.

The yields of various kinds of aberrations induced by a given dose of radiation in *Tradescantia* microspores is influenced by the temperature, as shown in Fig. 32 A, B.[1] The flower buds remained at the given temperature during the irradiation and for about an

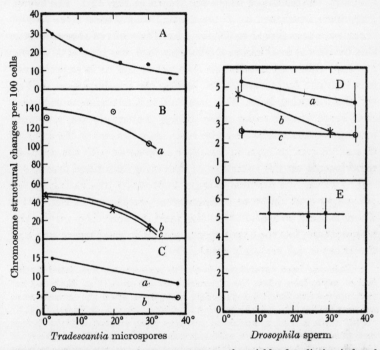

FIG. 32. The influence of temperature on the yield of radiation-induced chromosome structural changes. A, rings and dicentrics in unsplit chromosomes. B, split chromosomes: (*a*) chromatid breaks; (*b*) isochromatid breaks; (*c*) chromatid interchanges. C, rings and dicentrics in unsplit chromosomes: (*a*) irradiated at 3°, then transferred to 3 or 38°; (*b*) irradiated at 38°, then transferred to 3 or 38°. D, symmetrical interchanges between chromosomes II and III, irradiation at different temperatures. E, symmetrical interchange between chromosomes II and III, fertilization at different temperatures.

[1] The sources of the data are as follows: A. Sax, K. & Enzmann, E.V. (1939). The data of Tables 6, 7 and 8 of this paper have been grouped together. Doses 300–360 r. B. Catcheside, D.G., Lea, D.E. & Thoday, J.M. (unpublished). Dose 150 r. Rick, C.M. (1940) has irradiated unsplit chromosomes at 3° and 33°, confirming Sax and Enzmann's results with rings and dicentrics, and further showing that a similar effect occurs also with minute interstitial deletions.

hour following. The yield of aberrations at the higher temperatures is significantly lower than at the lower temperatures.[1]

If the view is correct that modifying factors such as temperature affect the joining of breakage ends rather than the primary production of breaks, then it should be possible to reduce the yield of structural changes by raising the temperature during the joining process. Sax and Enzmann carried out an experiment of this sort, the results of which are shown in Fig. 32c.[2] The flower buds were irradiated at 3° for 2 min., and 2 min. later half of them were transferred to 38° for 1 hr. The yield of aberrations in this batch was less than in the batch which was kept at 3° during the hour following irradiation, but not so low as in experiments in which the buds were at 38° during the irradiation as well as for the following hour. The result of this experiment indicates that some joining takes place during the irradiation, and that joining continues also for some time after the end of the irradiation. This conclusion is in complete agreement with the result of experiments on the variation of dose-rate (described later).

The yields of structural changes induced by irradiating *Drosophila* sperm at different temperatures are shown in Fig. 32D.[3] There is some indication that the yield is reduced by raising the temperature, but the experiments are not in good agreement and the effect is not certainly established.[4]

[1] This has been established both for microspores irradiated 24 hr. before metaphase when the chromosomes are split (Fig. 32B) and for microspores irradiated 5 days before metaphase when the chromosomes are unsplit (Fig. 32A; also Fabergé, A.C., 1940a). Sax, K. & Enzmann, E.V. (1939) made a trial in which microspores were irradiated 48 hr. before metaphase at a time when splitting was commencing. The majority of the structural changes were of the chromosome type, but a minority were chromatid changes. The number of chromatid changes per microspore *increased* at the higher temperatures in this experiment. In view of the contrary results obtained in experiments in which the microspores were irradiated either 24 hr. or 5 days before metaphase, it seems likely that the effect of raising the temperature at 48 hr. was to increase the proportion of chromosomes which were effectively split, rather than to raise the yield of aberrations per split chromosome.

[2] Sax, K. & Enzmann, E.V. (1939). Dose 160 r.

[3] The data for curves (a), (b), (c) are taken from the following sources: (a) Muller and Pontecorvo, reported in Muller, H.J. (1940); (b) Mickey, G.H. (1938); (c) Makhijani, dose 2000 r., reported in Muller, H.J. (1940).

[4] Muller, H.J. (1940), placing principal reliance on the data shown as curves (a) and (c) of Fig. 32D, concludes that the yield is independent of the temperature.

It is believed[1] that the joining in new arrangements of breaks induced by irradiating *Drosophila* sperm is delayed until fertilization of an ovum by the irradiated sperm occurs. In this event one might expect that the effect of temperature, if it exists, would be shown by the yield obtained being dependent upon the temperature at the time of fertilization. The data reproduced in Fig. 32 E[2] do not show any such dependence.

According to an experiment of Kaufmann and Hollaender,[3] irradiation of *Drosophila* male flies by either infra-red or ultra-violet radiation between the administration of two 2000 r. X-ray treatments reduces considerably the yield of chromosome structural changes induced by the X-rays. The infra-red or ultra-violet radiation alone had no effect. These results suggest that some restitution of breaks primarily induced by the X-rays is possible in the sperm, and is aided by the administration of ultra-violet or infra-red light. If the diminution of yield at higher temperatures suggested but not established by Fig. 32 D is confirmed, the explanation will presumably be that restitution in the sperm is aided by rise of temperature.

Marshak[4] has found that treating onion seedlings with a dilute solution of ammonia prior to irradiation reduces the proportion of abnormalities seen in anaphases 3 hr. after irradiation. It is not at all obvious what conclusion to draw from this observation. The treatment with ammonia delays the onset of prophase. Thus cells seen in anaphase 3 hr. after irradiation are not in the same stage at the time of irradiation in the ammonia-treated and untreated series, and this, rather than any more fundamental effect, may explain the observation. In any case, the chromosome aberrations studied are probably largely physiological changes (cp. p. 192).

The yield of chromosome structural changes obtained with a given dose in a given species is affected by the state of the chromosomes at the time of irradiation. This effect is best investigated in material in which it is possible to determine with certainty the stage of the cells at the time of irradiation. In the

[1] Muller, H.J. (1940).
[2] Muller and Pontecorvo, reported in Muller, H.J. (1940).
[3] Reported in Demerec, M., Kaufmann, B.P., Fano, U., Sutton. E. & Sansome, E. (1942).
[4] Marshak, A. (1938*a,b*).

experiments of Swanson,[1] the results of which are reproduced in Table 59, *Tradescantia* pollen grains were germinated on an artificial medium and irradiated at various times after germination, representing various stages of prophase of the pollen-tube mitosis.

The yield of aberrations is at a maximum 4 hr. after germination, when the cell is in mid-prophase and the chromosomes are spiralizing. The yield is small in the ungerminated pollen grain when the chromosomes are at rest. The results are consistent with the experiments already described in which chromosome movement was increased by centrifugation or decreased by colchicine or removal of the centromere, and from which it was concluded that movement of the chromosomes during irradiation favours permanent structural changes by hindering restitution.

TABLE 59. Yield of chromatid breaks and chromatid interchanges in *Tradescantia* pollen tubes irradiated at various times after germination (Swanson, 1943, 370 r.)

Hours after germination	Ungerminated	1	2	3	4	5
Chromatid breaks per 100 cells	2·0	25·6	28·3	30·4	38·0	34·0
Interchanges per 100 cells	3·0	6·8	11·1	14·1	15·2	6·0

Hours after germination	6	7	8	10	15
Chromatid breaks per 100 cells	18·0	4·6	1·9	1·4	0·8
Interchanges per 100 cells	7·7	0·0	0·0	0·0	0·8

The almost complete disappearance of aberrations when the chromosomes are in the fully condensed state 15 hr. after germination is attributed to the formation of a matrix round each chromosome which holds the chromosome together although breaks may be induced in the chromosome threads. In experiments on *Sciara* oocytes it has been established that irradiation during first meiotic metaphase and anaphase can produce chromosome structural changes (not detectable at the division concerned but observed in the salivaries of the F_1 larvae) actually with a greater frequency than during prophase.[2] The changes are, however, nearly all intrachromosomal, exchange between breaks in different chromosomes hardly ever occurring.[3] It is probable therefore that irradiation of metaphase and anaphase chromosomes can cause breaks which are not cytologically detectable

[1] Swanson, C.P. (1943).
[2] Reynolds, J.P. (1941).
[3] Bozeman, M.L. (1943).

at the division during which irradiation takes place, and which are less likely to give interchromosomal structural changes than are breaks induced in interphase or early prophase. If sister-union occurs at the breakage ends when the chromosomes split, such breaks induced at metaphase and anaphase may have a lethal effect at a subsequent division.

Experiments on a variety of materials have been described in which cells are irradiated, fixed after the lapse of varying intervals of time, and metaphase or anaphase figures examined for chromosome changes. These experiments therefore consist essentially in determining the sensitivity of chromosomes at various stages prior to metaphase. Their interpretation is complicated by the fact that radiation delays division, so that even though the time scale of the cell cycle may be known in the unirradiated material (which is not always the case), there is liable to be a doubt concerning the stage which has been reached at the time of irradiation by a cell which is found, for example, to be in metaphase 24 hr. later. The general result appears to be that cells become less sensitive as prophase advances,[1] in agreement with the data shown in Table 59. The sensitivity in interphase, prior to chromosome split, is rather lower than in early prophase, so that the highest sensitivity is reached in prophase.[2]

Dependence of the yield of structural changes on radiation intensity

Studies of the dependence of the yield of various types of structural change on the intensity, dose, and kind of radiation

---

[1] See, for example, Sax, K. & Swanson, C.P. (1941) on *Tradescantia* microspores; Marquardt, H. (1938) on *Bellevalia* microspores; Carlson, J.G. (1941a) on *Chortophaga* neuroblasts.

[2] Marshak, A. (1937, 1939b, 1942a, 1942b) has irradiated rat and mouse tumours, and the root tips of a variety of plants, and examined the proportion of anaphases which are abnormal at various times after irradiation. The maximum effect is shown by cells which are in anaphase 3 hr. after irradiation. It appears probable that the abnormalities at 3 hr. in these experiments are mainly physiological effects induced in cells sufficiently advanced in mitosis to escape the temporary inhibition of division experienced by cells less far advanced at the time of irradiation. When, instead of 'abnormal anaphases', an aberration definitely of the 'structural change' type was scored, the maximum yield was found in cells irradiated 18 hr. before metaphase (Marshak, A. 1939b scoring minute deletions in bean root tips).

have been of great value in elucidating the mechanism of the induction of structural changes by radiation, and these experimental results we now proceed to review.

A number of authors have investigated the manner in which the yield of chromosome structural changes induced in *Tradescantia* microspores by a given dose of radiation depends upon the time over which the irradiation is extended. The principal results are given in Fig. 33.[1] It may be seen that with both X-rays and neutrons the yield of chromatid and isochromatid breaks is independent of intensity. This is the result to be expected on the view that these aberrations are produced by the passage of a single ionizing particle through one or both respectively of the chromatids of a split chromosome.

It may be seen further that with both chromatid and chromosome interchanges, the yield produced by a given dose of X-rays diminishes with increase of the time over which the irradiation is extended. This result is readily explained on the view[2] that the two breaks which take part in an interchange are produced by separate ionizing particles. If the irradiation is extended over a prolonged time, a break in one chromosome has time to restitute before another break is produced in its vicinity with which exchange is possible. Experiments have also been made in which a given dose is either given in one concentrated exposure, or divided into fractions with rest periods between.[3] The yield is less with the fractionated dose, in agreement with the results of the intensity variation experiment.

In contrast to the X-ray results, it is found that with neutrons the yield of both chromosome exchanges (curve c, Fig. 33) and of chromatid exchanges (curve g, Fig. 33) is independent of the time over which a given dose is spread.[4] This suggests that a

[1] Derived from the experiments of Sax, K. (1939, 1940); Marinelli, L.D., Nebel, B.R., Giles, N.H. & Charles, D.R. (1942); Giles, N.H. (1943); Catcheside, D.G., Lea, D.E. & Thoday, J.M. (1946b). The results of Giles and of Catcheside, Lea and Thoday shown in curves e and f of Fig. 33 agree well. Those of Sax and of Marinelli, Nebel, Giles and Charles shown in curves a and b agree badly quantitatively, but both show a reduction of yield with prolongation of the time of exposure.

[2] Sax, K. (1939). A fuller discussion is given in Chapter VII.

[3] Sax, K. (1939, 1941b).

[4] Giles, N.H. (1943). It was also found that the effect of a given dose was independent of whether it was given as a single exposure or divided into two fractions separated by an interval of 15 or 32 min.

single ionizing particle usually causes both the breaks in the neutron-induced exchanges.[1]

Experiments have been made to test whether the yield of structural changes induced in *Drosophila* sperm by a given dose of X-rays or γ-rays is independent of the time over which the dose is spread. The experiments are made by irradiating male flies, or impregnated females, and detecting structural changes,

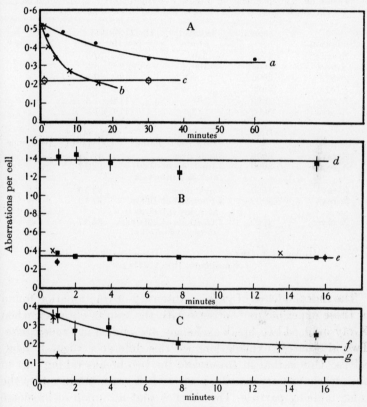

Fig. 33. Dependence upon duration of exposure of yield of aberrations per cell in *Tradescantia* microspores produced by a constant dose. A, chromosome exchanges: *a*, 300 r. X-rays (Marinelli, Nebel, Giles & Charles); *b*, 320 r. X-rays (Sax); *c*, 65v neutrons (Giles). B, chromatid and isochromatid breaks: *d*, chromatid breaks 150 r. X-rays (Catcheside, Lea & Thoday); *e*, isochromatid breaks (■ 150 r. X-rays (Catcheside, Lea & Thoday); × 130 r. X-rays (Giles); ● 26v neutrons (Giles)). C, chromatid exchanges: *f*, ■ 150 r. X-rays (Catcheside, Lea & Thoday), × 130 r. X-rays (Giles); *g*, ● 26v neutrons (Giles).

[1] Giles, N.H. (1940). A full discussion is deferred to Chapter VII.

either cytologically in the salivary chromosomes of the F_1 larvae,[1] or genetically by breeding tests.[2] The results are given in Table 60, and show on the whole no evidence for any effect of reduction of intensity or fractionation of dose. In either method only viable changes, i.e. symmetrical exchanges involving two or more breaks, are investigated.

TABLE 60. Independence of intensity or fractionation of the yield of chromosome structural changes induced in *Drosophila* sperm

Radiation	Dose r.	Intensity r./min.	II–III interchanges per 100 sperm	Reference
X-rays	2000	250	2.9 ± 0.3	1
X-rays	2000	100	3.8 ± 0.5	1
γ-rays	2000	0.8	3.4 ± 0.9	1
γ-rays	2000	0.05	4.2 ± 1.0	1
X-rays	5000	167	17.2 ± 1.5	2
X-rays	5000	12.5	17.0 ± 1.3	2

Radiation	Dose r.	How fractionated	Breaks per 100 sperm	Reference
X-rays	3000	Single dose	49.24	3
X-rays	3000	3 fractions of 1000 r. at 1 day intervals	52.17	3
X-rays	4000	Single dose	83.76	3
X-rays	4000	4 fractions of 1000 r. at 1 day intervals	81.82	3
X-rays	4000	2 fractions of 2000 r. at 16 days interval	83.57	3

1 Muller, H.J. (1940). 2 Dempster, E.R. (1941 b).
3 Kaufmann, B.P. (1941 b).

The independence of yield on intensity or fractionation found in these experiments contrasts with the results obtained when X-ray-induced two-break exchanges are studied in *Tradescantia*. Two alternative explanations for this difference suggest themselves. One is that in *Drosophila* the two breaks taking part in an interchange are usually, even with X-rays, produced by the same ionizing particle. The other is that no union of breakage ends takes place in the sperm, and that the breaks accumulate until opportunity for union occurs after fertilization. The first explanation would require the yield of aberrations to be directly proportional to the dose, and is ruled out by the experimental evidence (given below) that the yield increases more rapidly than

[1] Kaufmann, B.P. (1941 b).
[2] Experiments by Muller, H.J. (1940) in collaboration with Ray-Chaudhuri and Makhijani. Also Dempster, E.R. (1941 b).

the first power of the dose. The second explanation[1] is the accepted one. It is supported by the fact that cases have been reported of structural changes which apparently involve exchange between a paternal chromosome broken by irradiation of the sperm, and a maternal chromosome spontaneously broken, suggesting that joining of chromosomes broken by irradiation of the sperm is sometimes at any rate deferred until the sperm and egg chromosomes come into contact during the first cleavage of the fertilized egg.[2]

Dependence of yield on dose

The manner in which the yield of structural changes increases with increase of the dose of radiation has been extensively studied, and the results of these studies form the main basis on which theories of the mechanism of induction of these changes are built. The curves in Figs. 34 and 35 illustrate some of the principal results obtained.[3]

In some materials it is possible to observe simple breaks. These may be either chromosome breaks (affecting an unsplit chromosome) or chromatid breaks (affecting only one of the sister chromatids of a split chromosome) or isochromatid breaks (affecting both sister chromatids of a split chromosome at approximately the same locus). The yield of each of these types appears to increase linearly with increase of dose, as illustrated for X-rays and for neutrons in Fig. 34.[4] This result is consistent

[1] Muller, H.J. (1940).
[2] Sidky, A.R. (1940); Helfer, R.G. (1940). Experiments designed to secure interchange between maternal and paternal chromosomes by irradiation of both egg and sperm prior to fertilization have given negative results (Glass, H.B. 1940). Probably union of maternal broken chromosomes is not delayed until fertilization.
[3] Taken from the following papers: Bauer, H., Demerec, M. & Kaufmann, B.P. (1938), modified by addition of later work by Kaufmann, B.P. (1941b); Bauer, H. (1939, quoted by Kaufmann, B.P. 1941b); Sax, K. (1940, 1941b); Rick, C.M. (1940); Giles, N.H. (1940, 1943); Carlson, J.G. (1941a); Thoday, J.M. (1942); Newcombe, H.B. (1942b); Marinelli, L.D., Nebel, B.R., Giles, N.H. & Charles, D.R. (1942); Catcheside, D.G., Lea, D.E. & Thoday, J.M. (1946a); Kotval, J.P. (unpublished).
[4] In Fig. 34B (isochromatid breaks produced by X-rays) the results of Sax and of other workers do not agree exactly, but each separately gives a straight line. The estimates of dose given by Sax are probably too high (Giles, N.H. 1943). The linearity is not convincing in Figs. 34D,E (chromatid breaks produced by X-rays and neutrons). Chromatid breaks

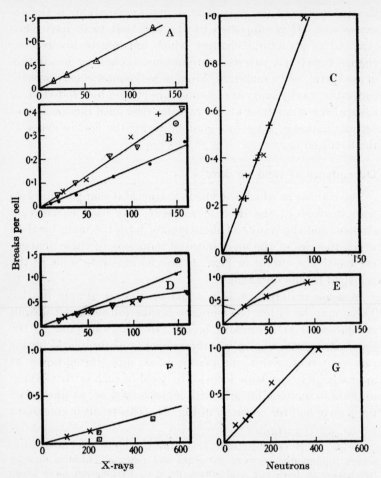

FIG. 34. Number of breaks per cell as a function of dose (X-ray doses in roentgens, neutron doses in v-units). A, chromatid and isochromatid breaks in *Chortophaga* neuroblasts. B, C, isochromatid breaks in *Tradescantia* microspores. D, E, chromatid breaks in *Tradescantia* microspores. F, G, chromosome breaks in *Tradescantia* microspores or mature pollen. ● Sax; △ Carlson; × Thoday; ⊡ Newcombe; + Giles; ⊙ Catcheside, Lea & Thoday; ▽ Kotval.

are difficult to see, and the departure from linearity may be subjective error (cp. discussion by Lea, D.E. & Catcheside, D.G. 1942). Newcombe, H.B. (1942b) found that the number of chromosome breaks produced by irradiating *Tradescantia* microspores increased considerably more rapidly than the first power of the dose. This result, obtained with doses of 240–960 r., he explains by supposing that the proportion of breaks primarily produced which restitute is reduced at high doses.

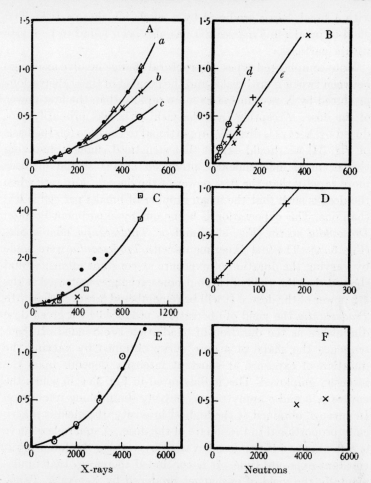

FIG. 35. Yield of two-break aberrations as a function of dose. Abscissae are doses in roentgens (X-rays) or v-units (neutrons). Ordinates are aberrations per cell (*Tradescantia*), or breaks per sperm (*Drosophila*). A, chromosome exchanges in *Tradescantia* microspores (Sax): △ constant time of 3 min.; • constant intensity of 160 r. per min.; × constant intensity of 20 r. per min.; ⊙ constant intensity of 2·7 r. per min. B, chromosome exchanges in *Tradescantia* microspores: × Thoday (Li+deuterons of 1eMV. energy); + Giles (Be+deuterons of 11eMV. energy); ⊕ Giles (Be+deuterons of 3eMV. energy). C, D, minute interstitial deletions in *Tradescantia* microspores and mature pollen: • Marinelli, Nebel, Giles & Charles; ⊡ Newcombe; × Rick; + Giles (Be+deuterons of 11eMV. energy). E, F, exchange breaks in *Drosophila* sperm: • Bauer, Demerec & Kaufmann; ⊙ Bauer; × Demerec, Kaufmann & Sutton.

with a break being produced by a single ionizing particle. The yield of breaks in *Tradescantia* has also been found to be linear with α-particles.[1]

More complicated types of structural change involve exchange between two or more breaks, and the number of these aberrations produced by X-rays increases more rapidly than the first power of the dose. Accepting that the yield of breaks primarily produced by X-rays is directly proportional to the dose (on the basis of Fig. 34) we should expect that structural changes involving two or more independently produced breaks should increase more rapidly than the first power of the dose, at any rate when the dose is such that the mean number of breaks per cell is less than one. This expectation is borne out by experiment both on *Drosophila* sperm (Fig. 35 E) and on *Tradescantia* microspores (Fig. 35 A).[2] The first experiments with *Tradescantia* were made by varying the duration of exposure at constant intensity, and showed that the yield of chromosome exchanges increased as the 3/2 power of the dose.[3] It will be remembered however that with *Tradescantia* the yield of aberrations produced by a given dose diminishes as the duration of irradiation is extended. In consequence the shape of a dose curve obtained by varying the duration of exposure at constant intensity depends upon the intensity employed. This is illustrated in Fig. 35 A, in which the curves a, b and c apply to successively diminishing intensities. In curve a, obtained at the highest intensity, the yield is practically proportional to the square of the dose. A square-law curve is also obtained if the dose is varied by varying the intensity at constant exposure time. It is concluded therefore[4] that fundamentally the yield of exchanges produced by X-rays in *Tradescantia* increases as the square of the dose, but that the results are distorted by restitution of breaks occurring unless it is arranged either that the irradiation is completed in a *short* time (making restitution negligible during the exposure) or that the irradiation extends over the *same* time at all doses (so equalizing the effects of restitution at the different doses).

[1] Kotval, J.P. & Gray, L.H. (1947).

[2] Fig. 35A refers to exchanges between unsplit chromosomes. The yield of chromatid exchanges obtained by irradiating split chromosomes also increases more rapidly than the first power of the dose (Sax, K. 1940; Thoday, J.M. 1942).

[3] Sax, K. (1938).   [4] Sax, K. (1940, 1941 *b*).

The yield of chromosome and chromatid exchanges induced in *Tradescantia* by neutrons is found to increase in direct proportion to the dose, as illustrated (for chromosome exchanges) in Fig. 35 B.[1] This is believed to be due to both the breaks taking part in the exchange being produced by the same ionizing particle in neutron experiments,[2] a hypothesis which also explains the fact (mentioned on p. 226) that the yield of exchanges is, with neutrons, found to be independent of the intensity. With α-rays also, the yield of chromatid exchanges has been found to be directly proportional to dose.[3]

Interstitial deletions form a special class of exchanges which have been separately studied in *Tradescantia*. As shown in Fig. 35 C and D, the yield of interstitial deletions is proportional to dose in neutron experiments, but in X-ray experiments increases with dose according to a power of the dose intermediate between the first and the second.

Extensive experiments have been made to study the yield of structural changes induced in *Drosophila* by irradiation of the sperm. The exchanges are detected either by examination of the salivary chromosomes of F_1 larvae, or by breeding tests. In either event study is limited to viable types involving structural changes which are all exchanges involving two or more breaks. The results by the cytological method are given in Fig. 35 E and F, and by breeding methods in Table 61.

It is seen that in all experiments on the production of gross structural changes by X-irradiation of *Drosophila* sperm, the yield increases in proportion to a power of the dose intermediate between the first power and the square, and is generally stated to be proportional to the 3/2 power of the dose. Muller[4] believes the power to approach the square more nearly at the lower doses, but the evidence is not at present convincing on this point.

When minute structural changes are studied, i.e. changes involving two breaks separated by a distance of the order of 1 % of the length of the chromosome or less, the yield is found to be proportional to the first power of the dose, as illustrated in

[1] The difference in gradients between curves *d* and *e* in Fig. 35B is believed by Giles, N.H. (1943) to be real, and due to the fact that the neutrons used in *d* were somewhat less energetic than those used in *e*.
[2] Giles, N.H. (1940).
[3] Kotval, J.P. & Gray, L.H. (1947).
[4] Muller, H.J. (1940).

Table 61. This suggests that a single ionizing particle usually causes both breaks in such rearrangements.

The evidence available at present on the variation with dose of the yield of gross structural changes induced by neutrons cannot be considered adequate to determine the shape of the curve. It is given in Fig. 35 F.

TABLE 61. X-ray-induced structural changes in *Drosophila* chromosomes investigated by genetical methods

Change studied	Dose r.	Aberrations per sperm	Power of dose	Reference
Gross structural changes				
Chromosome exchanges involving X chromosome	1500	0·0058 ± 0·002	1·88	1
	3000	0·0214 ± 0·005	1·68	
	6000	0·0735 ± 0·01		
Chromosome interchanges between chromosomes II and III	380	0·0021 ± 0·0008	1·80	2
	1500	0·0252 ± 0·003		
	1000	0·0116 ± 0·001	1·42	3
	4000	0·0830 ± 0·008		
Chromosome exchanges involving chromosome IV[7]	1000	0·00043 ± 0·00008	1·44	4
	4000	0·00318 ± 0·0004		
	1000	0·0007 ± 0·0002	1·03	5
	2000	0·0014 ± 0·0003	1·32	
	4000	0·0036 ± 0·0006	1·04	
	6000	0·0054 ± 0·0010		
Chromosome exchanges	1000	0·017 ± 0·003	1·24	5
	2000	0·041 ± 0·005	1·37	
	4000	0·105 ± 0·008	1·59	
	6000	0·200 ± 0·016		
Minute structural changes				
Minute rearrangements in X chromosome[8]	1000	0·00043 ± 0·00007	0·96	6
	4000	0·00162 ± 0·00022		
	1000	0·00087 ± 0·00014	0·93	4
	4000	0·00317 ± 0·0004		

1 Timoféeff-Ressovsky, N.W. (1939).
2 Muller, H.J. (1940). The yield quoted at 380r. is compounded of some X-ray data at 375r. and some γ-ray data at 400r.
3 Muller, H.J. (1940), quoting experiments of Muller and Sidky.
4 Muller, H.J. (1940), quoting experiments of Muller and Makki.
5 Khvostova, V.V. & Gavrilova, A.A. (1938).
6 Belgovsky, M.L. (1939).
7 Exchanges detected are those which transfer the IVth chromosome gene *cubitus interruptus* to a euchromatic region of any chromosome. Such exchanges have the effect of weakening the dominance of the gene, and permit a fly heterozygous for the wild type and the recessive to show the phenotype of the latter.
8 The X chromosome used is one in which a large inversion (sc^8) has brought the gene y^+ close to the heterochromatin. The apparent mutation rate of the y^+ gene is much greater than in an ordinary X chromosome, and is due mainly to minute rearrangements in the neighbourhood of the gene (possibly deletions) rather than to gene mutation proper.

Experiments are sometimes made in which a material is irradiated, fixed some hours or days later, and anaphase figures examined and classified as normal or abnormal. This procedure gives less insight into the mode of action of radiation than does the more laborious process of classifying the chromosome structural changes into the various types. It may, however, be the only procedure practicable when dealing with cells with a large number of small chromosomes, and even with more favourable materials may be adopted when it is desired to obtain rapidly a statistically significant amount of data. As a guide to the interpretation of such experiments we have plotted in this manner the results of experiments[1] on the irradiation of *Tradescantia* microspores, irradiated either when the chromosomes were split (Fig. 36 A, B) or when they were unsplit (Fig. 36 C, D). Plotted on a logarithmic scale, the points representing the percentages of normal division figures lie quite close to straight lines.

With neutrons the yields of all types of structural changes induced in *Tradescantia* chromosomes were found to be proportional to the dose, so that we should expect the percentage of normal division figures to be linear against dose when plotted on a logarithmic scale.[2] In the case of the X-ray experiments where the yields of some of the structural changes are proportional to dose, while the yields of others are more nearly proportional to the square of dose, we should expect the curve of the percentage of normal division figures against dose to be convex upwards when plotted on a logarithmic scale. The square law aberrations, however, being in a minority, the convexity is hardly noticeable.[3]

In Fig. 36 E and F we show the percentage of normal anaphase figures in a mouse lymphoma fixed 12 hr. after irradiation.[4]

[1] Thoday, J.M. (1942). [2] Cp. Chapter III, pp. 72–75.

[3] In Fig. 36, A and C approximate more closely, perhaps, to straight lines than is typical. In the case of Fig. 36A (irradiation 24 hr. before metaphase) there is believed to be some observational error leading to underestimation of the number of chromatid breaks at high doses (cp. footnote 4, p. 229).

As regards the experiment in which microspores were irradiated 5 days before metaphase (Fig. 36c), American authors (Sax, Rick, Giles) find a relatively higher proportion of two-break types of aberration than does Thoday, and curves of normal division figures derived from their material when plotted as in Fig. 36 would probably depart more markedly from a straight line.

[4] Marshak, A. (1942b).

Division figures were classified as abnormal when they showed lagging (acentric) fragments, or bridges (dicentric chromatids).

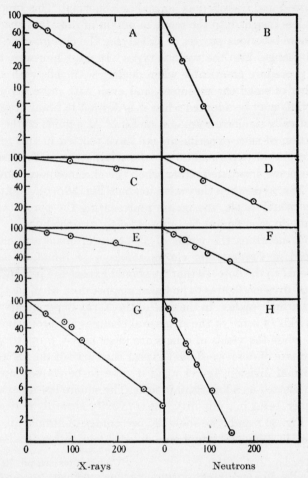

FIG. 36. The percentage of normal division figures as a function of dose. Abscissae are in roentgens (X-rays) or v-units (neutrons). A, B, *Tradescantia* microspore division, irradiation 24 hr. before metaphase (Thoday). C, D, *Tradescantia* microspore division, irradiation 5 days before metaphase (Thoday). E, F, mouse lymphoma mitosis, irradiation 12 hr. before anaphase (Marshak). G, H, bean root-tip mitosis, irradiation 3 hr. before anaphase (Marshak & Malloch).

Such fragments and bridges can be produced either by asymmetrical interchange between two chromosomes, a process expected to be approximately square-law with X-rays on the basis of *Tradescantia* experiments, or by a break in an unsplit chromosome, or an isochromatid break in a split chromosome, which are first-power-law processes. The curve shown in Fig. 36 E suggests that the single break process is predominant in these experiments, as in the *Tradescantia* experiments shown in Fig. 36 A and C.

Marshak has also made a number of experiments on rapidly dividing tissues in which the percentage of abnormal anaphases was determined in material fixed three hours after irradiation, at a time when the temporary reduction of mitotic activity caused by the radiation was most marked. The results of experiments of this type[1] are shown in Fig. 36 G and H. If, as suggested earlier, abnormalities observed at anaphase three hours after irradiation are a mixture of physiological changes and structural changes in the chromosomes, it appears from these results that the proportion of cells subject to the former type of effect as well as the latter increases approximately in linear proportion to the dose.

The relative efficiencies of different wave-lengths and types of ionizing radiation

Study of the relative efficiency of radiations of different iondensity in producing structural changes in chromosomes is an important method of attack on the mechanism of this process. The efficiencies of X-rays and neutrons have been compared with a number of materials. Providing that X-ray doses are measured in roentgens, and neutron doses in v-units,[2] units which represent approximately equal energy dissipations in tissue, the ratio of the yields for equal doses of the two radiations may be taken to be the ratio of the efficiencies per ionization of the densely ionizing particles (protons) in the neutron experiments, to the less densely ionizing particles (electrons) in the X-ray experiments.

Some experiments have been made to compare the efficiencies of X-rays and neutrons in producing structural changes by irradiation of *Drosophila* sperm. The results are not at present conclusive, the experiments shown in Fig. 35 E and F[3] suggesting

[1] Marshak, A. & Malloch, W.S. (1942).

[2] In reporting American experiments in which the neutron doses are given in n-units, we have throughout converted to v-units on the basis that 1 n-unit $= 2 \cdot 5$ v-units (cp. p. 20).

[3] X-ray experiments by Bauer, H., Demerec, M. & Kaufmann, B.P. (1938); Bauer, H. (1939); Kaufmann, B.P. (1941 b); neutron experiments by Demerec, M., Kaufmann, B.P. & Sutton, E. (1942).

that neutrons are slightly less effective, while other experiments¹ suggest that they are slightly more effective.

With all other materials which have been investigated, and for all types of chromosome aberrations, neutrons have been found to produce a greater yield of aberrations for the same dose than X-rays, as may be seen by comparing curves for the two

TABLE 62. Relative efficiencies of neutrons and X-rays

(A figure greater than 1·0 means that a greater yield of aberrations is produced by 1 v-unit of neutrons than by 1 r. of X-rays.)

Material studied	Type of aberration	Hours between irradiation and fixation	Relative efficiency	Reference
Tradescantia	Chromatid breaks	24	2·2	1
microspores	Isochromatid breaks	24	3·6	1
			4·1	2
	Chromosome breaks	120	4·3	1
Bean root tips	Abnormal anaphases	12	6·0	3
Mouse lymphoma	Abnormal anaphases	12	3·5	4
	Abnormal anaphases	3	2·3	4
Seedling root tips	Abnormal anaphases	3 Bean	2·6	3
		Pea	2·6	5
		Tomato	2·6	5

1 Thoday, J.M. (1942), as worked up by Lea, D.E. & Catcheside, D.G. (1942).
2 Giles, N.H. (1943).   3 Marshak, A. (1942a).
4 Marshak, A. (1942b).   5 Marshak, A. (1939b).

radiations in Figs. 34, 35 and 36. It is only possible to state a precise figure for the ratio of efficiencies when the dosage curves are of the same shape for both radiations, otherwise the ratio varies with dose. In Table 62 we give the ratios of efficiencies in cases where the yield of aberrations is a linear function of dose with both radiations.

In addition to ratios of efficiencies for aberrations of known types, for which data are only available for *Tradescantia*, ratios obtained by Marshak by recording abnormal anaphases in rapidly dividing tissues have also been included. In the cases where fixation was made 12 hr. after irradiation the anaphase abnormalities are presumably structural changes, probably mainly chromosome or isochromatid breaks, and the ratios obtained are comparable with the ratios for these types of aberrations obtained with *Tradescantia*. Where fixation was made 3 hr. after irradiation, the anaphase abnormalities are probably a mixture

1 Dempster, E.R. (1941a).

of physiological and structural changes in the chromosomes. The lower value of the ratio of efficiencies of neutrons and X-rays then obtained suggests that there is less difference in efficiency between the two radiations in producing the physiological effects than in producing structural changes.

In experiments on *Tradescantia* microspores, the yields of chromatid and isochromatid breaks have been found to be greater with α-rays than with neutrons, while the yield of chromatid exchanges has been found to be less with α-rays than with

TABLE 63. Relative efficiencies of different wave-lengths in inducing breaks

Aberration	Wave-length A.	Number of breaks per 100 cells per r.
A. *Tradescantia* pollen-tube mitosis[1]		
Chromatid breaks	~0.15	0.58 ± 0.04
	1.5	0.62 ± 0.12
	4.1	1.10 ± 0.11
	8.3	0.065 ± 0.015
Isochromatid breaks	~0.15	0.099 ± 0.018
	1.5	0.092 ± 0.046
	4.1	0.158 ± 0.040
	8.3	0.0034 ± 0.0034
B. *Tradescantia* pollen-grain mitosis[2]		
Chromatid breaks	~0.15	0.50 ± 0.015
	~0.015	0.47 ± 0.013
Isochromatid breaks	~0.15	0.25 ± 0.011
	~0.015	0.23 ± 0.009

[1] Catcheside, D.G. & Lea, D.E. (1943). It is doubtful if any isochromatid breaks are produced by X-rays of 8·3A. The yield given is based on the one isochromatid break observed in the chromosomes irradiated by this wavelength; it was probably a spontaneous aberration not caused by the radiation.
[2] Kotval, J.P. (unpublished).

neutrons.[1] The explanation of these results is discussed in Chapter VII. An important difference between α-rays and other radiations is the fact (pointed out on p. 210) that breakage ends produced in *Tradescantia* chromatids by α-rays appear to be unjoinable in a much higher proportion of cases than breakage ends produced by neutrons and X-rays.

The efficiencies of X-rays and γ-rays in producing structural changes by irradiation of *Drosophila* sperm have been compared by Muller and Ray-Chaudhuri.[2] The efficiencies per roentgen of

[1] Kotval, J.P. & Gray, L.H. (1947). The experiments were made by immersing the inflorescences in a solution of radon.
[2] Reported by Muller, H.J. (1940).

X-rays generated at 50 kV. and γ-rays were equal. Kotval found that γ-rays were slightly, but not significantly, less efficient than X-rays in producing chromatid and isochromatid breaks in *Tradescantia* microspores (Table 63B).

Fabergé,[2] irradiating *Tradescantia* microspores, found no difference in the yield of aberrations between X-rays of effective wave-lengths 0·077 and 0·44 A., generated at 400 and 50 kV. respectively. As explained in Chapter I, however, a considerable change in the wave-length of X-rays can be made in this region without causing a very marked change in the average ion-density of the electrons traversing the tissue, and greater significance therefore attaches to experiments made with still longer wave-lengths. By germinating pollen grains on artificial medium[3] and irradiating the pollen tubes it is practicable to work with very soft X-rays of long wave-length. The yields per roentgen of chromatid and isochromatid breaks produced by X-rays of different wave-length are given in Table 63A.[4] It is seen that a maximum efficiency is attained at a wave-length of about 4 A., and that a wave-length of 8·3 A. is very inefficient.[5] The explanation of the existence of a wave-length of maximum efficiency is discussed in Chapter VII.

Coefficients of aberration production in *Tradescantia*

To serve as the basis for the theory of structural change in *Tradescantia* given in Chapter VII, we have collected in Table 64 the coefficients of production of the various types of aberration. In the case of aberrations the yield of which is proportional to dose and independent of intensity (viz. chromatid, chromosome, and isochromatid breaks in X-ray experiments, and all types of aberration in neutron and α-ray experiments), the coefficients

[1] Kotval, J.P. (unpublished).
[2] Fabergé, A.C. (1940*b*). The number of extra chromosome bodies per cell was scored.
[3] According to the method of Swanson, C.P. (1940).
[4] Catcheside, D.G. & Lea, D.E. (1943).
[5] X-rays of this wave-length are very soft, and the explanation immediately suggests itself that the low efficiency is due to the dose received by the chromosomes being on this account much less than the estimated dose (despite allowance for absorption having been made in estimating the dose). This explanation is however discounted by the fact that α-rays, which are also very soft, were found to be highly efficient in the same series of experiments.

given are the values of m obtained by fitting the equation $y = mx$ to the experimental results by the least squares method. y is the number of aberrations per cell, and x the dose in roentgens or

TABLE 64. Coefficients of production of aberrations in *Tradescantia* microspores

Split chromosomes	Chromatid breaks per cell per r. $10^{-2} \times$	Isochromatid breaks per cell per r. $10^{-2} \times$	Chromatid exchanges per cell per r. $10^{-2} \times$	per r.^{2} $10^{-5} \times$	Reference
X-rays (0·15 A.)	0·725 ±0·08	0·271 ±0·02	—	1·81 ±0·21	1
(1·5 A.)	0·62 ±0·12	0·26 ±0·13	—	—	2
(4·1 A.)	1·10 ±0·11	0·44 ±0·11	—	—	2
(8·3 A.)	0·065 ±0·015	<0·009	—	—	2
Neutrons[6] (Li+D)	1·58 ±0·08	0·99 ±0·03	0·907 ±0·03	—	1
α-rays (Rn+RaA+RaC′)	1·96 ±0·09	2·10 ±0·09	0·59 ±0·05	—	3

Unsplit chromosomes	Chromosome breaks per cell per r. $10^{-2} \times$	Minute interstitial deletions per cell per r. $10^{-2} \times$	per r.^{2} $10^{-5} \times$	Asymmetrical exchanges[7] per cell per r. $10^{-2} \times$	per r.^{2} $10^{-5} \times$	Reference
X-rays (0·15 A.)	0·06 ±0·01	—	—	—	0·52 ±0·08	1
(0·1–1 A.)	—	—	0·40 ±	—	—	4
Neutrons[6] (Li+D)	0·26 ±0·02	—	—	0·32 ±0·02	—	1
(Be+D)	—	0·49 ±0·03	—	—	—	5

1 Thoday, J.M. (1942), as worked up by Lea, D.E. & Catcheside, D.G. (1942).
2 Catcheside, D.G. & Lea, D.E. (1943). The experiments with soft X-rays were made with pollen tubes, and have been converted to yields in microspores for the purpose of Table 64. The conversion is made possible by the fact that yields have been determined in both pollen tubes and microspores in the case of medium X-rays (0·15 A.) and α-rays. For a given type of aberration, the ratio of the yields in pollen tubes and microspores appears to be not very different for the two radiations. The yields of chromatid breaks in pollen tubes and microspores are equal within the error of the experiment. The yield of isochromatid breaks is definitely less in pollen tubes, the ratio of yields being 0·36 (for X-rays). This is believed to be due to the sister chromatids being farther apart in the pollen tubes 3–3½ hr. after sowing than in the microspores 24 hr. before metaphase. The yield of chromatid interchanges also is less in pollen tubes than in microspores, the ratio of yields (for X-rays) being 0·267. This is believed to be due to the chromosomes being less favourably disposed for interchange in the tube-shaped pollen tube than in the spherical microspore.
The yield of isochromatid breaks with 8·3 A. X-rays is described as <0·009 per cell per 100 r. for the following reason. 1 isochromatid break was obtained in 181 irradiated nuclei, which corresponds to the coefficient 0·009. Among 445 unirradiated nuclei however, 3 isochromatid breaks were observed, indicating a spontaneous rate of 1 per 148 nuclei. The 1 isochromatid break observed in the 181 irradiated nuclei was therefore probably spontaneous. The yield is therefore recorded as <0·009.
3 Kotval, J.P. & Gray, L.H. (1947).
4 The coefficient given is derived from Fig. 35c, p. 231, which includes the results of Rick, C.M. 1940); Newcombe, H.B. (1942b); Marinelli, L.D., Nebel, B.R., Giles, N.H. & Charles, D.R. (1942).
5 Giles, N.H. (1943). 6 Neutron doses are in v-units.
7 Only asymmetrical exchanges are recorded in unsplit chromosomes. To allow also for symmetrical exchanges, the yields of asymmetrical exchanges given in the table should be doubled.

v-units. In the case of X-ray-induced exchanges, in which the yield obtained with a given dose diminishes with increase of the duration of the exposure, the yields are first extrapolated to zero exposure time,[1] and the extrapolated yields then fitted to the equation $y = kx^2$ by the least squares method.

1 Extrapolation is made with the aid of the theoretical formula relating yield to exposure time derived in Chapter VII.

Structural changes induced by ultra-violet light

The induction of chromosome structural changes by the irradiation of maize pollen with ultra-violet light has already been discussed in Chapter VI. The principal result is that terminal deletions (i.e. simple breaks) are freely induced by ultra-violet light but that aberrations involving exchange between two breaks, namely, interstitial deletions and chromosome interchanges, occur with very low frequency. All of the very few interchanges detected are incomplete. It is also found that usually deletions affect only one chromatid and rarely both sister chromatids simultaneously; in other words that chromatid breaks, but rarely if ever isochromatid breaks, are caused by ultra-violet light.

Experiments in which *Drosophila* sperm have been irradiated by ultra-violet light[1] have shown that gene mutations are freely produced but that gross structural changes (which with X-rays involve breaks produced by separate ionizing particles) are rare, if they occur at all. Evidence regarding the induction of minute structural changes, involving exchange between two breaks which in X-ray experiments are produced by a single ionizing particle, is at present conflicting, salivary chromosome observations having been reported as showing that a proportion of ultra-violet-induced lethals are minute deficiencies,[2] while genetic tests designed to detect a particular type of minute rearrangement which with X-rays is comparatively frequent failed to reveal it in ultra-violet experiments.[3]

Direct observation of chromosome changes induced by ultra-violet light at the metaphase following irradiation have been made by Swanson,[4] who germinated *Tradescantia* pollen grains on a synthetic medium and irradiated at different intervals after germination. Chromatid breaks were the only structural changes observed, apart from a very few isochromatid breaks and inter-

[1] Muller, H.J. & Mackenzie, K. (1939); Mackenzie, K. & Muller, H.J. (1940); Demerec. M., Hollaender, A., Houlahan, M.B. & Bishop, M. (1942).

[2] Slizynski. B.M. (1942).

[3] Mackenzie, K. & Muller, H.J. (1940), looking for a minute rearrangement involving the y^+ locus in the sc^8 chromosome. See footnote 8, p. 234.

[4] Swanson, C.P. (1940, 1942, 1943). The principal results have been confirmed by Catcheside, D.G. & Lea, D.E. (unpublished).

changes, the frequency of which did not exceed the frequency of spontaneous aberrations in unirradiated material. These observations therefore are in agreement with the results with maize and *Drosophila* in showing that ultra-violet light can produce breaks but that the breaks do not take part in exchanges.

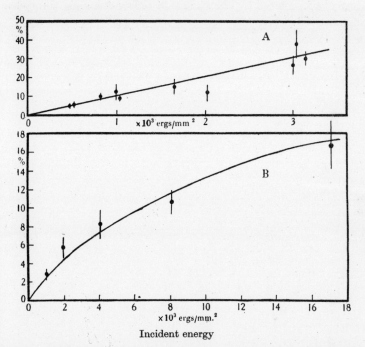

FIG. 37. Yield of ultra-violet-induced deletions per 100 nuclei as a function of incident energy for wave-length 2536A. A, chromatid breaks in *Tradescantia* pollen tubes (Swanson). B, endosperm deficiencies in maize pollen grains (Stadler & Uber).

The yields of breaks produced in *Tradescantia* pollen tubes and in maize pollen grains[1] by monochromatic light of wave-length 2536A. is plotted against dose in Fig. 37. It is seen that with *Tradescantia* the yield is proportional to dose. With maize the yield increases rather less rapidly than the first power of the dose. It has been shown that this non-linearity is due to the

[1] Detected by recording endosperm deficiencies, as described in Chapter V, p. 184. The method would only record deficiencies involving the three marker genes employed in these experiments, and thus record some but not all of the breaks in three of the ten chromosomes. The experiments and calculations are by Stadler, L.J. & Uber, F.M. (1942).

effect of unequal absorption of the ultra-violet light in differently orientated pollen grains. The curve in Fig. 37 B is a theoretical one calculated by Stadler and Uber on the basis that the probability of a chromosome break is strictly proportional to the dose delivered to the sperm nucleus, and making allowance for the absorption suffered by the radiation before reaching the sperm nucleus.

Chapter VII

THE MECHANISM OF INDUCTION OF CHROMOSOME STRUCTURAL CHANGES[1]

The production of structural changes in the chromosomes by irradiation of *Tradescantia* microspores and *Drosophila* sperm has been extensively studied, and a certain amount of progress has been made towards the elucidation of the mechanisms involved. Other materials have not yet been investigated sufficiently.

The process of structural change following irradiation of *Drosophila* sperm was discussed in Chapter v, in connexion with dominant lethals. The yields of dominant lethals and of viable structural changes could be correlated on the view that union among the breakage ends was random, but that a certain proportion of the breakage ends failed to unite before splitting, and sister-union occurred at these breakage ends. More experimental data, particularly concerning the yields of dominant lethals and viable structural changes with neutrons and α-rays, are required to test this theory, and to determine whether a single ionization or many ionizations are required to cause a break.

In the case of *Tradescantia*, experimental data of this sort are available, and the theory of the process of structural change can be carried further than has so far been possible with *Drosophila*. It appears that union of breakage ends in irradiated microspores is far from random, only breakage ends which at the time of production were close together having much chance of joining. The contrast of this standpoint with that provisionally adopted in the case of *Drosophila* does not necessarily weaken the latter, since union in *Tradescantia* microspores takes place within a few

[1] In this chapter citation of experimental data will usually be made by figure, table or page reference to Chapter vi, rather than to the original authors. The sources of the data are stated in Chapter vi. The theory of the process of structural change in *Tradescantia* which occupies the bulk of the chapter is based on the treatment of Lea, D.E. & Catcheside, D.G. (1942, and unpublished). A number of the basic principles were first stated by Sax, K. (1940). All calculations are based on the physical data tabulated in Chapter i, which fact necessitates certain numerical changes in the published theory of Lea and Catcheside, which used less accurate physical data.

minutes of breakage, and in the case of *Drosophila* is deferred until after the chromosomes of the irradiated sperm have entered the egg, when the spatial relations of breakage ends may bear little relationship to their spatial relations at the time of irradiation.

It further appears with *Tradescantia* that breaks are caused by single ionizing particles, but that a single ionization does not suffice. It is not yet known whether a single ionization suffices to break a *Drosophila* chromosome.

Reasons for believing that a break is caused by a single ionizing particle

The theories of chromosome structural change take as their starting point the assumption that a break in a chromosome is produced by the passage of an ionizing particle, and we shall begin the discussion therefore by assembling the arguments on which this assumption is based.

Several of the tests brought forward in Chapters III, IV and V in discussing the view that the inactivation of a virus particle or the mutation of a gene is caused by a single ionization may be expected to apply here also, and we should expect that if a chromosome break is produced by a single ionizing particle, the number of breaks produced by a given type of radiation will be proportional to the dose and independent of the intensity and temperature. However, we are not in a position to observe all the breaks primarily produced. In *Drosophila*, the aberrations detectable in the salivary glands, or by breeding methods, are not simple breaks, but are rearrangements involving two or more breaks. In *Tradescantia*, in which, in addition to more complicated rearrangements, simple breaks are observable, the latter do not constitute the total number primarily produced but the residue after some of the primary breaks have taken part in rearrangements and many others have *restituted*.[1] Since the proportion of breaks which undergo rearrangement or restitution may very well depend on intensity and temperature, it is evident that there are difficulties in applying the tests suggested above.

[1] *Restitution* means the joining of a pair of breakage ends to reform the original chromosome. We avoid the term *heal*, which is ambiguous, being used by some authors to mean restitution, and by others to mean a change in a breakage end as a result of which it is no longer capable of uniting with another breakage end.

In *Tradescantia*, for example, the yield of chromatid breaks diminishes with rise of temperature (Fig. 32 B, curve *a*, p. 221), which is probably to be explained on the view that rise of temperature promotes restitution. There is some evidence that rise of temperature (Fig. 32 D, p. 221) or exposure to ultra-violet or infra-red light (p. 223) promotes restitution in *Drosophila* sperm.

Thus the experiments on the variation of the yield of chromosome aberrations with temperature probably have more bearing on the restitution process than on the primary breakage process.

The intensity test can be applied to *Drosophila*, since it appears that rearrangement is deferred until after fertilization has occurred. It is found (Table 60, p. 228) that the yield of structural changes produced by a given dose is independent of the intensity and manner of fractionation. The inference is that the number of breaks in the sperm immediately prior to fertilization is independent of the manner in which the dose given to the sperm is distributed in time.

In *Tradescantia*, rearrangement takes place during and immediately following irradiation, and the yield of X-ray-induced chromosome and chromatid exchanges (i.e. aberrations involving two independently produced breaks) markedly diminishes with increase of the duration of exposure (Fig. 33, curves *a*, *b*, *f*, p. 227). In view of the fact that the observed simple breaks constitute the residue of the primary breaks which remains after some breaks take part in exchanges and others restitute, it would not have been surprising to have found that the yield of simple breaks observed *increased* somewhat with increase of the duration of exposure. However, the proportion of the breaks primarily produced which take part in exchanges being small, this increase will be very slight. Within the accuracy of the experiment, the observed yield of simple breaks is independent of the duration of exposure (Fig. 33, curve *d*). Thus the intensity experiments with *Tradescantia*, as well as with *Drosophila*, are consistent with the yield of breaks primarily produced being independent of the intensity.

The expectation that the number of primary breaks per sperm is proportional to dose cannot be applied directly to *Drosophila*, since simple breaks are not observable cytologically. If, however, the view is accepted that dominant lethals are largely due to simple breaks (p. 164), then the fact that the yield of dominant lethals is proportional to dose at small doses (Fig. 27) is evidence that the yield of primary breaks is proportional to dose.

In *Chortophaga* neuroblasts and *Tradescantia* microspores, in which simple breaks are cytologically observable, the yield is proportional to dose (Fig. 34, p. 230). Thus the experiments on the dependence of the yield of structural changes on dose and intensity are consistent with the primary breaks being produced by single ionizing particles. The processes of rearrangement and restitution which intervene between the production of the breaks and their observation, however, lead to complications which, it must be admitted, weaken the cogency of these experiments as positive evidence that single breaks are produced by single ionizing particles.

In the case of *Tradescantia*, these arguments can be supplemented by an additional piece of evidence which, in the opinion of the present writer, renders them conclusive. This is the fact that the yield of exchanges which in X-ray experiments increases more rapidly than the first power of the dose (Fig. 35 A, C, p. 231) and is dependent upon intensity (Fig. 33, curves a, b, f, p. 227), in neutron experiments increases linearly with the dose (Fig. 35 B, D) and is independent of intensity (Fig. 33, curves c, g). Now with the doses commonly used a microspore is traversed by a score or so of ionizing particles (protons) in neutron experiments or a few hundred ionizing particles (electrons) in X-ray experiments. The view we are advocating is that a particular break is produced by a particular one of these particles. The alternative hypothesis would be that a break cannot be ascribed to a particular ionizing particle but is in some way a cumulative effect due to them all (for example, caused by a change in the chemical composition of the nuclear sap). Now the yield of primary breaks is proportional to dose on the average, and subject to Poisson distribution in any individual nucleus (Table 56, p. 218), which conclusions are based on experimental evidence, and while favouring the first hypothesis, are not necessarily inconsistent with the alternative hypothesis. Granted this, we should then expect on either hypothesis that the probability of a pair of breaks being independently produced within a given short distance of each other would be proportional to the square of the dose. Exchanges, which involve the production of two breaks a short distance apart, are in fact produced by X-rays

SEPARATION OF BREAKS WHICH EXCHANGE 249

with a frequency approximately proportional to the square of the dose. This result therefore is not inconsistent with either hypothesis about the mechanism of production of the primary breaks. But the fact that with neutrons the yield of exchanges is proportional to the dose shows that the two breaks are not independently produced and can only be explained on the view that the *same* ionizing particle causes *both* breaks. The fact that the yield of neutron-induced exchanges is independent of intensity leads to the same conclusion. It is evident, therefore, that despite the fact that several ionizing particles traverse the nucleus, it is possible to ascribe the production of particular breaks to particular ionizing particles. The only reasonable interpretation is that a break is caused by the passage of an ionizing particle through the breakage point, or its immediate vicinity.

Distance apart at the moment of breakage of breaks which exchange

The theory of the process of structural change in *Tradescantia* which we develop in this and subsequent sections is founded upon the results of neutron and X-ray experiments. Most X-ray experiments have been made with inhomogeneous radiations of wave-length 0·1–1·0 A. Tissue irradiated by X-rays of this range of wave-lengths is traversed by electrons of a rather wide range of initial energies (Table 3, p. 12), which fact complicates calculations. Such experiments as have been made on the dependence of the yield of chromosome structural changes in *Tradescantia* on wave-length have indicated little variation of yield when the wave-length has been varied between 1·5 and 0·08 A. (p. 240, and Table 63, p. 239). We shall therefore simplify the calculations by using physical data for monochromatic X-rays of wave-length 1·5 A. even when the experiments were made with shorter wave-lengths. The advantage of the longer wave-length from the point of view of calculations is that all the electrons projected in tissue by this wave-length have the same initial energy, namely, 7·5 ekV. Their range (in tissue of density 1 g./cm.^{3}) is 1·5 μ. 6·7 of these photoelectrons are projected from each cubic micron of tissue by a dose of 1000 r. (Table 18, p. 32). Thus a dose of 50 r. causes the projection, on the average, of 303 photoelectrons in the nucleus of diameter 12 μ of the *Tradescantia* microspore.

The protons which traverse the tissue in a neutron experiment have for the most part ranges greater than the diameter of the *Tradescantia* nucleus, so that most of the protons which traverse the nucleus originate outside it. The mean path of a proton in the nucleus is 8μ (viz. two-thirds of the diameter), and with a dose of 1 v-unit the total range of all the protons in the nucleus is $3 \cdot 66\mu$ (being $4 \cdot 05\mu$ per 1000 v-units per cubic micron in tissue of unit density for Li + D neutrons, see Table 18, p. 32). We infer that with a dose of 50 v-units the *Tradescantia* nucleus of diameter 12μ is traversed on the average by 23 protons.

It is because the number of ionizing particles traversing the nucleus is much smaller in neutron experiments than in X-ray experiments that neutron-induced exchanges are predominantly 1-*hit* (i.e. both the breaks taking part in the exchange are caused by the same ionizing particle), while X-ray-induced exchanges are predominantly 2-*hit* (i.e. the two breaks are caused by separate ionizing particles). The difference is of course quantitative rather than qualitative. If it were practicable to use much larger doses, neutron exchanges would be predominantly 2-hit, while if it were practicable to use much smaller doses, X-ray exchanges would be predominantly single-hit.

The fact that with the doses commonly used neutron-induced exchanges are predominantly 1-hit means that exchanges between breaks is by no means random. For with 23 protons crossing the nucleus (with 50 v-units) the chance of a given break exchanging with a break produced by the same proton would, if exchange were random, be 22 times smaller than the chance of exchanging with a break produced by a different proton.[1] Evidently exchanges are in practice confined to breaks which (at the time of production) are close together. This argument can be made quantitative as follows. Suppose that a given break will only take part in exchange if a second break is produced within a distance h. For a 2-hit exchange to be produced it would therefore be necessary for a second proton to pass through a sphere of radius h centred at the given break. The fact that neutron exchanges with 50 v-units are not usually 2-hit means that a second proton will not usually pass through this sphere. The sphere of

[1] We suppose here, and elsewhere in this section, that the dose is administered in a short overall time, so that the breaks first formed have not time to restitute before the remaining breaks are formed.

SEPARATION OF BREAKS WHICH EXCHANGE

radius h must therefore have a cross-sectional area less than $\frac{1}{23}$ of the cross-sectional area of the nucleus of radius 6μ. We deduce that h is less than $1\cdot 3\mu$.

Turning now to X-ray-induced exchanges, the fact that these exchanges are usually 2-hit with 50 r. means that more than one electron is usually found within a radius h of a given break. Thus the sphere of radius h must have a volume exceeding $\frac{1}{303}$ of the volume of the nucleus, since with this dose 303 electrons are found in the nucleus. It follows that h exceeds $0\cdot 9\mu$.

These upper and lower limits for h are of course only valid as to order of magnitude, being based on the rather crude model that exchange is impossible outside the distance h, whereas no doubt the truth is that the probability of exchange diminishes smoothly with increasing separation of the breaks at the moment of formation. We may justifiably infer, however, that the distance apart at the moment of formation of breaks which take part in exchange is typically of the order 1μ, which means that exchange is far from being random in the nucleus.

A second method of estimating h comes from consideration of the relative frequencies of exchanges between breaks in the same chromosome (intrachanges) and between breaks in different chromosomes (interchanges). If for exchange to occur two breaks have to be very close together at the moment of breakage, then there is a high probability that pairs of breaks which exchange will be in the same chromosome. If, on the other hand, there is no such necessity, we shall expect there to be more interchanges than intrachanges (since there are six chromosomes). The relation between intrachanges and interchanges is geometrically analogous to the relation between 1-hit exchanges and 2-hit exchanges. The two breaks taking part in an intrachange are located on a single chromosome. The two breaks taking part in a 1-hit exchange are located on the path of a single ionizing particle. Suppose that in the nucleus of radius R the total (haploid) length of chromosome thread is L. In the sphere of radius h centred at a given break we shall expect to find a length of chromosome thread equal to $L(h^3/R^3)$ additional to the segment of chromosome thread in which the given break is located. Now if $L(h^3/R^3)$ exceeds $2h$, the second break with which the given break exchanges will usually be in an independent chromosome segment, and a gross structural change will result. If, however,

$L(h^3/R^3)$ is less than $2h$, then the second break is more likely to lie in the chromosome segment which contains the given break. (By chromosome segment we mean here the length $2h$ which is contained in the sphere of radius h.) In this event the result will be a minute intrachange, i.e. either inversion (which will escape notice) or a minute interstitial deletion (which is observable). Now the number of minute interstitial deletions observed in *Tradescantia* microspores irradiated at a time when the chromosomes are unsplit is roughly equal to the number of gross asymmetrical exchanges (Table 64, p. 241). We infer that $2h$ and $L(h^3/R^3)$ are approximately equal, so that $h = (2R^3/L)^{\frac{1}{3}}$. Taking L, the total haploid length of chromosome thread,[1] to be 486μ, and R, the radius of the nucleus, to be 6μ, we deduce that $h = 0.9\mu$, again indicating that exchanges usually occur between breaks separated by 1μ or less at the moment of breakage.

A further indication of the distance apart, at the moment of breakage, of breaks which exchange, is afforded by a study of the size distribution of minute interstitial deletions. It is found[2] that the most frequently occurring interstitial deletion is a sphere about 1μ diameter. This is consistent with the distance apart of the breaks at the moment of formation being of the order of 1μ.

To sum up: in *Tradescantia* exchange between breaks is far from random, but is for the most part confined to breaks which, at the moment of breakage, are separated by a distance of the order of 1μ or less. This conclusion has been separately reached from consideration of the following three independent lines of evidence:

(a) With doses of the order of 50 v-units or roentgens respectively, neutron-induced exchanges are predominantly 1-hit and X-ray-induced exchanges are predominantly 2-hit.

(b) Minute interstitial deletions and gross asymmetrical exchanges are approximately equally frequent.

(c) The most frequent size of minute interstitial deletion is about 1μ.

It is interesting that three such superficially unrelated facts should be determined by a single circumstance, namely, the value of h.

[1] Sax, H.J. & Sax, K. (1935).   [2] Rick, C.M. (1940).

The proportion of breaks which restitute in *Tradescantia*

It is generally supposed[1] that of the breaks primarily produced in chromosomes or chromatids of *Tradescantia* microspores, only a fraction, and probably a small fraction, survive to the time of fixation when they are observed either as simple breaks or in structural rearrangements. In developing theories of the mechanism of chromosome breakage by radiation it is necessary to be able to deduce from the observed aberrations the number of breaks primarily produced. It is particularly important to know whether the proportion of breaks which restitute is the same for different radiations, since if the proportion is not constant, it will be necessary to allow for this fact in deducing from the observations the relative efficiencies of different types of radiation in breaking chromosomes. We shall therefore devote this section to a discussion of several methods by which it is possible to estimate the proportion of breaks which restitute, and the number of breaks primarily produced per unit dose.

Method I.[2] The suggestion has been made that a connexion exists between the following facts: (*a*) a proportion of interchanges are incomplete; (*b*) a proportion of isochromatid breaks fail to show sister-union; (*c*) a proportion of chromatid breaks fail to restitute; and that experimental determinations of the proportions under headings (*a*) and (*b*) may be used to deduce the proportion under heading (*c*). Each of the three phenomena listed consists in the failure of a pair of breakage ends to unite. This failure might be attributed to a chance failure of the ends to come into sufficiently close proximity; alternatively, it might be attributed to one or both of the breakage ends being *unjoinable*, i.e. different in some way from a normal breakage end. Two considerations favour the latter alternative. One is that the frequency of the $NUpd$ type of isochromatid break greatly exceeds the product of the frequencies of the NUp and NUd types,[3] so suggesting that failure of a breakage end to join is not a chance phenomenon but is due to some cause which is likely to affect both breakage ends of a broken chromosome if it affects one. The second consideration is the fact (Table 50, p. 204, and

[1] E.g. Sax, K. (1938, 1940); Fabergé, A.C. (1940*a*).
[2] Catcheside, D.G., Lea, D.E. & Thoday, J.M. (1946*a*).
[3] Refer to Chapter VI, p. 204, for the terminology and Table 50 for the experimental data.

Table 53, p. 211) that the proportions of isochromatid breaks without sister-union, and of interchanges which are incomplete, are greater for α-rays than for less densely ionizing radiations, suggesting that a breakage end is likely to fail to join if the break has been produced by a densely ionizing particle. We shall therefore suppose that failure of a pair of breakage ends to join, whether in an incomplete interchange, an isochromatid break, or a chromatid break, is due to one or both of the ends concerned being unjoinable. We shall denote by f the proportion of primarily produced chromatid breaks which are unable to restitute owing to one or both breakage ends being unjoinable.

The relatively high frequency of occurrence of isochromatid breaks of the $NUpd$ type is evidence that if one of the four breakage ends is unjoinable, not only has the other breakage end of the same chromatid a high probability of being unjoinable, but so also have the breakage ends of the sister chromatid. For if this were not the case, the sister chromatid would restitute and a chromatid break and not an $NUpd$ configuration would be recorded.

It can readily be seen that according to the views we have been discussing, the following relations connecting the frequencies of different types of aberrations should hold:

$$\frac{\text{Incomplete interchanges}}{\text{Total interchanges}} = \frac{NUp + NUd}{SU + NUp + NUd}, \quad \text{(VII-1)}$$

$$f = \frac{NUp + NUd + NUpd}{2SU + NUp + NUd + NUpd}. \quad \text{(VII-2)}$$

The first equation is derived simply by considering that an isochromatid break is a particular case of interchange (but remembering that an 'interchange' in which both unions fail to occur is not an interchange at all but two chromatid breaks) In the second equation, the factor 2 in the denominator of the right-hand side is inserted under the belief that of the total number of isochromatid breaks having all four breakage ends joinable which are primarily produced, about one-half restitute or undergo symmetrical exchange, which is unobservable, and one-half undergo sister-union and are recorded as normal isochromatid breaks.[1] Table 65, which is based on the experimental

---

[1] If random union occurred between the four breakage ends, the probability (z) of sister-union occurring in place of an unobservable union would be $z = \tfrac{1}{3}$. If immediate separation of the two chromosome pieces

data given in Tables 50 (p. 204) and 53 (p. 211), shows that equation (1) is in fair agreement with the experimental data. Table 65 also lists the values of f calculated from equation (2). Combining all the X-ray data we obtain $f = 0.094$, while combining all the α-ray data we obtain $f = 0.667$.

While data concerning non-union isochromatid breaks are lacking for neutrons, since it is known that the proportion of neutron-induced interchanges which are incomplete is the same as with X-rays, the value $f = 0.094$ will be taken to apply also to neutrons.

TABLE 65. Correlation between non-union isochromatid breaks and incomplete interchanges

Radiation	Division	Isochromatid breaks						Interchanges			Reference
		SU	NUp	NUd	$NUpd$	$\dfrac{NUp+NUd}{SU+NUp+NUd}$	$\dfrac{NUp+NUd+NUpd}{2SU+NUp+NUd+NUpd}$	Complete	Incomplete	$\dfrac{\text{Incomplete}}{\text{Total}}$	
X-rays	p.g.	1027	98	68	33	0.1391	0.0883	848	132	0.1347	1
X-rays	p.g.	240	28	23	17	0.1753	0.1241	330	20	0.057	2
X-rays	p.t.	45	1	2	2	0.0625	0.0526	25	2	0.074	3
α-rays	p.g.	71	32	27	286	0.454	0.708	126	39	0.236	4
α-rays	p.t.	18	2	3	7	0.217	0.250	—	—	—	3

1 Catcheside, D.G., Lea, D.E. & Thoday, J.M. (1946a).
2 Kotval, J.P. (unpublished).
3 Catcheside, D.G. & Lea, D.E. (1943).
4 Kotval, J.P. & Gray, L.H. (1947).

The observed yield of chromatid breaks per roentgen being known (Table 64, p. 241), we are able from equation (2) to deduce the number of chromatid breaks per roentgen which restitute. Adding the numbers observed per roentgen as chromatid breaks, in isochromatid breaks, and in exchanges,[1] we deduce the total

occurred following the breakage, then sister-union would be favoured as compared with the unobservable types and the probability might approach unity. The assumption that $z = \tfrac{1}{2}$ is made as a compromise which is unlikely to be in error by a factor greater than 2.

[1] The number of exchanges per roentgen increases with increase of dose in the case of X-rays. The number of breaks in exchanges being, however, small compared with the number of breaks which restitute, it makes little difference to the final result. Twice the number of exchanges per roentgen at 100 r. was the number added to represent breaks appearing in exchanges.

number of chromatid breaks primarily produced per roentgen, which we denote by ξ. The values obtained are $\xi = 0{\cdot}086$ for X-rays, $\xi = 0{\cdot}206$ for neutrons, and $\xi = 0{\cdot}083$ for α-rays.

Some of the methods we are going to discuss lead directly to estimates of ξ. Evidently, by reversing the procedure just described, we can, by making use of the coefficients cited in Table 64, derive estimates of f from these estimates of ξ.

Method II. A second method of estimating ξ and f comes from the consideration that it is only the fact that the majority of the primarily induced breaks restitute which permits the yield of 2-hit exchanges to increase in proportion to the square of the dose. Clearly, if all the joinable breaks took part in exchange, the number of exchanges would be proportional to dose. We should expect therefore that at high doses the yield of exchanges will increase less rapidly than the square of the dose. The following approximate calculation will serve both as a demonstration of the (dose)^2 law at low doses, and as an indication of the manner in which the yield of exchanges may be expected to depart from this law at high doses.

$\xi D(1-f)$ being the number of joinable primary breaks produced in the nucleus of radius R by a dose D, the *mean* number (additional to the given break) in a sphere of radius h centre at a given break[1] will be $\xi D(1-f) h^3/R^3$, the actual number being distributed in a Poisson distribution about this mean. The probability of at least one break occurring in this sphere is therefore $1 - e^{-\xi D(1-f)h^3/R^3}$, which may be expanded in ascending powers as $\xi D(1-f) h^3/R^3 \{1 - \tfrac{1}{2}\xi D(1-f) h^3/R^3 + \ldots\}$. This expression is the probability that a given one of the $\xi D(1-f)$ joinable breaks in the nucleus shall exchange, and we deduce that the number of (complete) exchanges per nucleus, which is one-half of the number of breakage ends taking part in exchanges, is $\tfrac{1}{2}(\xi D)^2 (1-f)^2 h^3/R^3 \{1 - \tfrac{1}{2}\xi D(1-f) h^3/R^3 + \ldots\}$. For small doses all the terms in the bracket except the unity can be neglected, and we obtain for the yield of exchanges kD^2 per cell, where $k = \tfrac{1}{2}\xi^2(1-f)^2 h^3/R^3$. At higher doses the second term

[1] Assuming that the breaks are uniformly distributed in the nucleus. In view of this assumption the exchanges covered by this formula do not include the minute interstitial deletions, the high frequency of which depends on the higher than random probability of finding a second break in the same chromosome.

in the bracket is not negligible, and we obtain for the yield of exchanges

$$kD^2 \left\{ 1 - \frac{kD^2}{\xi D(1-f)} \right\} \text{ or approximately}$$

$$kD^2 \left\{ 1 - \frac{\text{Number of exchanges}}{\text{Total number of joinable breaks primarily produced}} \right\}.$$

(VII-3)

Sax[1] has described experiments, using large numbers of cells, in which the yield of exchanges was determined as a function of dose at a time when the chromosomes were unsplit, the time of irradiation being kept constant. The yield increased very slightly less rapidly than the square of the dose, the difference not being statistically significant. These experiments consequently do not enable ξ to be determined; they can, however, be used to fix a lower limit[2] for ξ, and hence an upper limit for f. The results obtained (for X-rays on unsplit chromosomes) are that $\xi > 0.014$, $f < 0.05$.

Newcombe,[3] employing higher doses, found that the yield of 2-hit exchanges increased with a power of the dose intermediate between the first and the second. At these high doses f is no longer constant, as is shown by the fact that the yield of simple breaks increases more rapidly than the first power of the dose. Apparently with high doses of X-rays the proportion of breakage ends which are joinable diminishes. Thus while the total number

[1] Sax, K. (1940, Table 2, and 1941b, Table 5). Only asymmetrical exchanges were scored (i.e. rings and dicentrics). The total number of exchanges is obtained by doubling the number of asymmetrical exchanges.

[2] The method adopted was to assume a value for ξ, to fit formula (3) by the least squares method, and to test the goodness of fit by the χ^2 method. By trial the value of ξ was found which gave a value of χ^2 corresponding to $P = 0.05$. This was taken to be the lowest value of ξ consistent with Sax's data.

[3] Newcombe, H.B. (1942b, Table 5). The time of irradiation was constant and equal to 12 min. We take as the number of asymmetrical exchanges:

centric rings + dicentric chromosomes + 2 × tricentric + 3 × tetracentric,

and double to allow for the unobserved symmetrical exchanges. Acentric rings are omitted to make the results comparable with Sax's, and to avoid the inclusion of any aberrations which from our point of view are minute interstitial deletions.

of primary breaks is proportional to the dose, the number of *joinable* breaks primarily produced increases less rapidly than the first power of the dose. This effect, as well as the fact that at high doses the number of breaks taking part in exchanges is no longer a negligible fraction of the total number of primary breaks, can be taken into account by a slight modification of the theory we have already given. Defining $k' = \frac{1}{2}\xi^2 h^3/R^3$, we obtain the following approximate expression for the yield of (complete) exchanges produced by a dose D:

$$k'D^2 \left\{ 1 - \frac{\text{Exchanges} + 2 \times \text{chromosome breaks at dose } D}{\text{Number of primary breaks at dose } D} \right\}.$$
(VII-4)

This formula was found to fit fairly well the manner of variation with dose of the yield of exchanges found by Newcombe at 240, 480 and 960 r. when the value $\xi = 0 \cdot 010$ was employed. The value of f at low doses (where f is constant) deduced by combining this value of ξ with the coefficients of aberration production listed in Table 64 (p. 241) is 0·078.

Two methods of estimating ξ and f somewhat similar to the method just described may be mentioned, though it has not proved possible to apply them in practice. They depend on the fact that the chromatid or chromosome breaks which are observed constitute a certain fraction of the primary breaks remaining after some have taken part in exchanges. Since the yield of X-ray-induced exchanges increases more rapidly than the first power of the dose when the dose is varied, we might expect the number of chromatid or chromosome breaks observed to increase rather *less* rapidly than the first power of the dose. The larger the number of breaks primarily produced relative to the number which take part in exchanges, the smaller will be the departure of the yield of breaks from a linear function of the dose, and analysis of the dose curve should thus enable the number of breaks primarily produced per roentgen to be determined. The experimental dose curves at present available (Fig. 34 D, F, p. 230) are not sufficiently accurate to permit of this method being applied.

In a similar manner, since the yield of X-ray-induced exchanges decreases when the duration of exposure is increased at constant dose, we should expect the yield of chromatid and chromosome breaks to increase slightly when the duration of exposure is increased at constant dose. The experimental points shown in Fig. 33, curve d (p. 227), do not indicate any increase of the yield of chromatid breaks with the duration of exposure at constant dose. They do not, however, rule out an increase of up to 10 % in the yield between the shortest and longest time used in the experiments, and when the calculation is made the corresponding limit set for f turns out to be $f < 0.75$. This, though doubtless true, is not helpful. The method is evidently insensitive, and data of very high precision would be needed to obtain a useful result from it.

Method III. On p. 256 it was shown that, apart from smaller terms which we need not now consider, the yield of 2-hit exchanges induced by X-rays was kD^2, where $k = \frac{1}{2}\xi^2(1-f)^2 h^3/R^3$. Now R, the radius of the nucleus, is 6μ, and h, the distance within which exchange occurs, was found earlier in this chapter to be 1μ. Thus from the experimental values of the coefficient of exchange production per cell per roentgen squared listed in Table 64, p. 241, we can deduce ξ and hence f. A similar method can be applied to the 1-hit exchanges induced by neutrons and α-rays. If d is the length of path of all the protons (or α-rays) which traverse the nucleus per roentgen,[1] and $\xi(1-f)$ the number of joinable breaks primarily produced per roentgen, then $\xi(1-f)/d$ is the number of joinable breaks primarily produced per micron of path length. Hence $2h\xi(1-f)/d$ is the probability that a second joinable break shall be produced within a distance h of a given break by the same ionizing particle. Thus $\xi^2(1-f)^2 hD/d$ is the mean number of (complete) exchanges per nucleus produced by dose D. Again taking $h = 1\mu$, we are able to deduce ξ, and hence f, by comparing this formula[2] with the experimental coefficients of exchange production listed in Table 64 (p. 241). The values of ξ and f obtained by the application of method III are listed in Table 67.

[1] Table 18, p. 32, gives the path length per cubic micron. Multiplication by the volume of the nucleus ($905\mu^3$) gives d.

[2] When applying this formula, or the formula for 2-hit exchanges, to split chromosomes, ξ should be replaced by ($\xi - 2c$), where c is the coefficient of production of isochromatid breaks per cell per roentgen, on the grounds that the breaks which constitute isochromatid breaks are not available for exchange.

Method IV.[1] Most of the interchanges observed in a nucleus irradiated at a time when the chromosomes are split are chromatid interchanges, in which one chromatid of each of two chromosomes has broken and exchange has taken place between the two breaks. A certain number of configurations are seen which can be diagnosed as interchanges between an isochromatid break and a chromatid break. These c/i interchanges have been found with X-rays to be 8·74 times less frequent than ordinary c/c interchanges, which indicates that 17·48 primary chromatid breaks are produced by X-rays to every primary isochromatid break. But the number of chromatid breaks scored in the fixed nuclei in these particular experiments was 3·843 times as great as the number of isochromatid breaks. Hence it is inferred that the proportion (f) of primary chromatid breaks which persist is only about $3·843/17·48 = 0·22$ time as great as the proportion (z) of isochromatid breaks which persist. Taking $z = 0·5$ (see footnote 1, p. 254) we deduce $f = 0·11$.

Method V. Estimates of ξ and f can be derived from consideration of the relative frequencies of chromatid and isochromatid breaks, if we are prepared to admit that the production of a primary chromatid break requires the passage of an ionizing particle through one of the chromatids, and the production of an isochromatid break requires the passage of the same ionizing particle through both of the chromatids. Considering the sister chromatids as a pair of parallel cylinders of radius r, separated by a distance (axis to axis) of s, it can be shown that the probability that an ionizing particle which passes through a specified chromatid shall pass also through the sister chromatid is

$$g = \frac{2}{\pi} \left\{ \phi - \frac{s}{2r} (1 - \cos \phi) \right\},$$

where $\sin \phi = 2r/s$, g taking the values given in Table 66, ranging from 0·363 for chromatids in contact to 0·053 for chromatids separated by a clear distance of 5 diameters.

We cannot be certain that an ionizing particle which passes through a chromatid inevitably breaks it; we denote by p the probability ($\leqslant 1$) that it does so. Denote by z the ratio of the number of isochromatid breaks observed to the number of iso-

[1] Catcheside, D.G., Lea, D.E. & Thoday, J.M. (1946a).

VALUES OF ξ AND f

chromatid breaks primarily produced (the remainder restituting). Then $\tfrac{1}{2}zpg$ is the ratio of the number of isochromatid breaks observed to the number of chromatid breaks primarily produced. (The factor $\tfrac{1}{2}$ is required since an isochromatid break

TABLE 66. *The geometrical factor for isochromatid breaks*

Clear separation of chromatids in diameters ...	0	0·5	1·0	1·5	2	3	5
$2r/s$	1·0	0·67	0·5	0·4	0·33	0·25	0·167
g	0·363	0·222	0·163	0·128	0·106	0·079	0·053

implies two chromatids broken.) Taking $z = 0.5$ as before (cp. footnote 1, p. 254), and noting that p cannot exceed unity and g cannot exceed 0·363, it follows that this ratio < 0.091.

Now the number of isochromatid breaks per cell per v-unit obtained in neutron experiments is 0.99×10^{-2} (Table 64, p. 241). Hence $\xi > 0.99 \times 10^{-2} \div 0.091$, i.e. $\xi > 0.11$, whence we deduce $f < 0.22$.

TABLE 67. *Estimates of the number of primary breaks per cell per roentgen*[1] *(ξ), and of the fraction of primary breaks which are unjoinable (f)*

		X-rays		Neutrons		α-rays	
		ξ	f	ξ	f	ξ	f
split chromosomes	Method I	0·086	0·094	0·206	0·094	0·083	0·67
	Method III	0·098	0·081	0·213	0·090	0·114	0·33
	Method IV	0·075	0·110	—	—	—	—
	Method V	—	—	>0·11	<0·22	—	—
	Adopted values	0·09	0·09	0·21	0·09	0·10	0·5
nsplit chromosomes	Method II:						
	Sax's data	>0·014	<0·050	—	—	—	—
	Newcombe's data	0·010	0·078	—	—	—	—
	Method III	0·066	0·010	0·150	0·019	—	—

We do not apply this method to α-ray and X-ray-induced isochromatid breaks owing to complications which we discuss later (p. 277).

The estimates of ξ and f derived from the five methods we have discussed are collected in Table 67. With unsplit chromosomes the estimates are too few in number and too divergent in numerical values to enable any precise quantitative conclusions to be drawn. There appears to be no doubt, however, that the number of breaks primarily produced greatly exceeds the number visible at the time of fixation.

[1] Per v-unit in the case of neutrons.

The agreement between the estimates of ξ obtained by the several methods when applied to split chromosomes is sufficient to give us some confidence in the mean values of ξ. The principal results are:

(a) The proportion of breaks which are unjoinable (i.e. f) is the same for neutrons and X-rays, but considerably greater in the case of α-ray-induced breaks.

(b) The number of breaks primarily produced per unit dose (i.e. ξ) increases in the order X-rays, α-rays, neutrons. Thus the efficiency per ionization increases with increase of ion-density in going from X-rays to neutrons, but decreases with further increase of ion-density in passing from neutrons to α-rays. The significance of this result is discussed later.

The agreement between the results given by the diverse methods employed in estimating ξ and f is indirect confirmation of the general soundness of the basic postulates of the theory.

Method V only gives an upper limit for f and a lower limit for ξ owing to the fact that *a priori* one is only able to ascribe upper limits to g (viz. 0·363, Table 65) and to p (viz. 1·0). The fact that the limit for ξ given by method V is within a factor of 2 of the probable values indicates that the actual values of g and p do not depart by more than a factor of 2 from these upper limits. In other words, if sister chromatids are not touching 24 hr. before metaphase they are not separated by more than about 1 diameter (which would make $g = 0·163$), and if a proton passing through a chromatid does not inevitably produce a break, the probability of its doing so is not less than one-half.

Dependence of the yield of aberrations in *Tradescantia* upon the duration of exposure

In considering exchanges hitherto we have supposed that all the primary breaks coexist in the nucleus. This is so if the dose is administered over a short overall time. If the dose is spread over a prolonged time, either by dividing it into fractions or by use of a low intensity, many of the breaks restitute before time has elapsed sufficient for other breaks with which they might exchange to be produced in their vicinity. Consequently, the yield of exchanges produced by a given dose of X-rays diminishes with increase of the duration of exposure. If we make assumptions about the time for which a break remains free, we can work out a theoretical formula[1] for the dependence of the yield of exchanges on the time of exposure. Conversely, by fitting such a formula to the experimental data we are able to deduce the mean time for which a break remains free before restitution or exchange occurs.

Since we have shown that the great majority of the breaks primarily produced in the nucleus by X-rays restitute, we can without great error make the simplifying assumption that the gradual diminution in the number of free breaks which occurs after the cessation of irradiation is solely due to restitution, and neglect the fact that exchange accounts for a small part of this diminution. Thus if n_0 breaks exist in the nucleus at time $t=0$, at time t the number (in the absence of further irradiation) will be $n_0 f(t)$, where $f(t)$ is a diminishing function of t.

The probability of a given break recombining in any short interval of time will be proportional to the number of other breaks within range (i.e. within distance h). The mean number within range is a certain fraction of the total number n of breaks existing in the nucleus at that time. Thus the number of breaks combining per unit time is proportional to n^2, $=\beta n^2$ say. This result is analogous to the equation for a bimolecular reaction in chemistry, or the equation for the combination of positive and negative ions in electricity. The total number of exchanges formed may be written

$$\int \beta n^2 \, dt = \beta n_0^2 \int \{f(t)\}^2 \, dt, \qquad (\text{VII-5})$$

where the integration extends over the period for which breaks exist in the cell.

To proceed further requires a knowledge of the function $f(t)$. The simplest assumption would be that all the breaks remain open for a constant time τ after the moment of their formation, and then rejoin in the original formation if they have not already taken part in exchanges. Such uniformity of behaviour is, however, not very plausible, and instead the calculation has been made assuming that $n/n_0 = f(t) = e^{-t/\tau}$, i.e. $dn/dt = -n/\tau$, implying an *average* time τ elapsing between breakage and restitution, with the actual times distributed in a skew manner about

---

[1] The formula given is that of Lea, D.E. & Catcheside, D.G. (1942). An alternative treatment is given by Marinelli, L.D., Nebel, B.R., Giles, N.H. & Charles, D.R. (1942).

this mean. Some sort of justification for the choice of this particular function is provided by the chemical reaction analogy we have already mentioned. If union between ends from different breaks is analogous to a second-order reaction, union between the ends of a single break may perhaps be regarded as a first-order reaction, which has an equation of the type chosen.

Suppose that while the cell is being irradiated at intensity I roentgens per minute the rate of formation of primary breaks[1] is ξI. Allowing for the rate of reunion we have

$$dn/dt = \xi I - n/\tau, \qquad \text{(VII-6}a\text{)}$$

whence $\qquad n = \xi I \tau (1 - e^{-t/\tau}).$

At time $t = T$ suppose the radiation to cease, n will then diminish according to the equation

$$n = \xi I \tau (1 - e^{-T/\tau}) e^{-(t-T)/\tau}. \qquad \text{(VII-6}b\text{)}$$

Substituting these values of n in equation (5) and integrating we obtain as our estimate of the total number of exchanges

$$\int \beta n^2 \, dt = \tfrac{1}{2} \xi^2 \beta \tau (IT)^2 G,$$

where $\qquad G = 2(\tau/T)^2 \{T/\tau - 1 + e^{-T/\tau}\}. \qquad \text{(VII-7)}$

Thus the number of exchanges produced in the cell by dose $D = IT$ roentgen is proportional to

$$(\text{dose})^2 \, G. \qquad \text{(VII-8)}$$

The function G is tabulated in Table 68 and is seen to have the value unity at $T = 0$ and to diminish as T increases. We see immediately that the theory predicts, in accordance with Sax's experiments (cp. p. 232), that if the dose is varied by varying the intensity at constant time, so keeping T and therefore G constant, the number of exchanges produced is proportional to $(\text{dose})^2$. Further, we see that if at constant dose the intensity is varied, the yield of exchanges should be simply proportional to G. In other words, the curve of number of exchanges against time over which the irradiation is extended should be identical in shape with a plot of the function G. This affords the simplest method of determining τ by comparison with experiment.

[1] Strictly, to be consistent with our previous definition of ξ, we should write $\xi(1-f)$ in place of ξ throughout this section. The omission of the factor $(1-f)$ has no influence on the function G.

TABLE 68. The function G of the time-intensity theory

(For example: $G = 0.438$ when the duration of the exposure (T) is such that $T/\tau = 3.2$.)

T/τ	0.0	0.1	0.2	0.3	0.4	0.5	0.6	0.7	0.8	0.9
0	1.000	0.967	0.937	0.907	0.879	0.852	0.827	0.802	0.779	0.757
1	0.736	0.715	0.696	0.678	0.660	0.643	0.626	0.611	0.596	0.581
2	0.568	0.554	0.542	0.529	0.518	0.506	0.495	0.485	0.475	0.465
3	0.456	0.446	0.438	0.429	0.421	0.413	0.405	0.398	0.391	0.384
4	0.377	0.371	0.365	0.358	0.353	0.347	0.341	0.336	0.331	0.325
5	0.321	0.316	0.311	0.307	0.302	0.298	0.294	0.290	0.286	0.282
6	0.278	0.274	0.271	0.267	0.264	0.260	0.257	0.254	0.251	0.248
7	0.245	0.242	0.239	0.236	0.234	0.231	0.229	0.226	0.224	0.221
8	0.219	0.216	0.214	0.212	0.210	0.208	0.206	0.203	0.201	0.199
9	0.198	0.196	0.194	0.192	0.190	0.188	0.187	0.185	0.183	0.182

T/τ	0.0	0.2	0.4	0.6	0.8	1.0	1.2	1.4	1.6	1.8
10	0.180	0.177	0.174	0.171	0.168	0.165	0.163	0.160	0.158	0.155
12	0.153	0.150	0.148	0.146	0.144	0.142	0.140	0.138	0.136	0.134
14	0.133	0.131	0.129	0.128	0.126	0.124	0.123	0.121	0.120	0.119
16	0.117	0.116	0.115	0.113	0.112	0.111	0.110	0.108	0.107	0.106
18	0.105	0.104	0.103	0.102	0.101	0.100	0.0987	0.0978	0.0968	0.0959

T/τ	0	1	2	3	4	5	6	7	8	9
20	0.0950	0.0907	0.0868	0.0832	0.0799	0.0768	0.0740	0.0713	0.0689	0.0666
30	0.0644	0.0624	0.0605	0.0588	0.0571	0.0555	0.0540	0.0526	0.0512	0.0500

$G(x) = \dfrac{2}{x^2} \{x - 1 + e^{-x}\}.$

Approximations: Large x: $G(x) = 2/x - 2/x^2$.
Small x: $G(x) = 1 - x/3 + x^2/12 - x^3/60$.

For Sax's experiments[1] at room temperature we find that $\tau = 4$ min., both for chromatid and chromosome aberrations. Thus Fig. 38A shows that the theoretical curve fits Sax's data (Fig. 4 of his 1940 paper) of the production of chromosome exchanges by a constant dose of 320 r. spread over various times

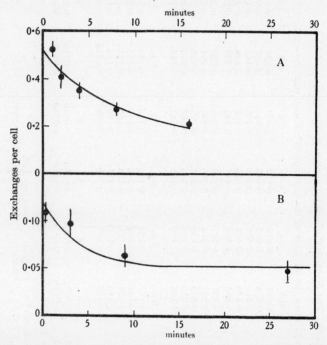

FIG. 38. Diminution with increased duration of exposure of the yield of exchanges produced by a constant dose of X-rays in *Tradescantia* microspores. A, chromosome exchanges, continuous irradiation over the time indicated, 320 r. B, chromatid exchanges, 150 r. divided into three 2 min. fractions with rest periods of 0, 3, 9, or 27 min. between fractions. Curves theoretical, points experiments of Sax.

by varying the intensity. The agreement is seen to be satisfactory. Repetitions of this experiment made in different laboratories have given different experimental curves and correspondingly different values of τ (cp. curves a and b of Fig. 33, p. 227). It also appears that τ is influenced by temperature.[2]

[1] Sax, K. (1939, 1940). We have multiplied the ordinate scale by 0·03 to convert it to exchanges per cell, and inserted estimates of error taking the standard deviation of a figure based on the recording of n exchanges to be a fraction $1/\sqrt{n}$.

[2] Sax, K. & Enzmann. E.V. (1939); Catcheside, D.G., Lea, D.E. & Thoday, J.M. (1946a).

For chromatid exchanges Sax's data refer to experiments of a slightly different type, in which the radiation was given at constant intensity, but split up into three 2 min. fractions with rest intervals between the fractions, which in the different experiments took the values 0, 3, 9 or 27 min. The application of the theory to this type of experiment presents no difficulties, but is a little tedious. The first stage is to calculate the number of free breaks existing at any given time. During the first fraction of duration T_1 the number of breaks increases according to equation (6a), during the rest period T_2 it decreases according to equation (6b); thus at time $T_1 + T_2$ the number of free breaks is

$$n_2 = \xi I \tau (1 - e^{-T_1/\tau}) e^{-T_2/\tau}.$$

During the next fraction T_3 the number of breaks again increases, following the differential equation $dn/dt = \xi I - n/\tau$, with initial value n_2; the solution of the differential equation is

$$n = \xi I \tau (1 - e^{-t/\tau}) + n_2 e^{-t/\tau}.$$

Thus at the end of the second rest interval T_4 the number of free breaks is

$$n_4 = \xi I \tau (1 - e^{-T_3/\tau}) e^{-T_4/\tau} + n_2 e^{-(T_3+T_4)/\tau},$$

and so on. Evaluating $\int \beta n^2 dt$ stage by stage gives the number of exchanges. Fig. 38 B shows the agreement of theory and experiment obtained using the same value $\tau = 4$ min. as was used for chromosome exchanges. There is no evidence in Sax's experiments of different rates of rejoining of breaks in split and unsplit chromosomes.

If the dose is varied by varying the time of exposure at constant intensity, then with increase of dose G diminishes and equation (8) shows that the yield of exchanges increases less rapidly than the square of the dose. Sax showed experimentally[1] that the yield of chromosome exchanges obtained at a constant intensity of 25 r./min. increased as the 1·5 power of the dose. His results are shown in Fig. 39, in which both dose and yield of exchanges are plotted on a logarithmic scale. With this method of plotting, a yield proportional to a power of the dose is represented by a straight line. The experimental points lie between lines a, representing (dose)^2 and line d, representing (dose)^1, and

[1] Sax, K. (1938). Centric rings and dicentric chromosomes were scored.

approximate closely to line *b*, representing (dose)$^{1\cdot5}$. Curve *c* is the function (dose)^2 G, G being calculated using the value $\tau = 4$ min. derived from Sax's experiments on the dependence of yield on duration of exposure at constant dose (Fig. 38A). It is

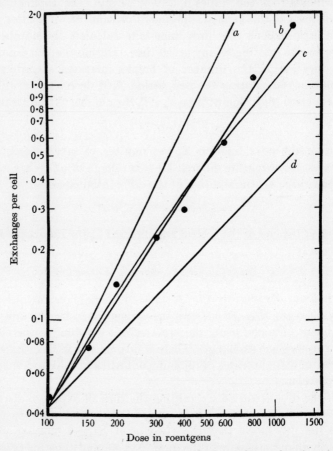

FIG. 39. Yield of chromosome exchanges in *Tradescantia* as a function of dose at a constant intensity of 25 r. per minute (X-rays). Points experimental (Sax), curves: *a* (dose)^2; *b* (dose)$^{1\cdot5}$; *c* (dose)^2 G; *d* (dose)^1.

seen to represent the experimental results satisfactorily except for some deviation at the highest doses (corresponding to exposure times exceeding 30 min.).

No special theoretical significance attaches to the power 1·5 in the (dose)$^{1\cdot5}$ law which fits these observations. If a lower intensity had been used over the same dose range, a lower power would have been obtained. If a higher intensity had been used, a higher power, approximating to (dose)^2 at sufficiently high intensity, would have been obtained. The experimental results shown in Fig. 35A (p. 231) illustrate the dependence of the shape of the dose curve on the intensity.

With the aid of the G function a satisfactory account can be given of the manner in which the yield of exchanges depends on dose and intensity in experiments having a duration up to about 30 min. In experiments extending beyond 30 min., the theoretical curve falls below the experimental curve for the longer exposure times. Fig. 39 illustrates this for an experiment on the variation of yield with exposure time at constant intensity, and it has also been demonstrated in experiments on the variation of yield with exposure time at constant dose.[1] The explanation may be that the number of breaks in the nucleus falls off on account of restitution more gradually than indicated by the formula $-dn/dt = n/\tau$, or it may be that an appreciable proportion of the exchanges produced by X-rays are 1-hit exchanges, the number of which produced by a given dose is, of course, independent of exposure time. This point has not yet been elucidated.

The relative efficiencies of different radiations

By the methods described in an earlier section it was possible to deduce from the experimentally observed yields of aberrations the numbers (ξ) of chromatid breaks primarily produced per cell per roentgen (or v-unit) by X-rays, neutrons, and α-rays in *Tradescantia* microspores irradiated at a time when the chromosomes were split (i.e. 24 hr. before metaphase of the first haploid mitosis). The average values of ξ are reproduced in Table 69. In addition, values of ξ are given for soft X-rays of wave-lengths 1·5, 4·1 and 8·3 A. These figures are obtained on the assumption that the values of ξ with different wave-lengths are proportional to the respective yields of chromatid breaks per roentgen given in Table 63 (p. 239). Since it was found that the proportion of joinable breaks was not detectably different for medium X-rays and neutrons (Table 67, p. 261), it seems likely that it will not

[1] Catcheside, D.G., Lea, D.E. & Thoday, J.M. (1946b).

be different for different wave-lengths of X-rays,[1] and that the procedure adopted for deducing ξ for soft X-rays will therefore be correct.

TABLE 69. Numbers of primary chromatid breaks per cell per roentgen (ξ) produced in *Tradescantia*[2]

Radiation	ξ
X-rays (0·15 or 1·5 A.)	0·09
(4·15 A.)	0·17
(8·3 A.)	0·010
Neutrons (Li+0·9eMV. deuterons)	0·21
α-rays (Rn+Ra A+Ra C′)	0·10

Reasons have already been given for supposing that a chromatid is broken by a single ionizing particle. We are now in a position to answer the question whether a single ionization causes the break or whether a number of ionizations are necessary. The tests by which actions of radiation caused by a single ionization can be recognized were discussed at length in Chapter III. One test is that the efficiency of different radiations (i.e. yield per unit dose, here ξ) should decrease in the order of increasing ion-density, i.e. in the order X-rays, neutrons, α-rays. Table 69 shows, however, that neutrons are more efficient, not less efficient, than X-rays. It is concluded, therefore, that a single ionization is not able to cause a break and that a number of ionizations are necessary. More exactly, if the more densely ionizing particles (protons) in neutron experiments produce n times as many ionizations per micron as the less densely ionizing particles (electrons) in X-ray experiments, the fact that neutrons are more efficient per ionization shows that the passage of a proton has a probability of causing a break *more than* n times as great. If we draw hypothetical curves (Fig. 40) relating the probability of an ionizing particle causing a break to its ion-density, a curve such as B and not such as C or D must be assumed. The essential feature required by the fact that neutrons are more efficient per ionization than X-rays is that the curve must have a steeply rising portion in which the probability of

[1] The ion-densities of the ionizing particles which traverse a tissue irradiated by soft X-rays are intermediate between the ion-densities of the ionizing particles when the tissue is irradiated by medium X-rays and neutrons respectively.

[2] Per v-unit in the case of neutrons.

EFFECT OF ION-DENSITY

breakage is increasing more rapidly than the first power of the ion-density. An extreme form of curve B is the square-cornered curve A, and while this is less plausible than a smooth curve such as B, we shall for the sake of the simplification thereby achieved,

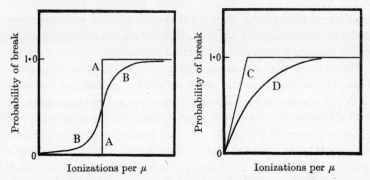

FIG. 40. Hypothetical relations between the ion-density of a particle, and the probability of its causing a break when traversing a chromatid.

assume the extreme form A in some calculations. The approximation involved is not an absurdly crude one as it would be if the true curve had a shape such as D, Fig. 40.

α-rays, which have an ion-density still greater than that of protons, produce fewer primary breaks per ionization (Table 69). It follows that the point representing α-ray ion-density is at a part of curve B where the probability of breakage is increasing less rapidly than the first power of the dose. It is therefore beyond the shoulder of the curve, and the probability of an α-particle breaking a chromatid through which it passes is evidently close to unity.

We can obtain further evidence leading to the same conclusion by taking account of the dimensions of the chromatids. The haploid length of the chromosome thread in *Tradescantia* is 486μ, and the diameter of the thread is said to be 0.1μ.[1] The volume of all the 12 chromatids in the prophase nucleus will therefore be $7.63\mu^3$. Since the total length of α-ray track produced per roentgen per μ^3 is $0.56 \times 10^{-3}\mu$ (Table 18, p. 32), it follows that the total length of α-ray track contained within the chromatid thread is $4.3 \times 10^{-3}\mu$ per roentgen, and that the number of inter-

[1] Sax, H.J. & Sax, K. (1935); Sax, K. (1938). The dimensions are those observed at pachytene, a stage when the chromosomes are greatly extended.

sections of α-rays with the chromatid thread is 0·043 per roentgen.[1] If δ-rays are allowed for (data also from Table 18), the total number of intersections of α-rays and δ-rays with the chromatid thread is 0·113 per roentgen.

The number of chromatid breaks primarily produced per roentgen is 0·10 (Table 69), which is compatible with the view that the probability of an α-particle which traverses the chromatid causing a break is approximately unity.

Making a similar calculation for neutrons (physical data from Table 18) we find that the mean number of chromatid intersections per v-unit is 0·31, allowing for proton intersections only, or 0·40 allowing also for δ-ray intersections. Comparing with the value of ξ given in Table 69, viz. 0·21 primary chromatid breaks per v-unit, we infer that the probability of a proton causing a break is a little less than unity. Actually the protons which traverse the tissue in these experiments cover a rather wide range of energies, and consequently of ion-densities. It is likely that the less energetic (i.e. more densely ionizing) protons have a probability of practically unity of breaking a chromatid through which they pass, and that the more energetic (i.e. less densely ionizing) protons have considerably lower probability. This is suggested by Giles's experiment (Fig. 35B, p. 231) in which a decrease of the mean neutron energy (i.e. increase in ion-density) caused some increase in the yield of aberrations per ionization.

On these grounds it appears likely that the rising part of curve B, Fig. 40 corresponds roughly to the range of ion-densities of the protons in neutron experiments.

We have now to attempt to interpret the variation with wavelength of the number of primary breaks produced per roentgen by X-rays. For equal doses of X-rays and neutrons many more chromatids are traversed by ionizing particles in X-ray experiments than in neutron experiments (as may be seen by comparing the figures of the total range in the tissue of the ionizing particles given in Table 18, p. 32). Yet X-rays produce fewer breaks. It follows that the probability of a chromatid being broken by the passage of an electron through it is rather low, a

[1] If a straight line intersects a long cylinder at random, the mean length of straight line lying within the cylinder is equal to the diameter of the cylinder, viz. 0·1 μ for the chromatid thread.

fact which is to be ascribed to the ion-density in an electron track being on the average considerably lower than in a proton track. The number of ionizations per micron produced by an electron increases rapidly towards the end of the track,[1] and for the last few tenths of a micron it is of the same order as in a proton track of a few electron-megavolts energy. If this part of an electron track traverses a chromatid, the probability of causing a break must be quite high. However, the ion-density towards the beginning of an electron track of, for example, 20 ekV. energy, is some ten times lower than in the last tenth of a micron. In view of the rapidity with which the probability of an ionizing particle causing a break diminishes with decrease of ion-density (cp. curve B, Fig. 40), it is evident that if this part of the electron track traverses a chromatid, the probability of a break being caused is very small. We are thus led to think that an electron is practically ineffective in causing breakage unless the densely ionizing 'tail' of the track traverses the chromatid, in which event the probability of causing a break may be quite high. The following considerations support this view. Soft X-rays of wave-lengths 1·5 and 4·15 A. dissipate their energy by means of photoelectrons of 7·5 and 2·5 ekV. respectively. For equal numbers of roentgens, the numbers of photoelectrons ejected by the two wave-lengths are in the ratio of 1 : 3 (Table 18, p. 32). The numbers of chromatid breaks primarily produced per roentgen are in the ratio of 1 : 1·9 (Table 69, p. 270). It follows that the efficiencies per photoelectron are in the ratio 1·57 : 1. Now the last 2·5 ekV. of path of the 7·5 ekV. photoelectron (which has a range of $1·5\mu$)[2] is of course indistinguishable from the whole track of the 2·5 ekV. photoelectron (of range $0·23\mu$). It follows that the probability of a break being caused by the first $1·27\mu$ of a 7·5 ekV. electron is little more than half as great as the probability of a break being caused by the last $0·23\mu$. The notion of an effective tail at the end of an electron track which is otherwise practically ineffective is thus supported by the experimental determination of the relative efficiencies of wave-lengths 1·5 and 4·1 A.

As a simplified model for the purpose of calculation, we shall suppose the probability of a chromatid being broken to be unity

[1] See Plate I c.
[2] Ranges read off from Table 10, p. 24.

if the densely ionizing 'tail' of the electron track traverses it, and to be zero if the earlier part of the track traverses it. On the basis of this model we can calculate the length l of the 'tail' and predict the variation of efficiency with wave-length.

In Fig. 41A we represent diagrammatically the passage of an electron through a chromatid thread. The effective 'tail' is the portion PQ of the electron track, the remaining ineffective portion being represented by the interrupted line. It is clear that

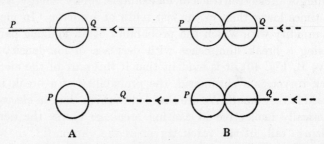

Fig. 41. The passage of an electron through a chromatid, or through sister-chromatids.

for PQ to traverse the chromatid, while remaining parallel to the direction in which it is drawn, P may occupy any point in a volume equal to $A(l-2r)$, where l is the length PQ, A is the area presented by the chromatid to the electron, and $2r$ is the mean path of the electron in the chromatid of radius r. The area presented by a chromatid to an electron which is perpendicular to the chromatid axis is *length × diameter*, but allowing for random inclination of electron path and chromatid axis, this is reduced by the factor $\frac{1}{4}\pi$. The total area of all the twelve chromatids (haploid length 486μ, diameter $0\cdot1\mu$) is thus $A = 76\cdot3\mu^2$. If the X-rays liberate n electrons per μ^3 per roentgen, the expected yield of primary chromatid breaks per roentgen is thus

$$\xi = nA(l-2r), \qquad (\text{VII-9})$$

where $A = 76\cdot3\mu^2$ and $2r = 0\cdot1\mu$.

In this formula l is the length of the effective 'tail' of the electron track, or the whole length of the electron track if this is so short that there is no ineffective portion. For X-rays of $4\cdot15$A., $n = 0\cdot0201$ electrons per μ^3 per roentgen (Table 18, p. 32), and $\xi = 0\cdot17$ (Table 69). Inserting these numerical values in equation (9) we deduce that $l = 0\cdot211\mu$. Now the photoelectrons projected by X-rays of this wave-length have an energy of 2·5 ekV. and a range of 0·23 μ. The approximate equality of this figure and the value of l just deduced suggests that probably the whole of a photoelectron track of this energy is effective. The length of effective 'tail' of a more energetic photoelectron must therefore be at least 0·2 μ.

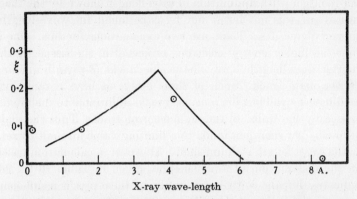

FIG. 42. The number of primary chromatid breaks per roentgen (ξ) as a function of wave-length in the soft X-ray region. Curve theoretical, points based on experiment.

The length of 'tail' may be deduced by making use of the data available for X-rays of shorter wave-length. For X-rays of wave-length 1·54 A., $n = 0·0067$ electron per μ^3 per roentgen (Table 18, p. 32) and $\xi = 0·09$ (Table 69, p. 270). Inserting these numerical values into equation (9) we deduce that the length l of the effective 'tail' is 0·28 μ, and its energy (Table 10, p. 24) is therefore 2·8 ekV. We are now able to use equation (9) to construct a theoretical curve of variation of ξ with wave-length over the soft X-ray region.[1] This calculated curve is shown in Fig. 42, together with the values of ξ given in Table 69, which are based on experiment. It is evident that the theory outlined gives a good general representation of the experimental facts.

Taking the length of the effective 'tail' to be 0·28 μ, it can be deduced with the aid of Table 10 (p. 24) that the energy

[1] Values of n are obtained by interpolation in Table 18, p. 32. l is 0·28 μ, or the length of the photoelectron track, whichever is the smaller. For a wave-length λ we take the photoelectron energy to be

$$12·4/\lambda - 0·5 \text{ ekV}.$$

(cp. p. 10), and read off the corresponding range in Table 10 (p. 24).

dissipated in a chromatid which intercepts a length $0 \cdot 1 \mu$ of the less densely ionizing end of the 'tail' is about $0 \cdot 6$ ekV., corresponding to the production of 15–20 ionizations. This is therefore the minimum amount of energy which, dissipated in a chromatid, is sufficient for the probability of breakage to approach unity.

The continued rapid decrease of the number of primary breaks per roentgen with diminution of wave-length shown by the theoretical curve is not borne out by experiment for wave-lengths shorter than $1 \cdot 5$ A.[1] There are two explanations for this. One is that the higher energy electrons projected in the tissue by the shorter wave-length X-rays have side tracks (δ-rays) branching off the main track. Such of these δ-rays as have a range exceeding $0 \cdot 1 \mu$ will be able to cause breaks additional to the breaks caused by the 'tails' of the main electron tracks. Thus the yield of breaks per roentgen tends to a limiting value and not to zero as the wave-length is diminished.[2] However, a calculation based on the distribution of δ-ray energies given in Table 16 (p. 28) indicates that the contribution made by the δ-rays is insufficient to account for the whole of the yield of $0 \cdot 09$ primary break per roentgen at a wave-length of $0 \cdot 15$ A., which is the experimental result (Table 69). A second factor must therefore enter, which may be the following. The theoretical curve of Fig. 42 is based on the rather crude model that parts of the electron track less densely ionizing than the 'tail' are completely ineffective in causing breaks. The adequate manner in which the calculated curve fits the observations down to a wave-length of $1 \cdot 5$ A. (i.e. a photoelectron energy of $7 \cdot 5$ ekV.) shows that the probability of a break being caused by the first $4 \cdot 7$ ekV. of a $7 \cdot 5$ ekV. electron track is indeed negligible by comparison with the probability of a break being caused by the 'tail', i.e. the last $2 \cdot 8$ ekV. It is, however, pushing the model to extremes to imagine that with, for example, a 50 ekV. electron track it is still true that the probability of a break being produced by the first 47 ekV. is completely negligible compared with the probability of a break being produced by the last $2 \cdot 8$ ekV. of track. The application of

[1] Some decrease with diminution of wave-length has been found. Thus X-rays of $0 \cdot 15$ A. produce slightly fewer breaks than X-rays of $1 \cdot 5$ A., and γ-rays produce slightly fewer breaks than X-rays of $0 \cdot 15$ A. (Table 63, p. 239). But this decrease is much less rapid than shown by Fig. 42.

[2] Cp. a calculation by Fano, U. (1943c).

equation (9) should therefore be limited to the soft X-ray region (i.e. wave-lengths exceeding 1 A.). A quantitative theory covering shorter wave-lengths is not yet available. To develop such a theory it will probably be necessary to make assumptions about the form at low ion-densities of the curve relating probability of breakage to ion-density of ionizing particle (curve B, Fig. 40). We have been able to develop theories for soft X-rays without such assumptions, since for these wave-lengths it was satisfactory to use the square-shaped curve A, Fig. 40.

Isochromatid breaks

The values of ξ derive mainly from the experimental yields of chromatid breaks per roentgen, the yields of other types of aberration entering into the calculation of ξ only in subsidiary degree. The agreement between the theoretical curve and experimental points in Fig. 42, and the correlation (discussed on pp. 271 and 272) between the values of ξ for neutrons and α-rays

TABLE 70. Yields of isochromatid breaks per cell per roentgen (or v-unit)

Radiation	Isochromatid breaks per cell per roentgen (c)	Primary chromatid breaks per cell per roentgen (ξ)	Ratio (c/ξ)
X-rays (0·15 or 1·5 A.)	0·0027	0·09	0·030
(4·15 A.)	0·0044	0·17	0·026
(8·3 A.)	< 0·00009	0·010	< 0·009
Neutrons (Li + D)	0·0099	0·21	0·047
α-rays (Rn + Ra A + Ra C')	0·0210	0·10	0·21

and the numbers of ionizing particles crossing the chromatids, thus serve as tests of the capacity of the theory to explain the experimental yields of chromatid breaks. We turn now to a discussion of isochromatid breaks.

In Table 70 we list the numbers (c) of isochromatid breaks obtained per cell per roentgen in *Tradescantia* microspores (taken from Table 64, p. 241), and in the last column give the values of c/ξ. c/ξ is z times the ratio of the numbers of isochromatid and chromatid breaks primarily produced,[1] and is thus a measure of the relative frequency of primary production of isochromatid and chromatid breaks. We proceed to discuss

[1] z being the ratio of the numbers of isochromatid breaks observed and isochromatid breaks primarily produced (cp. footnote 1, p. 254).

the explanation of the marked variation of c/ξ with different radiations.

According to the treatment developed on p. 261, which applies to neutron-induced isochromatid breaks, $c/\xi = \frac{1}{2}zpg$, where p is the probability of an ionizing particle which traverses a chromatid causing a break, and g is a geometrical factor depending on the distance apart of the chromatids. A theoretical expression for the number of isochromatid breaks per cell per roentgen (c) can be derived for X-rays by a combination of the arguments of pp. 261 and 274. It is

$$c = \tfrac{1}{2}zpg \, nA \, (l-4r). \qquad (\text{VII-10})$$

Comparison with equation (9) for the number of primary breaks per cell per roentgen, viz. $\xi = nA\,(l-2r)$, shows that for X-rays

$$c/\xi = \tfrac{1}{2}zpg \, (l-4r)/(l-2r).$$

This is less than the value of c/ξ for neutrons by the factor $(l-4r)/(l-2r)$. The term $(l-4r)$ in equation (10) takes the place of the term $(l-2r)$ in equation (9) owing to the presumption that to cause an isochromatid break it is necessary for the electron to traverse both chromatids and not merely one, as illustrated diagrammatically in Fig. 41 B. $2r = 0.1\mu$, the chromatid diameter, and for X-rays of wave-length less than 3·8 A., $l = 0.28\mu$ (the length of the effective tail). Thus the factor $(l-4r)/(l-2r)$ is 0·44. According to Table 70, the experimental ratio of the c/ξ values for X-rays of wave-length less than 3·8 A., and neutrons, is 0·64.

For X-rays of wave-length exceeding 3·8 A., the photoelectrons projected by which have a range less than 0.28μ, the value of l required is the photoelectron length. $(l-0.2)/(l-0.1)$ thus decreases, eventually to zero, with increase of wave-length (and consequent diminution of photoelectron range) beyond 3·8 A. This is in qualitative agreement with the behaviour of the experimental values of c/ξ in Table 70.

c/ξ, and consequently $\tfrac{1}{2}zpg$, is considerably higher for α-rays than for neutrons. Three factors contribute to this. One is that the proportion of isochromatid breaks having unjoinable breakage ends is higher in the case of α-rays. For such isochromatid breaks $z = 1$, and is probably less for isochromatid breaks having joinable breakage ends. The second factor is that in neutron experiments p is probably less than unity for the higher energy protons (cp. p. 272). p will therefore be higher for α-rays than for protons. The third factor invokes δ-rays. Numerous short electron tracks branch from an α-ray track (see Plate IA, p. 10; and numerical data in Table 17A., p. 30). It may happen sometimes that when an α-ray passes through one chromatid but not through the sister chromatid, the latter is nevertheless broken owing to a δ-ray from the α-ray track passing through it. Some indication of the probable magnitude of this effect is afforded by the fact (taken from Table 17A) that the number of δ-rays emitted from a length of α-ray track equal to a chromatid diameter $(0{\cdot}1\mu)$ and having a range exceeding a chromatid diameter (i.e. an energy exceeding 1·5 ekV.) is about 0·3. The addition of a term of this order to g in the formula $c/\xi = \tfrac{1}{2}zpg$ will result in an appreciable increase in c/ξ, since g is between 0·18 and 0·363 (p. 262).

We conclude that the theory affords a qualitative explanation of the experimental result that the ratios of the frequencies of primary production of isochromatid and chromatid breaks increase in the order soft X-rays, harder X-rays, neutrons, α-rays.

Recapitulation

We conclude the discussion of *Tradescantia* by recapitulating, in non-mathematical language, the description of the process of structural change which has been built up in the preceding sections. When *Tradescantia* microspores are irradiated (in the stage in which the chromosomes are split) by X-rays, neutrons or radioactive radiations, the number of chromatid breaks primarily induced greatly exceeds the total number of breaks observed at the time of fixation. The majority of the breaks restitute after being open for a period of a few minutes. Exchange between breaks is possible during the time the breaks are open. If the dose of radiation is given in a short overall time, so that all the primary breaks are open in the nucleus at the same time, the yield of X-ray-induced exchanges is proportional to the square of the dose. If the dose is spread over a longer time, restitution of the breaks first produced can take place before the later breaks are formed and the yield of X-ray-induced exchanges under these conditions is reduced.

For there to be an appreciable probability of exchange

occurring between two breaks, it is necessary for them to be produced at an initial separation of not more than about 1μ.

In α-ray and neutron experiments, in which a comparatively small number of ionizing particles traverse the nucleus with the doses commonly given, the majority of the exchanges are exchanges between pairs of breaks produced simultaneously by the same ionizing particle. Consequently the yield of these exchanges (unlike the 2-hit X-ray-induced exchanges) is proportional to the first power of the dose and independent of the intensity.

If the ionizing particle which breaks a chromatid also breaks the sister chromatid at approximately the same locus, then an isochromatid break usually results.

A proportion of the chromatid breaks primarily produced have unjoinable breakage ends. As a result a certain proportion of the interchanges are incomplete, a certain proportion of isochromatid breaks fail to show sister-union, and a certain proportion of the breaks not taking part in exchanges or isochromatid unions remain as visible chromatid breaks instead of restituting. The proportion of breaks which are not joinable is the same for X-rays and neutrons, but higher for α-rays.

The number of chromatid breaks primarily produced per cell per roentgen can be inferred from the experimental yields of aberrations by allowing for the chromatid breaks which restitute. It is found that this number is different for different radiations. Correlation of the number of primary breaks per cell with the numbers and ranges of the ionizing particles which traverse the nucleus, enables the following conclusions to be drawn. A proton (of not too high energy) or an α-ray traversing a chromatid has a probability almost unity of causing a break. An electron, however, is only likely to cause a break if it is the 'tail' end, where the ion-density is highest, which traverses the chromatid. The length of this effective 'tail' is $0\cdot3\mu$,[1] and the minimum number of ionizations necessary to be produced in the chromatid thread of diameter $0\cdot1\mu$ for the probability of causing a break to approach unity is 15–20. In this way the relative efficiencies of different wave-lengths and types of radiation in breaking chromatids can be explained. Among X-rays, a wave-length of about 4 A. has a higher efficiency per roentgen than either longer

[1] Cp. Plate I c.

or shorter wave-lengths; neutrons are more effective than the most effective X-rays, and α-rays are less effective.

With materials other than *Tradescantia*, the experimental data are not adequate for an analysis of this sort to be carried out. According to experiments cited in Table 62 (p. 238), neutrons are more effective than X-rays in inducing chromosome structural changes in bean roots and in mouse tumours. If this result is confirmed it is evidence that in these materials, as in *Tradescantia*, several ionizations and not a single ionization are needed to cause a break. To test the applicability to these materials of other of the conclusions drawn from the *Tradescantia* experiments, it will be necessary to have additional data of ion-density variation (e.g. α-ray or soft X-ray experiments), and data on the dependence of the yield of aberrations on intensity. It will be highly desirable to classify structural changes into chromatid breaks, isochromatid breaks, and exchanges, and to determine the dependence on dose and intensity of the yields of these individual types of aberration. Observations classified simply as abnormal anaphases yield much less information.

Chapter VIII
DELAYED DIVISION
Introduction

A temporary inhibition of division appears to be a general action of radiation, having been demonstrated in a great variety of cells. The duration of the delay increases with increasing dose, so that this effect of radiation differs from those we have discussed in previous chapters in being a graded action rather than an all-or-none action. It is therefore usually more appropriate to consider the mean delay produced in a batch of cells, rather than the proportion of the cells which are delayed, as a measure of the effect produced by a given dose.

In the actions of radiation which we have considered previously, to produce the effect studied (e.g. mutation or chromosome breakage), it is necessary to produce a single ionization, or a certain concentration of ionization, in a very limited locality. In consequence the effect is always produced by a single ionizing particle,[1] since the chance is slight of more than one ionizing particle passing through the same locality. The region within which ionization must be produced to delay division is probably not so small, and many ionizations are probably necessary to cause appreciable delay. Consequently, in X-ray experiments, many ionizing particles contribute to the effect. Thus the non-random spatial distribution of ionization in irradiated tissue, and the 'target-theory' type of calculation, which play a large part in the interpretation of the actions of radiation considered in previous chapters, are much less important here.[2]

The fact that it is a delay in division, and not a permanent inhibition of division which is being studied, means that the action of radiation concerned is one from which recovery occurs, and any interpretation of experiments on delayed division must incorporate some mechanism of recovery.

It seems likely that the effect of the radiation in inhibiting division is due to some chemical change in the cell, but it is not

[1] X-ray-induced chromosome exchanges are, it is true, caused by two ionizing particles, but this is simply because each exchange involves two breaks. The initial effect of the radiation, the production of a break, is caused by a single ionizing particle.

[2] See, however, p. 304.

yet clear whether the change consists essentially in the destruction of some component needed for division to occur, or the production of some substance having an inhibitory effect. The recovery, which eventually permits division to proceed, presumably implies the re-formation of the component destroyed by the radiation (on the former view), or the removal of the inhibitory substance (on the latter view). The state of the cell at a given moment is thus determined partly by the dose which it has received, and partly by the extent to which recovery has taken place. If the chemical change we have postulated were understood we could doubtless express the state of the cell in terms of the concentrations of the compounds concerned. It is, of course, highly desirable that the chemical reactions should be investigated.[1] Pending the acquisition of this information, a certain amount of progress can be made in the formal interpretation of the results by introducing the concept of *cumulative dose*. The cumulative dose is a measure of the state of the cell allowing for the radiation it has received and the recovery which has occurred. It can be defined for our purpose as that dose of radiation in roentgens which given instantaneously to the cell would bring it to the same state (as regards the chemical change which is responsible for the delay of division) as that in which the cell finds itself at the given moment. The concept of cumulative dose has been used by several authors[2] in developing theories of actions of radiation from which recovery is possible, and in which the chemical changes involved are at present unknown. If it should eventually appear that the chemical change involved is the production by the radiation of some inhibitory substance, then the cumulative dose will be proportional to the concentration of this substance. If, on the other hand, the essential change is the destruction of some component in the cell, and recovery is its re-formation, then the cumulative dose will be proportional to the deficit of this component existing at any given time.

[1] The suggestion has been made (Mitchell, J.S. 1942) that the delay in division is due to an interference in the nucleic acid cycle, resulting in a failure of the conversion of ribonucleic acid to desoxyribonucleic acid. Evidence in support of this view is provided by the observation that ribonucleotides accumulate in the cytoplasm of cells after irradiation.

[2] Cf. Hoffmann, J.G. & Reinhard, M.C. (1934); Quimby, E.H. & MacComb, W.S. (1937); Lea, D.E. (1938a,b).

Different aspects of the phenomenon of delayed division have been studied in widely different experimental materials, including bacteria, invertebrate eggs, and various rapidly dividing plant and animal tissues, and in this chapter we give an account of some of these researches. It will be realized that delay in division is not the only effect of radiation on a cell. If appreciable delay can be produced by doses which do not kill the cell, then delayed division can be studied. If, however, the dose needed to cause appreciable delay kills a large proportion of the cells irradiated before division occurs, then it may prove impossible to investigate delayed division. It will be realized further that the term 'recovery' is used here to mean recovery from whatever change is inhibiting division, and does not imply that the 'recovered' cell is in all respects normal. It may, for example, have chromosome structural changes or other changes which lead to its eventual death. It appears that delayed division is a phenomenon sufficiently distinct from lethal action to justify its separate discussion.[1]

The delay of first cleavage in sea-urchin eggs

An illuminating series of researches has been carried out by Henshaw and his colleagues[2] on the delay of first cleavage induced by X-rays in *Arbacia*. The general procedure is as follows. Separate suspensions are prepared of the sperm and the eggs. When the suspensions are mixed, a sperm enters each egg, the male and female pronuclei fuse, the diploid nucleus undergoes mitosis, and a cleavage furrow appears across the cell. The time of first cleavage can be determined by microscopic examination of the living cells, and is very nearly uniform among all the eggs of a batch. The time when 50 % of the eggs have cleaved is taken as the time of cleavage, and can be determined to about 1 min., the whole time elapsing between the mixing of the egg and sperm suspensions and cleavage being about 45 min. (at 25°). If either the egg or the sperm or both are irradiated before fertilization, or if the zygote is irradiated after fertilization, the time elapsing

[1] This point is further discussed in Chapter IX, p. 311.

[2] Henshaw, P.S. (1932); Henshaw, P.S., Henshaw, C.T. & Francis, D.S. (1933); Henshaw, P.S. & Francis, D.S. (1936); Henshaw, P.S. (1938); Henshaw, P.S. (1940); Henshaw, P.S. & Cohen, I. (1940). The interpretation of these experiments in terms of cumulative dose follows the treatment of Lea, D.E. (1938b).

between fertilization and cleavage is increased, the amount of the increase over the controls being referred to as the *cleavage delay*. The cleavage delay produced by a given dose to the sperm is independent of their concentration, and of the fluid in which the sperm are suspended, indicating that the effect is a direct action on the gamete and not an indirect action due to the irradiation of the suspending fluid.[1]

Henshaw's experiments consisted in determining the cleavage delay under various conditions of irradiation. Cleavage delay can be caused by irradiation of either egg or sperm, and for a given dose is approximately the same whichever gamete is irradiated.[2] In view of the very much greater quantity of cytoplasmic material in the egg, this observation suggests that the action of the radiation in delaying division is on the nucleus and not on the cytoplasm. This has been convincingly demonstrated by experiments on half-eggs.[3] If eggs are centrifuged at high speed they break into two halves, one half containing the intact nucleus, the other half having no nucleus. Either half when fertilized will eventually cleave, though the time required is abnormally long for the non-nucleated half-egg. Irradiation of a nucleated half-egg prior to fertilization causes the same delay as irradiation of a whole egg, but no delay is caused by irradiation of a non-nucleated half-egg.

The validity of the concept of cumulative dose is supported by the experimental results[4] shown in Fig. 43. Prior to fertilization either the eggs, or the sperm, or both, are irradiated by various doses of X-rays spread over a constant time of 40 min. The cleavage delays obtained when both eggs and sperm are

[1] Evans, T.C., Slaughter, J.C., Little, E.P. & Failla, G. (1942). Some other actions of radiation on the sperm were found by these authors to be indirect.

[2] Henshaw's earlier experiments (Henshaw, P.S. & Francis, D.S. 1936), and also the experiments of Mavor, J.W. & de Forest, D.M. (1924), suggested that irradiation of the sperm produces a greater effect than irradiation of the unfertilized eggs. Henshaw subsequently showed (1940) that this apparent difference in sensitivity is really due to partial recovery occurring in the eggs prior to fertilization. If a sufficiently short time elapses between the commencement of the irradiation and fertilization, so that recovery is avoided, there is little difference in sensitivity between egg and sperm, the egg being slightly more sensitive.

[3] Henshaw, P.S. (1938).
[4] Henshaw, P.S. & Francis, D.S. (1936).

irradiated (curve C) are greater than when either eggs alone (curve A) or sperm alone (curve B) are irradiated, but less than the sum of these two delays (curve D). The delays are therefore not additive. However, what we should expect to be additive is not the delay but the cumulative dose. Inspection of curves A and B shows that x roentgens to the egg produces the same delay as $\frac{1}{2}x$ roentgens to the sperm.[1] On the cumulative dose basis we should

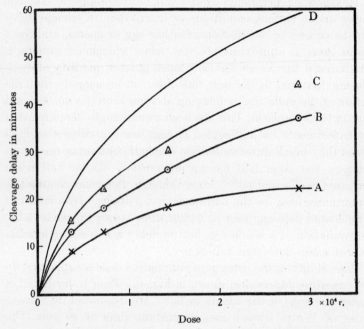

FIG. 43. Cleavage delay in *Arbacia* caused by X-rays (Henshaw & Francis). A, eggs irradiated; B, sperm irradiated; C, eggs and sperm irradiated.

therefore expect that x roentgens each to egg and sperm should produce the same delay as $3x/2$ roentgens to sperm alone. Curve C is calculated on this basis from the experimental curves A and B, and is seen to fit the observations satisfactorily.

Henshaw's observations on fixed material[2] have shown that the delay in cleavage caused by irradiation is due mainly to a great prolongation of prophase. The stages prior to prophase,

[1] The duration of 40 min. was long enough for appreciable recovery to occur during the irradiation of the eggs (cp. footnote 2, p. 285). Sperm do not recover.

[2] Henshaw, P.S. (1940).

namely, fertilization, the approach of sperm and egg pronuclei, and their fusion, are not delayed, while the later stages of division (metaphase, anaphase and telophase) are somewhat prolonged but to a much less marked degree than in the case of prophase.[1] We presume therefore that the effect of the radiation is

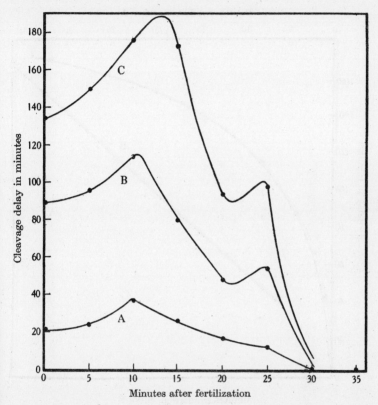

FIG. 44. Cleavage delay when the fertilized egg is irradiated at various times after fertilization (*Arbacia*, Henshaw & Cohen). A, 1 min.; B, 2 min.; C, 4 min.; exposure at 7800 r./min.

to hinder the condensation of the chromosomes which occurs during prophase, and that the completion of prophase signifies that recovery is nearing completion. The experimental results shown in Fig. 44[2] are in agreement with this view. In these experiments fertilized eggs were exposed to X-rays at various

[1] For the purpose of this description we are including in early prophase the stage which, in the notation of Fry, H.J. (1936), is stage 4.
[2] Henshaw, P.S. & Cohen, I. (1940).

times ranging from 0 to 35 min. after fertilization, thus irradiating the eggs in various stages up to the end of prophase. The maximum effect of a given dose is seen to be produced when the dose is given 10–15 min. after fertilization, i.e. when fusion of the pronuclei is complete and immediately before the visible

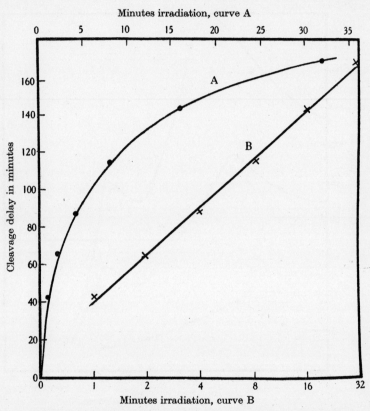

Fig. 45. Cleavage delay as a function of the dose received by the sperm (7800r./min., *Arbacia*, Henshaw).

onset of prophase. The effect is reduced if the irradiation is made after prophase has started, and is negligible if made after the completion of prophase.

Fig. 45, curve A,[1] shows how the cleavage delay increases with increase of the dose given to the sperm prior to fertilization. A straight line is obtained if the dose is plotted on a logarithmic scale as in Fig. 45, curve B, the cleavage delay increasing by 25

[1] Henshaw, P.S. (1940).

min. each time the dose is doubled.[1] On our interpretation, according to which the duration of the delay is the time required for recovery, we deduce that, in the absence of further irradiation, the cumulative dose decays to half value in 25 min. The decay being exponential, we represent it by the function $e^{-t/\tau}$. The half-value time of 25 min. corresponds to τ, the characteristic time being 36 min., and the rate of decay of the cumulative dose in the fertilized egg being a fraction $1/\tau = 2\cdot 8\%$ per min.

The fact, shown by Fig. 44, that the cleavage delay caused by a given dose is greatest if the dose is given 10 min. after fertilization is to be expected on the present view. The cleavage delay depends on the cumulative dose existing immediately prior to the commencement of prophase. If the irradiation is made immediately after fertilization, there is time for some decay of cumulative dose to occur during the time elapsing between the irradiation and the onset of prophase. This will be shown by the effect of a given dose given immediately after fertilization being less than the effect of an equal dose given some minutes later. The gradients of the curves of Fig. 44 during the first 10 min. are consistent with recovery occurring at about the rate ($2\cdot 8\%$ per min.) already found.

The decay of cumulative dose can be followed over a longer period in experiments[2] in which unfertilized eggs are irradiated, and are then left (in sea water) for varying periods before fertilization. Under these conditions the cleavage delay produced by a given dose is found to diminish with increase of the rest period. With the aid of an experimental curve showing cleavage delay as a function of dose to the unfertilized egg, we are able to derive from the experimental results, which show cleavage delay as a function of rest period, curves showing cumulative dose as a function of rest period.[3] When the cumulative dose is plotted on a logarithmic scale, a straight line is obtained, as illustrated in Fig. 46B, where the four lines refer to four different doses. The cumulative dose thus decays according to the formula $e^{-t/\tau}$. The value of τ at 20–25° is deduced from the

[1] The data of curve B, Fig. 43, obtained some years earlier, indicate that the delay is increased by about 10 min. when the dose is doubled.
[2] Henshaw, P.S. (1932); Henshaw, P.S., Henshaw, C.T. & Francis, D.S. (1933).
[3] See Lea, D.E. (1938b) for details.

gradients in Fig. 46B to be 104 min., and to be independent of the dose.

In an experiment in which the unfertilized eggs were kept at 0° after irradiation,[1] it was found that the recovery was slower than

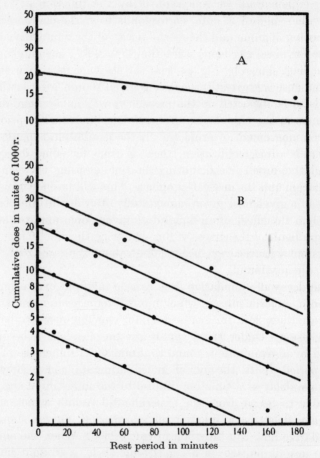

FIG. 46. Decay of cumulative dose in unfertilized *Arbacia* eggs.
A, at 0°; B, at 20–25°.

at room temperature. Converting cleavage delay into cumulative dose in the same fashion as before, Fig. 46A is obtained. The gradient corresponds to $\tau = 360$ min.

There appears to be a correlation between the differences in the τ values found for fertilized eggs at 20–25°, unfertilized eggs

[1] Henshaw, P.S. (1940).

at 20–25°, and unfertilized eggs at 0°, and the differences in metabolic rate as measured by the rate of oxygen uptake.[1] It is seen in Table 71 that $100/\tau$, which is the percentage decay of cumulative dose per minute, is approximately proportional to the rate of oxygen uptake.

TABLE 71. Correlation of recovery rate with rate of oxygen uptake

System	Temperature °C.	τ min.	Rate of decay of cumulative dose $(100/\tau)$ % per min.	Relative rate of oxygen uptake
Fertilized egg	20–25	36	2·8	3–5
Unfertilized egg	20–25	104	1·0	1·0
Unfertilized egg	0	360	0·3	0·2
Sperm	20–25	Large	0·0	0·01

In experiments in which sperm were irradiated Henshaw found that a rest period intervening between irradiation and fertilization did not reduce the cleavage delay. Thus the correlation between rate of recovery and rate of oxygen uptake extends also to sperm, since it is known that the rate of oxygen uptake by a sperm is negligible compared with that by an egg.

The fact that sperm left in sea water after irradiation do not show recovery makes it improbable that the recovery of irradiated unfertilized eggs left in sea water is due to diffusion out of the egg of inhibitory substances. The correlation between recovery rate and oxygen uptake favours the interpretation that the effect of the radiation is to destroy some nuclear constituent necessary for the condensation of the chromosomes, and that recovery consists in the re-formation of this constituent as a result of the metabolic activity of the cell.

It would be unjustifiable to deduce from Table 71 that the recovery process is one which directly requires oxygen. The rate of oxygen uptake is presumably an indication of the general level of metabolic activity, and in *Arbacia* eggs appears to vary in different states of the egg in much the same way as does whatever reaction is responsible for recovery. It does not necessarily follow that the rates of recovery in different organisms will be proportional to their respective rates of oxygen uptake. No re-

[1] Experimental data on the rate of oxygen uptake are given by Loeb, J. (1910); Loeb, J. & Wasteneys, H. (1911); Needham, J. (1931); Tang, P.S. (1931); Whitaker, D.M. (1933).

covery from the delay of cleavage produced by irradiation was found in the unfertilized eggs of the clam *Cumingia*,[1] although the oxygen consumption[2] is rather higher than in the unfertilized eggs of *Arbacia*.

In *Arbacia* recovery occurs if a rest period elapses between the irradiation of unfertilized eggs and fertilization. It is to be presumed that recovery will also be occurring during the irradiation if this is extended over a prolonged time, so that the effect of a given dose will be less if it is administered at low intensity than if it is administered at high intensity. This has, in fact, been shown to be the case.[3] These considerations can easily be put into a quantitative form.[4]

If the cumulative dose existing at any time t is D, then the rate of decay due to the recovery process is D/τ, and the rate of increase of cumulative dose due to the radiation being administered is I (where I is the dose rate in roentgens per minute). Thus

$$dD/dt = I - D/\tau, \qquad (\text{VIII-1})$$

integrating to $\qquad D = I\tau(1 - e^{-t/\tau}). \qquad (\text{VIII-2})$

Two batches of eggs were irradiated by equal doses spread over 30 and 150 min. respectively by the use of intensities in the ratio of 5 : 1. They were fertilized immediately after the completion of the irradiation and the cleavage delays were found to be 21 and 12 min. respectively. According to equation (2), the cumulative doses at the time of fertilization should be in the ratio of

$$\frac{5(1 - e^{-30/104})}{1(1 - e^{-150/104})} = 1 \cdot 64 : 1,$$

since $\tau = 104$ min. for unfertilized eggs at room temperature. Since the actual dose used in this experiment is not stated, one proceeds to calculate it from the fact that the 30 min. exposure resulted in a cleavage delay of 21 min., and thence calculates the delay to be expected for the 150 min. exposure. The result of this

[1] Henshaw, P.S., Henshaw, C.T. & Francis, D.S. (1933).
[2] Whitaker, D.M. (1931).
[3] Henshaw, P.S., Henshaw, C.T. & Francis, D.S. (1933). In *Cumingia*, however, in which no recovery occurs when a rest period intervenes between irradiation and fertilization, the delay produced by a given dose was found to be independent of the intensity.
[4] Lea, D.E. (1938 *a,b*).

calculation is 14 min., in fair agreement with the experimental figure of 12 min.

It appears therefore that the dependence of the effect of a given dose upon the time over which it is spread can, in these experiments, be explained on the basis of essentially the same recovery process which permits the eventual completion of the division.

The stage of division which is subject to delay

When division is delayed by irradiation, it is of interest to determine in which stage or stages the retardation occurs, and also at which stage the cells should be irradiated to give the greatest delay. The answers to these questions can most readily be obtained in the case of materials in which the cells develop synchronously, so that all the cells can be irradiated at the same stage. The irradiation being made at a chosen stage, samples are fixed at intervals and the stages reached determined. Comparison with a control series enables the delay suffered in each stage to be measured. Henshaw[1] has made experiments in this fashion to find which stages are subject to delay in the cleavage division in *Arbacia*.

In many experimental materials cells are present at all stages of mitosis and interphase, so that it is not possible to choose a single stage to be irradiated. Experiments made by fixing such materials at intervals after irradiation and determining the numbers of cells in the various stages are therefore more difficult to interpret. The interpretation is facilitated if the material chosen is one in which it is possible to follow the course of mitosis in the living cells. A cell can be selected, and irradiated when it is in a known stage, and its subsequent behaviour studied. The examination of living cells in this way has been used to supplement observations on fixed preparations in the case of embryonic chick tissue growing in culture,[2] and whole grasshopper embryos.[3] In the latter material it has been found that irradiation produces the greatest delay if it is made at a stage in prophase when the chromosomes are already discrete, but prior to the breakdown of the nuclear membrane. A dose of 10–20r.

[1] Henshaw, P.S. (1940).
[2] Canti, R.G. & Donaldson, M. (1926).
[3] Carlson, J.G. (1941*b*). The cells examined were neuroblasts.

at this critical stage causes the cell to remain unchanged for some hours. Cells irradiated at an earlier stage of prophase are also held up in middle prophase, but for a shorter time, so that they may reach metaphase actually before the cells which were irradiated in the critical stage, although they were less advanced at the time of irradiation. A larger dose (250r.) causes the cells to be held up in early prophase (a stage when the chromosome threads are still barely visible) instead of in middle prophase, and cells somewhat more advanced but not yet passed the critical stage appear to revert to the condition of early prophase, where they remain some hours before resuming mitosis.

Irradiation after the breakdown of the nuclear membrane causes no delay. Irradiation at a stage intermediate between the critical stage and the breakdown of the nuclear membrane causes a short delay, which is shorter the further removed the cell is from the critical stage.

These conclusions are in general agreement with those of Henshaw on *Arbacia* eggs, already mentioned (p. 286), and with observations on other materials. The main point is that moderate doses of radiation given late in prophase or at a subsequent stage cause little retardation. Somewhere in prophase, or shortly before the visible onset of prophase, there is a stage where the delay produced by a given dose is a maximum. This is well shown in Fig. 44. Irradiation at or immediately prior to this critical stage causes the cells to be held up. The duration of the delay depends upon the cumulative dose at this time. If irradiation is made some time previously, the cumulative dose has partly decayed by the time the critical stage is reached, and hence the effect is less than if the same dose had been given nearer the critical stage. This critical stage, at which the effect of a given dose is a maximum, and near which cells whose division is held up accumulate, is described by Carlson as being in prophase. In other materials[1] it has been regarded as late interphase (i.e. immediately prior to prophase). This difference is probably not fundamental, but reflects a difference between different materials in the ease with which early prophase can be distinguished from the resting stage. Thus in Carlson's material (grasshopper neuroblasts)[2] the stage described as prophase lasts twice as long as metaphase, anaphase and telophase combined, and in fact

[1] Cp. Spear, F.G. (1935).   [2] Carlson, J.G. (1941b).

occupies half the time interval between the telophase of one division and the metaphase of the next. In mammalian[1] and chick[2] tissue, however, the stage described as prophase lasts only about one-third as long as metaphase, anaphase and telophase combined.

Bearing this in mind, the observations of Carlson which we have described are consistent with the conclusion reached by previous authors investigating a variety of tissues[3] that a dose of up to a few hundred roentgens temporarily prevents cells from entering mitosis, but does not prevent cells already in mitosis completing division.

Rapidly dividing tissues

The manner in which the mitotic count in a rapidly dividing tissue varies during the hours following irradiation is illustrated in Fig. 47B, which shows the result of an experiment[4] in which chick tissue growing in culture was irradiated by γ-rays at an intensity of 25 roentgens per minute for $1\frac{1}{2}$, $2\frac{1}{2}$, 9 or 30 min., and fixed at various intervals after exposure. With small doses (cp. the $2\frac{1}{2}$ min. curve) the retardation of mitosis at a stage just previous to that recognized as prophase results in an initial diminution of mitotic count. When mitotic activity returns, there are two classes of cells entering division. One class consists of those cells which were so far from division at the time of irradiation that they had recovered from the dose given by the time they had reached the stage in which cells which are delayed accumulate. These cells therefore divide at the normal time. The second class consists of those cells which were in early prophase at the time of irradiation, and whose further development was held up for some hours. As these cells recover (i.e. the cumulative dose decays) they slowly pass through prophase.[5]

In consequence of the fact that these two classes of cells are

[1] Tansley, K., Spear, F.G. & Glücksmann, A. (1937) using rat retina.
[2] Juul, J. & Kemp, T. (1933); Lasnitzki, I. (1940).
[3] Alberti, W. & Politzer, G. (1923, 1924) using corneal epithelium of amphibian larvae; Strangeways, T.S.P. & Oakley, H.E.H. (1923); Strangeways, T.S.P. & Hopwood, F.L. (1926), and later workers using chick tissues grown in culture; Mottram, J.C., Scott, G.M. & Russ, S. (1926) using Jensen rat sarcoma.
[4] Canti, R.G. & Spear, F.G. (1929).
[5] Cp. Carlson, J.G. (1941b).

entering mitosis together, the mitotic activity during this period is higher than in unirradiated material. Since the cells of the second class are recovering as they pass through prophase, and therefore are spending an abnormally long time in this phase, the

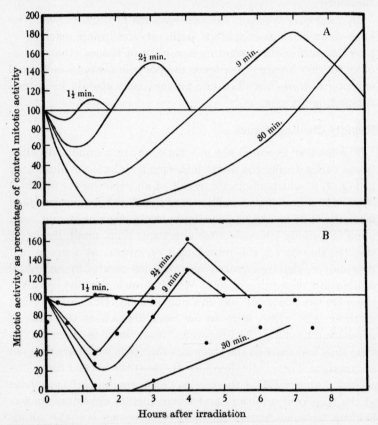

FIG. 47. Mitotic activity at various times after irradiation. Exposures of 1½, 2½, 9, and 30 min. at 25 r./min. A, calculated curves; B, experimental results (Canti & Spear, γ-rays on chick tissue in culture).

proportion of mitotic cells which are in prophase is higher than normal during the recovery wave.[1]

Larger doses of radiation cause the death of many of the cells entering division, and such cells are recorded as degenerating

[1] Demonstrated in bean root tips by Pekarek, J. (1927); in rat tissue by Tansley, K., Spear, F.G. & Glücksmann, A. (1937); in tadpole tissue by Spear, F.G. & Glücksmann, A. (1938); in chick tissue in culture by Lasnitzki, I. (1940).

cells and not as dividing cells. In consequence, with larger doses the mitotic count fails to exceed that in unirradiated material during the period of recovery.

Curves such as those shown in Fig. 47 can be constructed theoretically if the relation between delay in division and dose

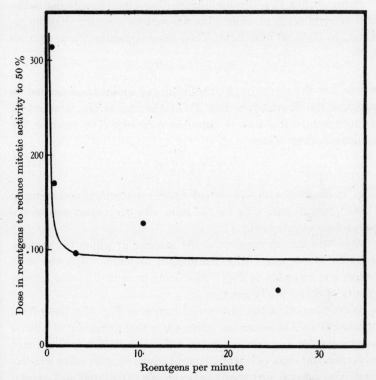

FIG. 48. Dependence upon the intensity of the dose required to produce a given delay of division. Curve theoretical, points experiments of Canti & Spear.

is known. As explained in the discussion of cleavage delay in *Arbacia*, this is essentially the same as the relation between the cumulative dose remaining at a given moment and the time which has elapsed since irradiation. But the rate of decay of cumulative dose can be deduced from experiments on the variation of the effect of a given dose with the intensity at which it is given.

Canti and Spear[1] irradiated chick tissue in culture and counted the number of cells in division 80 min. after the irradiation

[1] Canti, R.G. & Spear, F.G. (1927).

finished. They determined the dose needed to reduce the mitotic count to half that in unirradiated cultures at several different intensities, obtaining the results shown in Fig. 48. At low intensities the dose required to produce a given effect is several times greater than at high intensities.

Suppose that D_0 is the cumulative dose which must exist at the end of irradiation in order that the mitotic activity shall be reduced to 50% 80 min. later. Then from equation (2), p. 292, we have

$$D_0 = I\tau(1 - e^{-t/\tau}),$$

where t is the duration of irradiation at intensity I needed to produce the cumulative dose D_0 at the end of the irradiation. It is therefore the dose required at intensity I to produce the observed effect. Thus

$$It = D_0 \frac{(t/\tau)}{(1 - e^{-t/\tau})}. \qquad \text{(VIII-3)}$$

Fig. 48 shows that a theoretical curve constructed according to this equation, and with $\tau = 217$ min., fits the experimental observations satisfactorily.[1]

A second datum required is the manner in which the mitotic activity at 80 min. after irradiation depends upon the dose (when the intensity is high), this being provided by the experiments of Spear and Grimmett.[2]

With these data the theoretical curves of Fig. 47A have been constructed.[3] These curves, which show the variation of mitotic activity over a period of 8 hr. after irradiation utilize only experimental data of mitotic counts made 80 min. after irradiation. The general agreement between the calculated and experimental curves shows that, as in the case of *Arbacia* eggs, the recovery during irradiation which results in the effect of a given dose being diminished by spreading it over a prolonged time, is essentially the same process as the recovery after irradiation which permits the eventual return of mitotic activity.

That the experimental curves in Fig. 47 show a lower recovery wave than the theoretical curves is due to the circumstance that

[1] Cp. Lea, D.E. (1938*a*).

[2] Spear, F.G. & Grimmett, L.G. (1933).

[3] See Lea, D.E. (1938*b*) for details of the calculation, and for a comparison of theoretical and experimental curves obtained when a second irradiation is made 80 or 160 min. after the first.

cells which degenerate when they enter division are not included in the experimental curves.

The duration of mitotic delay as a function of dose has been investigated in a number of tissues, and Fig. 49 illustrates some of the results obtained.[1] A precise figure cannot be given to the

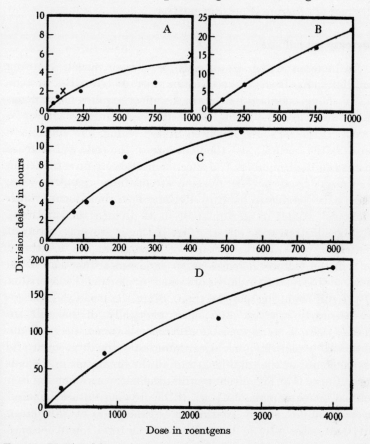

FIG. 49. Division delay in hours as a function of dose in roentgens. A, chick tissue in culture (● Canti & Spear; × Lasnitzki). B, *Chortophaga* neuroblasts (Carlson). C, rat retina (Tansley, Spear & Glücksmann). D, *Triton* cornea (Alberti & Politzer).

[1] Taken from the experiments of the following authors: Alberti, W. & Politzer, G. (1924), using the cornea of the newt *Triton*; Canti, R.G. & Spear, F.G. (1929), Lasnitzki, I. (1943a), using chick tissue in culture; Tansley, K., Spear, F.G. & Glücksmann, A. (1937), using the retina of the rat; Carlson, J.G. (1938a), using neuroblasts of the grasshopper *Chortophaga*.

duration of the delay produced by irradiation of a rapidly dividing tissue, since cells are irradiated at different stages and suffer different delays. In constructing Fig. 49 the delay has been taken to be the period elapsing between the irradiation and the commencement of the rise in mitotic activity which follows the period of reduced mitotic activity.

Increase of cell size

An increase in the size of cells, or of their nuclei, following irradiation has been observed in a number of materials.[1] In some cases, though perhaps not in all, the effect requires no explanation other than the inhibition of division which is known to occur. If the cells are prevented from dividing, but continue to grow at the normal rate, then the size of the cells will increase instead of their number. The effect is most easily investigated in the case of bacteria. The division rate of these organisms being high, it is possible to hold up division for a time during which each cell would, if unirradiated, have divided several times. Fig. 50 illustrates the effect of irradiating a growing culture of *Bacterium coli*.[2] In the absence of irradiation, the number of cells in the culture increases as an exponential function of the time, so that a straight line is obtained by plotting the logarithm of the cell count against the time. Since the mean cell size remains nearly constant, and since practically all the cells are viable, there is no appreciable difference between the gradients obtained by plotting (on a logarithmic scale) the total count, the viable count, or the total length of all the organisms in the culture. Curve C of Fig. 50 shows this gradient, corresponding to a tenfold increase in about $4\frac{1}{2}$ hr. If the growing culture is placed over a radium source (continuous γ-irradiation at 35r./min.), division ceases after a short time and the total count becomes practically constant as shown by curve A. The viable count diminishes (curve B) owing to the fact that in addition to

[1] E.g. bacteria, Spencer, R.R. (1935), Lea, D.E., Haines, R.B. & Coulson, C.A. (1937); yeast, Holweck, F. & Lacassagne, A. (1930), Wyckoff, R.W.G. & Luyet, B.J. (1931); Protozoa, Mottram, J.C. (1926), Robertson, M. (1932); fungal spores, Luyet, B.J. (1932); algae, Luyet, B.J. (1934); bean root tips, Pekarek, J. (1927), Mottram, J.C. (1933a); mouse tumours, Mottram, J.C. (1927), Ludford, R.J. (1932); human tumours, Glücksmann, A. (1941).

[2] Lea, D.E., Haines, R.B. & Coulson, C.A. (1937).

inhibiting the division of all the organisms, the radiation kills some of them. Meanwhile the length of the cells increases very greatly (see Plate IV A, B). A measurement of the total length of all the cells in the culture would underestimate the growth rate, since dead cells as well as growing ones would be included.

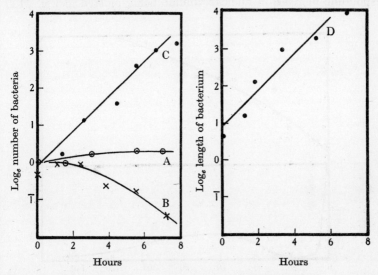

FIG. 50. Inhibition of division in *Bact. coli* (Lea, Haines & Coulson). A, total count in irradiated culture. B, viable count in irradiated culture. C, total length in unirradiated culture. D, length in irradiated culture. γ-rays at 35 r./min.

Curve D therefore shows the manner in which the length of the *longest* organism increases during the irradiation. It is seen that the gradient of curve D, i.e. the rate of increase of length of the non-dividing cells in the irradiated culture, is practically the same as the gradient of curve C, i.e. the rate of increase of total length of all the cells in the unirradiated culture. Thus the growth rate (meaning rate of volume increase) is the same for unirradiated cells as for (living) irradiated cells, but in the former case division occurs so that the cells maintain a nearly constant size but increase in number, while in the latter case division is inhibited so that the size of the individual cells increases but their number remains constant.

Fig. 51 shows the results of an experiment on the irradiation of bean root-tips.[1] Normally about 10 % of the cells in the

[1] Pekarek, J. (1927).

dividing region of the root-tip are in mitosis at any given moment, and the average nuclear diameter is about $8{\cdot}5\mu$. After irradiation (by about 1600 r.) the mitotic activity is greatly reduced as shown in Fig. 51 B. Meanwhile the mean nuclear diameter increases, rising to about $11{\cdot}5\mu$ after 100 hr., the corresponding increase in nuclear volume being plotted in Fig. 51 A.

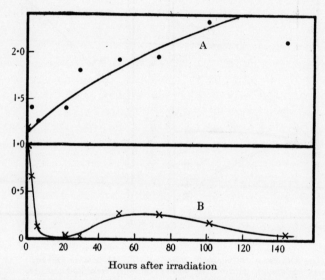

FIG. 51. Inhibition of division in bean root-tips (Pekarek). A, nuclear volume relative to control nuclear volume. B, mitotic activity relative to control mitotic activity.

The fact that the nuclear volume doubles in about 3 days is compatible with the rate of growth being unchanged by the irradiation, but resulting in increase in cell size instead of increase in cell number.

The increase of cell or nuclear size in plant and animal tissues after irradiation cannot be expected to be so striking as the increase of size of bacteria illustrated in Plate IV A, B. The inhibition of division often persists for a time which is only a fraction of the intermitotic period. Thus if the intermitotic period is 3 days, and the inhibition lasts for 1 day, we shall expect a 30 % increase of cell or nuclear volume, which corresponds to an increase of only 10 % in diameter. Careful measurement of many cells will be required to establish such an increase.

The relative efficiencies of different radiations in causing delay of division

Probably the most convenient material to use in comparing the relative efficiencies of different radiations in causing delay in division would be marine invertebrate eggs, but experiments with radiations other than X-rays have not yet been made using this material. A few experiments have however been made using rapidly dividing plant and animal tissues.

With a rapidly dividing tissue two methods of experiment can be used. One is to irradiate the tissue and to determine the mitotic activity at intervals after irradiation, and to measure the duration of the delay as a function of the dose in the manner of Fig. 49. The efficiencies of different radiations can then be obtained from comparison of the doses needed to cause equal delays.

The second method is to determine the mitotic count at a definite time after irradiation, preferably when mitotic activity is at a minimum (which is at a time after irradiation somewhat in excess of the duration of mitosis). The relative efficiencies of different radiations in inhibiting division are then deduced from measurement of the doses needed to produce equal reduction of mitotic count.

Of the few experiments at present available, the best have been made on the bean root-tip,[1] using the second method. The results are shown in Fig. 52, in which the mitotic count 3 hours after irradiation is expressed as a fraction of the mitotic count in unirradiated controls, and plotted on a logarithmic scale against the dose. Plotted in this way, the points lie approximately on straight lines, and the relative efficiencies can be taken to be the ratios of the gradients. The efficiencies of the three radiations, neutrons, γ-rays, and α-rays for equal ionization in the tissue are approximately in the proportions 4 : 2 : 1 respectively.

The explanation suggested by the authors for their results is that for inhibition of division to occur a certain number of ionizations must be produced, not in the nucleus as a whole, but

[1] Gray, L.H., Mottram, J.C., Read, J. & Spear, F.G. (1940); Gray, L.H., Mottram, J.C. & Read, J. (unpublished). I am indebted to Dr Gray for communicating the results of these experiments, and his interpretation of them, prior to publication.

anywhere within a smaller volume estimated to be about 3μ in diameter.[1] It is supposed that a single α-particle traversing this region produces several times more than the minimum amount of ionization (a few thousand ionizations) needed for inhibition of division to occur. α-rays are thus less efficient per ionization

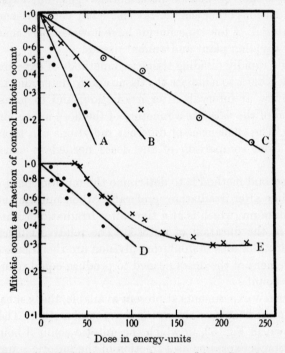

FIG. 52. Reduction of mitotic count as a function of dose (Gray, Mottram, Read & Spear). A, neutrons; B, γ-rays; C, α-rays (bean root-tips fixed 3 hr. after irradiation). D, neutrons; E, γ-rays (chick tissue in culture fixed 80 min. after irradiation).

than less densely ionizing radiations, for the same reason as in the case of chromosome breakage and virus inactivation, namely, that the extra ionization contributes to the dose without increasing the proportion of cells affected.

One would expect on the basis of this argument that neutrons, being more densely ionizing than γ-rays (though less densely

[1] The authors point out that the nucleolus is of about this size. If the cause of the delayed division is interference with the nucleic acid cycle, it is not unplausible that the nucleolus should be concerned, since it is a reservoir for ribonucleic acid during the resting stage. Cp. also Mottram, J.C. (1932).

ionizing than α-rays) would be less effective, or at any rate not more effective, than γ-rays. The authors have put forward an ingenious mechanism by which the higher efficiency of neutrons can be explained, but it involves some assumptions which appear to be rather unplausible.

An alternative explanation of the lower efficiency of α-rays compared to less densely ionizing radiations would be that the change produced in the nucleus leading to the inhibition of division is a chemical reaction involving 'activated water'. α-rays would then be less efficient for the same reason as that put forward in Chapter II to explain their low efficiency in decomposing proteins in aqueous solution (cp. p. 60), namely, that owing to the high concentration of active radicals in the neighbourhood of the α-ray track, some radicals disappear by recombination before reacting with the solute concerned. This mechanism would not provide an explanation of the fact that neutrons are more effective per ionization than γ-rays.

The relative efficiencies of the different radiations in causing inhibition of division cannot therefore be considered entirely understood. It is desirable that experiments should also be made by the first method mentioned, namely, the determination of the doses of the different radiations needed to produce the same interval between the initial fall of mitosis and its subsequent rise, since this experiment should decide between the alternative mechanisms which have been suggested for the lower efficiency of α-rays compared to γ-rays.[1]

No special significance is attached to the approximately exponential shape of the γ-ray curve of Fig. 52B. The cells, whose delay in division causes the reduction in mitotic count at fixation, three hours after irradiation, are the cells which are due to enter division any time between 0 and 3 hr. after irradiation.[2]

[1] As pointed out to me by Dr Gray. If the low efficiency of α-rays in reducing mitotic activity three hours after irradiation is due to an α-ray producing more than the minimum amount of ionization in the sensitive region, then α-rays should be no less effective than γ-rays when the doses needed to cause inhibition of division for an equal time (longer than 3 hr.) are compared. If however the low efficiency of α-rays is explained as a low ionic yield in some chemical reaction, then the efficiency should be low by whichever way it is measured.

[2] Taking the duration of mitosis to be approximately 3 hr. in the bean root-tip.

and the minimum delays needed to prevent them from being in division at the time of fixation therefore range from 3 to 0 hr. respectively. In view of this heterogeneity in the stages of the cells concerned and in the delays needed to be induced in them, it is not surprising that the number of cells in division at the time of fixation diminishes gradually with increase of dose.

Neutrons are also more effective per ionization than γ-rays in causing a reduction of mitotic count shortly after irradiation in the case of chick tissue grown in culture, as shown in Fig. 52, curves D, E.[1] In these experiments an exponential curve is obtained with neutrons (Fig. 52 D), but the γ-ray curve departs from the exponential form in having an initial flat portion (Fig. 52 E). This difference is interpreted by the authors[2] to mean that a single ionizing particle causes the delay in the neutron experiments and that many are required in the γ-ray experiments.

[1] Gray, L.H., Mottram, J.C., Read, J. & Spear, F.G. (1940). The γ-ray curve is quoted from Spear, F.G. & Grimmett, L.G. (1933).

[2] Spear, F.G., Gray, L.H. & Read, J. (1938); Gray, L.H., Mottram, J.C., Read, J. & Spear, F.G. (1940).

Chapter IX

LETHAL EFFECTS

Lethal effects of radiation have been studied by a great number of authors on a great variety of experimental materials, partly because of the ease with which experiments can be made in which the criterion of effect is the death of an organism or of a cell, and partly because of the practical importance of the lethal action of radiations on cells in the treatment of cancer.

No attempt is made in this chapter to review the whole of this very considerable literature. Attention is confined to the killing of single-celled organisms and of individual cells of multicellular organisms, and is directed mainly to lethal actions which may be understood in terms of the mechanisms described in the earlier chapters. The killing of multicellular organisms as distinct from the killing of their individual cells is not discussed, nor is the killing of a cell discussed when this is due, as it probably is in some circumstances, to a change produced by the radiation in the surrounding tissues or fluids rather than to the dissipation of energy by the radiation in the cell itself.

Death precipitated by division

It has been shown in many experimental materials that when a cell is killed as a result of irradiation, death does not occur immediately, but at or following the next division that the cell undergoes. To kill a cell immediately requires a much larger dose than to cause a cell to die at or following its next division. Thus, 10,000 r. of X-rays delivered to yeast does not cause immediate death. The irradiated cells divide once, but the daughter cells are usually unable to divide further and eventually die. A much larger dose is required to kill the cells without division, 30,000 r. sufficing only to kill 50 % in this way.[1]

Similarly, when chick tissue growing in culture is irradiated,[2] doses of 2500 r. and upwards are required to cause the death of an appreciable proportion of resting cells without the intervention of division, but a dose of 100 r. suffices to cause the death of

[1] Holweck, F. & Lacassagne, A. (1930). Cp. also Wyckoff, R.W.G. & Luyet, B.J. (1931).
[2] Lasnitzki, I. (1943 a,b).

a considerable proportion of those cells which attempt division in the hours following irradiation.

Fig. 53 shows the percentage of the pollen of *Tradescantia* which germinates, the anthers having received a dose of 800 r. at various times before dehiscing.[1] If the dose of radiation is

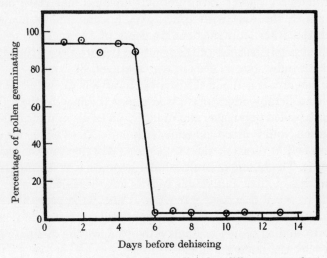

FIG. 53. Germination of pollen irradiated at different stages by a constant dose of 800 r. (Newcombe).

given one to five days prior to the time of dehiscing, the percentage of pollen germination is as high as in unirradiated pollen. If the same dose is given 6 days or more prior to dehiscing, only 3 % of the pollen grains germinate. The discontinuity is due to the occurrence of the pollen grain mitosis about 6 days before dehiscing. The radiation is evidently lethal if nuclear division intervenes between the irradiation and the germination test, but not if no nuclear division intervenes.

If the change which leads to death at division is one from which recovery can occur, it may be possible to reduce the lethal effect of a given dose by adopting some means of increasing the time interval between irradiation and the next division. Thus if eggs of the parasitic worm *Ascaris* are irradiated by 5000 r. and allowed to develop at 25°, only 1 or 2 % develop into normal

[1] Newcombe, H.B. (1942a). See also experiments by Poddubnaja-Arnoldi, V. (1936) and Koller, P.C. (1943).

embryos. If, however, the eggs are kept for 7 weeks at 5° after irradiation, so preventing cell division during this time, and are then restored to 25°, 40% of the eggs develop into normal embryos.[1]

That cells irradiated by small doses during the resting stage break down at division has been demonstrated by Spear and his

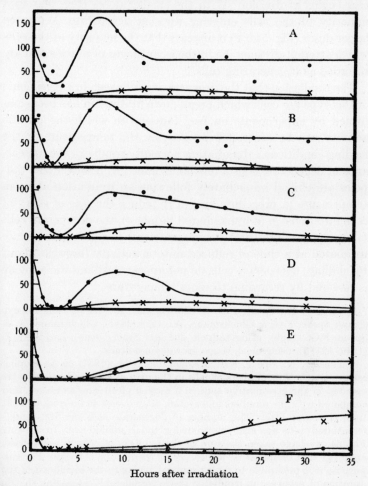

Fig. 54. Numbers of dividing and degenerating cells at various times after irradiation (rat eye, Tansley, Spear & Glücksmann). • mitotic count as percentage of control mitotic count; × percentage of cells which are degenerate. A, 2 min.; B, 4 min.; C, 6 min.; D, 10 min.; E, 12 min.; F, 30 min. exposure at 18 r./min. (γ-rays).

[1] Cook, E.V. (1939).

co-workers[1] in a number of rapidly dividing tissues. Fig. 54 illustrates the type of result obtained.[2] After irradiation mitotic activity is reduced, and may remain practically at zero for some hours after the larger doses (Fig. 54 E, F). The return of mitotic activity is found to coincide with the appearance of degenerating cells in the tissue, strongly suggesting that the degenerative process begins at division. With the smaller doses (Fig. 54 A, B) a minority of the cells entering division degenerate. With the larger doses (Fig. 54 E, F) it appears that the majority of the cells which attempt division die in the early stages of mitosis and are recorded as degenerating cells.

The conclusion that the degenerating cells seen in experiments of this type are cells which break down at division has been confirmed by experiments on frog tadpoles in which the mitotic activity can be controlled by varying the temperature or the feeding conditions.[3] Irradiation normally results in the appearance of degenerate cells a few hours later. If, however, the tadpoles are chilled immediately following an irradiation at room temperature (a procedure known to reduce the rate of entry of cells into division in unirradiated tadpoles), the degenerate cells fail to appear. Conversely, in experiments in which tadpoles are irradiated at a time of reduced mitotic activity (brought about by chilling) degenerate cells do not appear until mitotic activity is restored by returning to room temperature.

[1] Tansley, K., Spear, F.G. & Glücksmann, A. (1937), using rat eye tissue; Spear, F.G. & Glücksmann, A. (1938, 1941), Glücksmann, A. & Spear, F.G. (1939), using tadpole eye and brain tissue; Lasnitzki, I. (1940, 1943a), using chick tissue growing in culture.

[2] Tansley, K., Spear, F.G. & Glücksmann, A. (1937) on developing rat eye tissue. The numbers of cells in the various stages of mitosis were counted in the germinative zone in a section of an irradiated eye, and compared with the numbers in a section of an eye of an unirradiated rat from the same litter. The number of degenerate cells per 100 (undifferentiated) cells was also determined in the irradiated eye. In drawing Fig. 54, we have taken the sum of the numbers of cells in metaphase, anaphase and telophase as a measure of the mitotic activity. For this purpose it is preferable to omit prophase, owing to the likelihood of the duration of prophase in irradiated tissue considerably exceeding that in unirradiated tissue, in which event the numbers of cells in prophase will not be a valid measure of the rate of entry of cells into division.

[3] Glücksmann, A. & Spear, F.G. (1939). Dividing and degenerating cells were counted in the eye and brain.

The cause of death at division

The fact that in many experimental materials radiation delays division, and that a proportion of the cells die when they enter division after this delay, suggests the possibility that delay of division and death at division may be essentially similar phenomena differing in degree rather than in kind. On this supposition a moderate amount of damage to the cell causes some delay in entering division or a prolongation of the early stages, while a greater amount of damage (produced by a larger dose of radiation) causes increased delay in entering division and breakdown during division. This interpretation is not however inevitable, and appears to be definitely wrong in a number of instances, in which evidence has been obtained that delay of division, and death at the resumption of division, are independent phenomena. The general method of testing whether they are independent or related phenomena is to vary the conditions of the experiment in some way, e.g. the type of radiation, dose-rate, or temperature, and to see whether the two effects vary with the conditions in the same way or in different ways.

If marine invertebrate eggs are irradiated and then fertilized, it is found[1] in some species (e.g. in the sea-urchin *Arbacia* and the clam *Cumingia*) that a moderate dose causes a prolongation of the time elapsing between fertilization and the first cleavage of the fertilized egg, while larger doses completely prevent cleavage from occurring. This is in keeping with the view that the inhibition of division and the lethal action are actions differing in degree rather than in kind. But in other species (the marine worms *Chaetopterus* and *Nereis*) the largest dose which can be given without completely preventing cleavage in all the eggs causes little or no delay in cleavage in those which survive.

Again, if *Arbacia* sperm are irradiated and then mixed with unfertilized eggs, a moderate dose causes a prolongation of the time elapsing between fertilization and the first cleavage of the fertilized egg, while larger doses completely prevent cleavage. These two effects are caused by quite different mechanisms. Cleavage delay is due to the dissipation of energy by the radiation in the sperm itself, and the effect of a given dose is independent of the concentration of the sperm suspension and of the

[1] Henshaw, P.S., Henshaw, C.T. & Francis, D.S. (1933).

nature of the suspending fluid. Failure of the sperms to fertilize effectively is, however, an indirect action due to the production of some change in the suspending fluid, as is shown by the fact that the effect of a given dose is greater in dilute than in concentrated suspensions.[1]

As has already been mentioned, recovery from the lethal effect of radiation on the eggs of the parasitic worm *Ascaris* occurs if the eggs are stored for a prolonged time at 5° after irradiation.

TABLE 72. Relative doses of γ-rays, neutrons, and α-rays which produce equal effects in the bean root-tip

Criterion of effect	γ-rays	Neutrons	α-rays
Delayed division	1·0	0·5	2·0
Lethal effect	1·0	0·12	0·11

However, this storage leads to no recovery from whatever change is responsible for the cleavage delay also caused by radiation, for on returning to 25° after prolonged storage at 5°, the cleavage delay induced by the radiation is undiminished.[2]

In the bean root, the efficiencies of γ-rays, neutrons, and α-rays have been compared in causing delay in division in the rapidly dividing cells of the root-tip.[3] The efficiencies of the same radiations have also been compared in causing cessation of growth of the bean root, a phenomenon presumably due to the death of cells in the proliferating region of the root-tip.[4] In the upper line of Table 72 are listed the relative doses of the three radiations needed to reduce mitotic activity to the same fraction of the normal mitotic activity three hours after irradiation. In the lower line are listed the relative doses needed to cause cessation of growth in 50% of the root-tips. It is seen that the relative efficiencies of the three radiations depend upon which criterion of effect is chosen, a result which strongly suggests that the delay of division and the lethal action are due to independent mechanisms.

[1] Evans, T.C., Slaughter, J.C., Little, E.P. & Failla, G. (1942). Dilute suspensions can be protected by the addition of proteins. It appears likely that an 'activated water' reaction is concerned as in the inactivation of viruses in dilute suspension (p. 108).

[2] Cook, E.V. (1939).

[3] Gray, L.H., Mottram, J.C., Read, J. & Spear, F.G. (1940); Gray, L.H., Mottram, J.C. & Read, J. (unpublished). Cp. Chapter VIII, p. 303.

[4] Gray, L.H. & Read, J. (1942a,b,c,d); Gray, L.H., Read, J. & Poynter, M. (1943).

INACTIVATION OF VIRUSES

These examples show that it is not in general safe to assume that the same change in the cell is responsible for delay in division and for the breakdown of a cell which attempts to divide after the expiration of the delay, though the assumption may of course be correct in some materials.

To account for the fact that death does in many cases occur during or following division, we must seek for some action of radiation which is lethal at or following division but not lethal before division. Genetical changes, or changes in the chromosomes, suggest themselves as possibly fulfilling these requirements, and these types of lethal action of radiation we now proceed to discuss.

LETHAL MUTATIONS

The inactivation of viruses

The inactivation of the viruses, which has been separately discussed in Chapter IV, is the simplest example of a lethal action. It will be recalled that in the case of the small, crystallizable (macromolecular) viruses, an ionization anywhere in the virus particle suffices to cause inactivation. In the case of the larger viruses, it appears that a single ionization is sufficient to inactivate a virus particle, but that it is necessary for the ionization to be produced, not anywhere in the virus, but in the 'radiosensitive' part, which constitutes only a fraction of the total volume of the virus particle. Reasons were given in Chapter IV for believing that this radiosensitive part is to be identified with the genetically important part of the virus, the larger viruses being supposed to have genetical and non-genetical parts differentiated as in the higher cells, while in the macromolecular viruses no such differentiation is suspected.

An estimate of the size of the radiosensitive region can be deduced from the radiation experiments. As explained in Chapter III, from the measurement of the inactivation dose with α-rays, an estimate can be made of the area presented to the radiation by the sensitive material in the virus. The inactivation dose of vaccinia irradiated by α-rays of about 5 eMV. was found to be $2 \cdot 11 \times 10^5$ r.[1] Now for a dose of 1000 r. of α-rays of this energy, 0·71 α-particles cross each square micron of irradiated tissue (Table 18, p. 32). Thus with a dose of $2 \cdot 11 \times 10^5$ r. an average

[1] Lea, D.E. & Salaman, M.H. (1942). The inactivation curve is reproduced as Fig. 16B, p. 113.

of 1 α-particle crosses each area of $1/(211 \times 0.71) = 0.0067\mu^2$. This, then, is our estimate of the area presented to an α-ray beam by the sensitive part of vaccinia virus. It is less than the cross-sectional area of the whole virus, which is about $0.23 \times 0.17 = 0.04\mu^2$. Electron micrographs have however shown internal structure in this virus. The micrograph reproduced in Plate IV D[1] shows a centrally placed body of diameter ranging from 0.08 to 0.12μ in different virus particles (and possibly additional smaller bodies), the corresponding area being 0.005 to $0.011\mu^2$. These areas are in fair agreement with the figure $0.0067\mu^2$ deduced from the α-ray inactivation experiments.

It is a plausible conjecture to identify the radiosensitive region, the existence of which is inferred from the radiation experiments, with the internal structure (or structures) revealed by the electron microscope, and to suppose that the genetical apparatus of the virus is contained in this structure. A first speculation would be that a single ionization anywhere within this structure of diameter about 0.1μ causes inactivation of the virus. However, this appears not to be the case. For if the target diameter is calculated on the basis that there is a single spherical target, ionization anywhere within which leads to inactivation of the virus, we find values of $31\,m\mu$ for γ-rays, $41\,m\mu$ for X-rays, and $70\,m\mu$ for α-rays (see Table 36, p. 123). As explained in Chapter III, a systematic increase of calculated target diameter with increase of ion-density of the radiation is evidence that there is not a single spherical target, but either a target which is very far from spherical, or a multiplicity of spherical targets ionization in any one of which leads to inactivation of the virus.

This last (multi-target) model is plausible if we are prepared to regard inactivation of vaccinia virus as a lethal mutation, produced by the ionization of any one of a large number of genes. Since nothing is known, as yet, about the genetical apparatus of viruses, this explanation is necessarily speculative. We shall, however, pursue it to the extent of showing that it affords an explanation of the manner in which the inactivation doses of the different radiations vary with the ion-density of the radiations. It is also possible to estimate the size and number of the genes in

[1] Green, R.H., Anderson, T.F. & Smadel, J.E. (1942). Similar internal structure is shown in electron micrographs obtained by Salaman, M.H. & Preston, G.D. (unpublished).

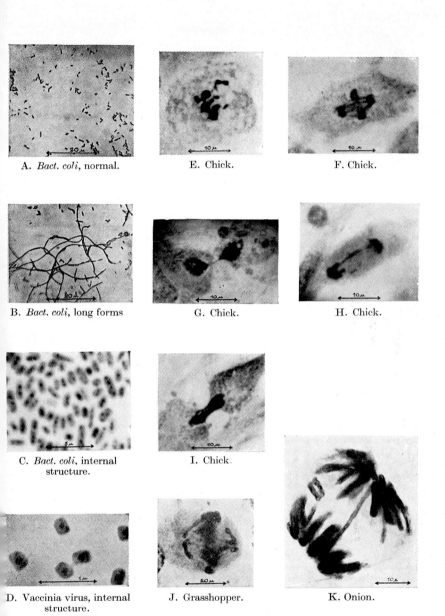

PLATE IV. Bacteria; viruses; abnormal division figures

vaccinia in the same way that estimates can be obtained of the size and number of the genes in *Drosophila* by study of radiation-induced lethal mutations (p. 179).

The calculation is facilitated by the graphs provided in Chapter III, and is made as follows. Experimentally, the inactivation doses of α-rays of 5eMV. and γ-rays are $2 \cdot 11 \times 10^5$ and $0 \cdot 80 \times 10^5$ r., which are in the ratio $2 \cdot 64 : 1$. Interpolating in Fig. 10 (p. 90) between the curves for 3eMV. and 6eMV. α-rays, we deduce that this ratio of inactivation doses corresponds to the target diameter $2r$ and density ρ, satisfying the relation $2r\rho = 8m\mu$. Taking the density to be $1 \cdot 35$ (this being about the density of virus protein) we deduce that the diameter of the target is $6m\mu$.

The inactivation doses of the three radiations to be expected for a single spherical target of diameter $6m\mu$ can be read off from Fig. 8A, and are seen to be 100–120 times greater than the experimental inactivation doses. It follows that there are about

TABLE 73. *Vaccinia virus, calculated and experimental inactivation doses*

Radiation	γ-rays	X-rays (1·5A.)	α-rays (5eMV.)
Experimental	0·80	1·04	2·11
Calculated	0·76	1·14	2·18
	All $\times 10^5$ r.		

110 targets in one virus particle, any one of which ionized leads to inactivation of the virus (since the inactivation dose with N targets is $1/N$ times the inactivation dose with one target (cp. p. 90)). We conclude that in vaccinia there are 110 genes each of diameter (or strictly target diameter) $6m\mu$. The theoretical inactivation doses of the three radiations calculated on this basis[1] are set out in Table 73, and are seen to be in good agreement with the experimental inactivation doses.

[1] Fig. 8A shows that the inactivation dose of γ-rays expected with a single target of $6m\mu$ diameter is $0 \cdot 84 \times 10^7$ r. With 110 targets the expected inactivation dose is $0 \cdot 84 \times 10^7/110 = 0 \cdot 76 \times 10^5$ r. The calculation is made similarly for other radiations. For α-rays of 5eMV. it is necessary to interpolate between the curves given for 3 and 6eMV. The number and size of the genes given here are somewhat different from the number and size inferred by Lea, D.E. & Salaman, M.H. (1942) from the same experimental inactivation doses owing to their use of less accurate physical data and methods of calculation.

Undue weight should not be attached to the estimate of the number of genes, since this is very sensitive to any error in the experimental inactivation doses or in the physical data and method of calculation. We can safely say, however, that if we are correct in interpreting the inactivation of vaccinia virus as a lethal gene mutation, then the number of genes is certainly greater than one, though a great deal smaller than in an organism such as *Drosophila* in which there are believed to be several thousand.

The killing of bacteria

It is convenient to consider the bactericidal action of radiations at this point, since this can also be interpreted as lethal mutation.

Many authors have studied the bactericidal action of X-rays, neutrons, radioactive radiations, and ultra-violet light.[1] The general method is to irradiate the bacteria either dry, or in aqueous suspension, or spread on the surface of a nutrient agar medium. If the last-mentioned procedure is adopted the ob-

[1] X-rays: Lacassagne, A. & Holweck, F. (1928, 1929a,b); Wyckoff, R.W.G. (1930a,b); Ellinger, P. & Gruhn, E. (1930); Claus, W.D. (1933); Levin, B.S. & Lominski, I. (1935); Pugsley, A.T., Oddie, T.H. & Eddy, C.E. (1935); Luria, S.E. (1939); Zirkle, R.E. (1940); Lorenz, K.P. & Henshaw, P.S. (1941); Lea, D.E., Haines, R.B. & Bretscher, E. (1941).

α-rays: Chambers, H. & Russ, S. (1912); Hercik, F. (1933, 1934a,b); Bruynoghe, R. & Mund, W. (1935); Lea, D.E., Haines, R.B. & Coulson, C.A. (1936); Luria, S.E. (1939); Bonét-Maury, P. & Olivier, H.R. (1939); Zirkle, R.E. (1940).

β-rays and cathode rays: Chambers, H. & Russ, S. (1912); Wyckoff, R.W.G. & Rivers, T.M. (1930); Knorr, M. & Ruff, H. (1934); Baker, S.L. (1935); Spencer, R.R. (1935); Lea, D.E., Haines, R.B. & Coulson, C.A. (1936).

γ-rays: Spencer, R.R. (1935); Dozois, K.P., Ward, G.E. & Hachtel, F.W. (1936); Lea, D.E., Haines, R.B. & Coulson, C.A. (1937).

Neutrons: Lea, D.E., Haines, R.B. & Bretscher, E. (1941); Spear, F.G. (1944).

Ultra-violet light: Coblentz, W.W. & Fulton, H.R. (1924); Baker, S.L. & Nanavutty, S.H. (1929); Gates, F.L. (1929a,b, 1930); Ehrismann, O. & Noethling, W. (1932); Wyckoff, R.W.G. (1932); Dushkin, M.A. & Bachem, A. (1933); Duggar, B.M. & Hollaender, A. (1934a,b); Dreyer, G. & Campbell-Renton, M.L. (1936); Hollaender, A. & Claus, W.D. (1936); Hercik, F. (1936); Koller, L.R. (1939); Lea, D.E. & Haines, R.B. (1940); Wells, W.F. (1940); Rentschler, H.C., Nagy, R. & Mouromseff, G. (1941).

servation consists simply in incubating the plate for a few hours until the colonies reach a size allowing them to be counted by naked eye or under low magnification, and comparing the number of colonies on the irradiated area with the number of colonies on an equal unirradiated area. If the bacteria have been irradiated in the absence of nutriment, suitable dilutions are inoculated onto nutrient agar plates after irradiation, and incubated until the colonies are countable by naked eye. The criterion of death in these experiments is thus inability to give rise to a colony visible to the naked eye or under the low power of the microscope. The usual method of reporting the result of an experiment is to plot a survival curve showing the fraction of the organisms capable of giving rise to colonies as a function of the dose.

It has been shown that the fraction of the organisms in an aqueous suspension which are killed by a given dose of radiation is independent of the concentration of the organisms in the suspension,[1] indicating that the death of a bacterium is due to energy dissipated by the radiation in the bacterium itself, and is not an indirect action due to the dissipation of energy in the water.

Some caution is required in the use of the method in which bacteria are irradiated while spread on the surface of nutrient agar, since it has been found that large doses of radiation produce poisons in the nutrient medium which kill bacteria sown on it.[2]

In quantitative experiments precautions should be taken to ensure that the organisms in a suspension are thoroughly dispersed before being plated. For if the organisms are clumped, each colony will be produced not from a single organism but from a clump of organisms. A dose of radiation which renders half the organisms incapable of colony formation will not, under these circumstances, reduce the number of colonies observed to half, since a clump of several organisms half of which are killed will still give rise to a visible colony. To avoid trouble of this sort the aqueous suspension should be shaken mechanically for about

---

[1] Lea, D.E. & Haines, R.B. (unpublished). *Bact. coli* was irradiated by γ-rays in aqueous suspensions of concentrations from 10^4 to 10^9 organisms per ml.

[2] Pugsley, A.T., Oddie, T.H. & Eddy, C.E. (1935), using X-rays; Coblentz, W.W. & Fulton, H.R. (1924), using ultra-violet light.

half an hour before use, and organisms such as *Staphylococci* having a strong tendency to clump are best avoided.

Most workers studying the bactericidal action of radiations have obtained exponential survival curves, indicating that the

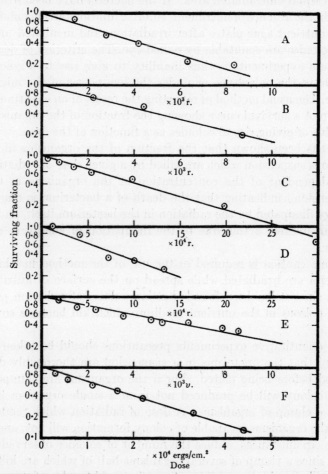

FIG. 55. Exponential survival curves of irradiated bacteria. A, X-rays (1·5A.) on *S. aertryke* (Wyckoff). B, α-rays on *Bact. coli* (Zirkle). C, β-rays on spores of *B. mesentericus* (Lea, Haines & Coulson). D, γ-rays on spores of *B. megatherium* (Lea, Haines & Coulson). E, neutrons on *Bact. coli* (Lea, Haines & Bretscher). F, ultra-violet light (2803A.) on *B. megatherium* (Hercik).

fraction of the organisms killed by a given increment of dose remains constant throughout the irradiation. When the surviving fraction is plotted on a logarithmic scale against the dose, a straight line is obtained.[1] Examples of exponential survival curves obtained with various organisms and radiations are plotted in this manner in Figs. 55 and 57.

Occasionally survival curves deviating systematically from the exponential shape have been reported. The deviation takes the form of the fraction of organisms killed by a given increment of dose being less at small doses than at large doses, so that a *sigmoid* survival curve is obtained. A typical sigmoid survival curve, plotted on linear and logarithmic scales respectively, is shown in Fig. 56, curves c and d, curves a and b being exponential survival curves.

It is probable that exponential survival is the typical result, and that the occasional finding of a sigmoid curve is due to some disturbing factor. One such disturbing factor is clumping of the organisms. If the proportion of organisms which survive a given dose is the exponential function e^{-x}, where x is proportional to the dose, then the probability that an individual organism shall be killed by this dose is $1-e^{-x}$. If the organisms are sown in clumps of n individuals, the probability of all n organisms of a clump being killed will be $(1-e^{-x})^n$. Hence the proportion of the clumps which produce colonies after a dose proportional to x will be $1-(1-e^{-x})^n$. This function represents, not an exponential but a sigmoid survival curve. Curves c and d of Fig. 56 have been calculated by this formula with $n=4$, and illustrate the survival curves to be expected when the organisms are in clumps of 4. Curves a and b of Fig. 56 are the survival curves e^{-x}, i.e. the survival curves to be expected for the same organisms when not clumped. Sigmoid survival curves which have been obtained with organisms such as *Staphylococci* having a strong tendency to clump are probably to be explained in this fashion.[2]

A second disturbing factor resulting in a sigmoid survival curve being obtained may be met with when the organisms are irradiated while spread on the surface of a nutrient medium in which toxic substances are produced by the radiation. At small doses the concentrations of the poisons will be insufficient to kill

[1] See Chapter III, p. 72, for a fuller discussion of methods of plotting survival curves, and the implications of exponential survival.

[2] Gates, F.L. (1929a) obtained sigmoid survival curves with *Staph. aureus* irradiated by ultra-violet light.

any bacteria, and the rate of death observed will be that due to the direct action of the radiation on the bacteria. With larger doses the concentration of poisons may be large enough to cause appreciable mortality, and the rate of death will then be greater than the rate due to the direct action of the radiation alone.[1]

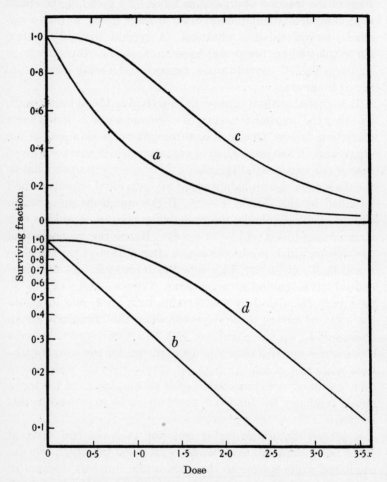

FIG. 56. Survival curves (theoretical). a and b, exponential survival curves e^{-x}. c and d, sigmoid survival curves $1-(1-e^{-x})^4$.

[1] This may be the explanation of the distinctly sigmoid survival curve obtained by Lacassagne, A. & Holweck, F. (1929a) irradiating '*Pyocyanique S*' with X-rays of wave-length 8·3 A. Lea, D.E., Haines, R.B. & Coulson, C.A. (unpublished) irradiating the same organisms with the

The most convenient way of summarizing the result of an experiment in which an exponential survival curve is obtained is to state the dose which reduces the surviving fraction to $e^{-1} = 37\%$.[1] This dose, which in the case of virus and enzyme inactivation we referred to as the inactivation dose, is more appropriately referred to as the *mean lethal dose* when bacteria are concerned.

Table 74 summarizes the results of experiments which show that the mean lethal dose is the same, within the error of the

TABLE 74. Independence of mean lethal dose on intensity [2]

Radiation	Organism	Intensity r./min.	M.L.D. r.	Reference
α-rays	Spores of *B. mesentericus*	9.84×10^3 6.12×10^4	2.3×10^4 2.6×10^4	1
X-rays (8 A.)	Spores of *B. mesentericus*	6.24×10^3 6.02×10^4 4.70×10^5	1.5×10^5 1.2×10^5 1.7×10^5	1
X-rays (0·15 A.)	*Bact. coli*	65 209	5.6×10^3 5.7×10^3	2
		ergs/cm.²/sec.	ergs/cm.²	
Ultra-violet light (2536 A.)	*Bact. coli*	1.2×10^2 3.1×10^3 6.4×10^4	7.5×10^3 8.5×10^3 8.5×10^3	3

1 Lea, D.E., Haines, R.B. & Bretscher, E. (1941).
2 Lea, D.E. & Haines, R.B. (unpublished).
3 Lea, D.E. & Haines, R.B. (1940). The exposure time required to deliver the mean lethal dose in these experiments ranged from about 1 min. to about 0·1 sec. Rentschler, H.C., Nagy, R. & Mouromseff, G. (1941) have shown that the mean lethal dose is still unchanged when it is delivered in a time of the order of 10^{-6} sec.

experiment, whether the irradiation is made at a low intensity and spread over a prolonged time, or at a high intensity and concentrated in a short time.

The effect of a given dose has also been found to be independent of the temperature at which the bacteria are maintained during the irradiation. Table 75 shows[3] the fractions of spores of *B. mesentericus* which survive a constant dose of 2×10^4 r. of α-rays given at various temperatures. The mean lethal dose has

same wave-length of X-rays obtained an exponential survival curve, the fraction of the organisms surviving a given dose being greater than in Lacassagne and Holweck's experiment, especially at large doses.

1 Cp. Chapter III, p. 74.
2 See also Table 86, p. 374, for additional data.
3 Lea, D.E., Haines, R.B. & Coulson, C.A. (1936).

also been found to be independent of the temperature when spores of *B. mesentericus* are irradiated by β-rays[1] at 41, 20 and $-20°$; when *Bact. coli* is irradiated by γ-rays[2] at 0 and 37°; when *B. prodigiosus* is irradiated by α-rays[3] at various temperatures from 2·5 to 36°; and when *Bact. coli* is irradiated by ultra-violet light[4] at 5 and 37°.

TABLE 75. Independence of temperature

(α-rays, 2×10^4 r., on spores of *B. mesentericus*.)

Temperature °C.	Fraction surviving
+50	0·51 ± 0·06
+20	0·52 ± 0·03
0	0·56 ± 0·03
−20	0·56 ± 0·03

A limited number of researches have been made to compare the efficiencies of different radiations in killing bacteria. It is highly desirable that the experiments with the different radiations should be made by a single group of workers under conditions as nearly as possible alike. Misleading conclusions are likely to be reached if results obtained with the different radiations by workers in different laboratories are collated, even though the organisms they are using are nominally the same.

Lacassagne and Holweck,[5] irradiating '*Pyocyanique* S' with two wave-lengths of soft X-rays, obtained an exponential survival curve with X-rays of 4A., and a sigmoid survival curve with X-rays of 8A. They interpreted the exponential survival curve to mean that a single quantum of the 4A. radiation sufficed to kill the organism. The sigmoid survival curve was interpreted on the multi-hit target theory (cp. Chapter III, p. 71) to mean that four quanta of the soft radiation were needed to kill the organisms. The sigmoid curve has however not been reproduced by other workers irradiating the same organisms with X-rays of the same wave-length,[6] so that this interpretation is probably incorrect.

1 Lea, D.E., Haines, R.B. & Coulson, C.A. (1936).
2 Lea, D.E., Haines, R.B. & Coulson, C.A. (1937).
3 Hercik, F. (1934b).
4 Rentschler, H.C., Nagy, R. & Mouromseff, G. (1941).
5 Lacassagne, A. & Holweck, F. (1929a).
6 See p. 320, footnote 1.

Wyckoff irradiated *Bact. coli* with soft X-rays of several wavelengths, and obtained the survival curves shown in Fig. 57.[1] The mean lethal doses read off from these curves[2] are given in Table 76, and are seen to increase on the whole as the wave-length of

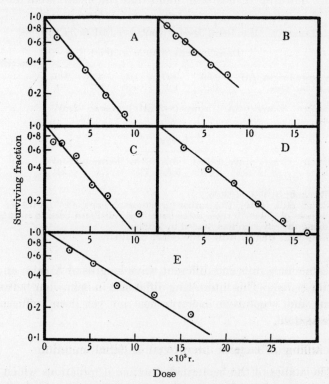

Fig. 57. Killing of *Bact. coli* by soft X-rays (Wyckoff). A, 0·6 A.; B, 0·7 A.; C, 1·5 A.; D, 2·3 A.; E, 4·0 A.

the radiation increases, i.e. as the ion-density of the photo-electrons which dissipate the energy in the bacteria increases. Results obtained by other workers studying the relative efficiency of different radiations in killing *Bact. coli* are also given in Table 76. It is seen that, while the results of the three series of experiments do not agree with each other numerically, in each

[1] Wyckoff, R.W.G. (1930b).
[2] Except in the case of wave-length 0·71 A., where the mean lethal dose given in the table is the mean of two experiments, only one of which is shown in Fig. 57.

series the mean lethal doses increase from left to right in the table, which is the order of increasing ion-density.

A similar increase of mean lethal dose with increase of ion-density has been found with other vegetative bacteria.[1] With spores, however, it has been found that the mean lethal dose is

TABLE 76. Mean lethal doses of various radiations on *Bact. coli*

(Doses in units of 10^3 r. or 10^3 v.)

						Reference
X-ray wave-length (A.)	0·56	0·71	1·5	2·3	4·0	1
Mean lethal dose	4·2	4·6	4·3	6·7	8·4	
Radiation	X-rays (0·3A.)	α-rays (∼5eMV.)	α-rays (∼2eMV.)			2
M.L.D.	3·9	5·7	6·4			

			X-rays					
Radiation	β-rays	γ-rays	0·15A.	1·5A.	8·3A.	Neutrons	α-rays	3
M.L.D.	4	5·2	6·0	6·5	7·5	7·1	24	

1 Wyckoff, R.W.G. (1930b).
2 Zirkle, R.E. (1940). This author gives doses in units of eV./μ^3 tissue. For the purpose of Table 76 these doses have been calculated back to roentgens, employing the physical data quoted by Zirkle.
3 Lea, D.E., Haines, R.B. & Bretscher, E. (1941).

the same for γ-rays and different wave-lengths of X-rays, and is less for α-rays.[2] This interesting difference in behaviour between spores and vegetative bacteria has not yet been sufficiently investigated.

The killing of bacteria interpreted as lethal mutation

The studies of the bactericidal action of radiations which we have reviewed in the preceding section have established the following results:

(i) The survival curves are exponential.

(ii) The mean lethal dose is independent of the intensity and of the temperature at which the irradiation is made.

(iii) In the case of vegetative bacteria, the mean lethal doses of different radiations increase in the order of increasing ion-density.

1 Lea, D.E., Haines, R.B. & Coulson, C.A. (1936), using *S. aureus*; Lea, D.E., Haines, R.B. & Coulson, C.A. (unpublished), using *B. pyocyaneus* and *B. proteus*.
2 Lea, D.E., Haines, R.B. & Coulson, C.A. (1936), Lea, D.E., Haines, R.B. & Bretscher, E. (1941), using spores of *B. mesentericus*.

These results parallel those obtained in the study of the inactivation of viruses and the production of mutations in *Drosophila*, and suggest (cp. Chapter III) that a single ionization (in the right place) suffices to kill a bacterium. This conclusion is only plausible biologically if we suppose the effect to be on the genetical apparatus of the bacterium. For the typical effect of one ionization is chemical change in one molecule, and it is unlikely that the concentration of any cytoplasmic constituent is so nicely balanced that change in a single molecule of it leads to the death of the bacterium. On the other hand, in organisms which have been investigated genetically, lethal effects due to losses of, or changes in, single genes are common.[1]

The idea that the killing of bacteria by radiation is to be interpreted as lethal mutation is on other grounds plausible. In the researches on which is based the conclusion that death is due to a single ionization, the criterion of death has invariably been the inability of the irradiated bacterium to give rise to a colony visible to the naked eye or under low magnification. The few microscopical studies which have been made of the behaviour of individual bacteria after irradiation have shown that a bacterium exposed to a dose sufficient to prevent its giving rise to a colony does not suffer rapid disintegration, as happens for example when bacteria are killed by strong chemical disinfectants. A motile bacterium retains its motility,[2] and though failing to give rise to a visible colony, a bacterium which has been 'killed' by radiation may divide once or twice when sown onto a nutrient medium.[3,4] A spore which has received a dose large enough to prevent its giving rise to a visible colony will nevertheless germinate when sown onto a nutrient medium.[4] These results suggest that, as is often the case with higher cells, the death of an irradiated bacterium occurs at or following division. The cause may very well be an effect of radiation on the genetical apparatus. We shall continue the discussion on the assumption that the effect in question is of the nature of a lethal gene mutation. Since practically nothing is known about the genetical apparatus of bacteria, this standpoint is provisional. We adopt

[1] At any rate in *Drosophila*, where losses of single genes usually behave as recessive lethals.
[2] Bruynoghe, R. & Mund, W. (1935). [3] Luria, S.E. (1939).
[4] Robinow, C.F. & Lea, D.E. (unpublished).

it since a good representation of the radiation results can be given on the assumption that we are dealing with lethal gene mutation.

According to the bottom line of Table 76, the inactivation dose of *Bact. coli* irradiated by α-rays of about 5 eMV. is $2 \cdot 4 \times 10^4$ r. With α-rays of this energy, $0 \cdot 71$ α-particles cross each square micron of irradiated tissue per 1000 r. (Table 18, p. 32). Thus with a dose of $2 \cdot 4 \times 10^4$ r. an average of 1 α-particle crosses each area of $1/(24 \times 0 \cdot 71) = 0 \cdot 06 \mu^2$. Now while the ordinary bacteriological staining methods do not show internal structure, by the use of suitable methods internal structures have been demonstrated in bacteria. Plate IV c shows a micrograph of *Bact. coli*[1] taken from an 18 hr. culture such as was used in the radiation experiments. Internal structures are clearly visible having a diameter of about $0 \cdot 3 \mu$ and an area therefore of about $0 \cdot 07 \mu^2$. In view of this coincidence of areas, it is plausible to suppose that for an α-particle to produce a lethal mutation it is necessary for it to pass through this structure. The internal structures are believed on observational grounds to be chromosomes.[2]

Further calculation proceeds exactly as in the case of vaccinia virus which was discussed earlier (p. 315). It is apparent to begin with that the hypothesis that a single ionization *anywhere* in the body of diameter $0 \cdot 3 \mu$ causes lethal mutation is not satisfactory. For, if we make the assumption that there is a single spherical target ionization anywhere within which causes lethal mutation, we obtain a target diameter which increases with increase of iondensity from $0 \cdot 08 \mu$ for γ-rays to $0 \cdot 2 \mu$ for α-rays. We can however obtain a consistent representation of the data by assuming that there are a large number of genes, ionization of any one of which causes lethal mutation. By proceeding as described in the case of vaccinia virus it is found that the assumption of 250 genes having an average diameter of $12 \, m\mu$ satisfactorily fits the experimental data. The comparison of experimental mean lethal doses with the values calculated[3] on this assumption is made in Table 77.

[1] Robinow, C.F.
[2] Badian, J. (1933); Robinow, C.F. (1942, 1944).
[3] Somewhat different estimates of the size and number of the genes were obtained by Lea, D.E., Haines, R.B. & Bretscher, E. (1941) analysing the same experimental data, owing to the use of less accurate physical data and methods of calculation.

LETHAL MUTATIONS IN BACTERIA

Studies of the killing of bacteria by monochromatic ultra-violet light have shown that the survival curves are exponential and that the effect of a given dose is independent of the intensity.[1] These results suggest that, as with ionizing radiations,

TABLE 77. *Bact. coli*: calculated and experimental mean lethal doses

Radiation	β-rays	γ-rays	X-rays			Neutrons	α-rays
			0·15A.	1·5A.	8·3A.		
Experimental	4	5·2	6·0	6·5	7·5	7·1	24
Calculated	4·4	4·4	5·7	7·8	17·6	8·8	23

All $\times 10^3$ r. or $10^3 v$.

the effect is caused by a single unit action, which with ultra-violet light would mean the absorption of a single quantum. No third test, corresponding to the ion-density variation test with ionizing radiations, is applicable with ultra-violet light, so that the interpretation that the lethal effect of ultra-violet light is due to the absorption of a single quantum (in a particular region) is less securely established than is the interpretation that the lethal effect of an ionizing radiation is due to the production of a single ionization.

Quantitative experiments usually give the incident energy in ergs/cm.² necessary to reduce the surviving fraction to a given fraction, conveniently chosen to be 37 %. The absorption coefficient of bacterial protoplasm being known,[2] one can deduce the mean lethal dose in units of ergs per gram. Doses of ionizing radiations measured in roentgens can be converted to ergs per gram by means of Table 2 (p. 8). In this way it is found[3] that about a hundred times as much energy (in ergs) is dissipated in a bacterium killed by ultra-violet light as in a bacterium killed by X-rays. If one supposes the lethal effect to be a gene mutation, the quantum yield (i.e. the probability that a quantum absorbed in the gene shall cause lethal mutation) is very much less than the ionic yield. A similar conclusion was reached in discussing the inactivation of viruses (p. 125).

[1] E.g. Wyckoff, R.W.G. (1932); Lea, D.E. & Haines, R.B. (1940).
[2] Gates, F.L. (1930), quoted in Table 1, p. 5.
[3] See Lea, D.E. & Haines, R.B. (1940) for details.

LETHAL CHROMOSOME STRUCTURAL CHANGES

Incidental reference to types of chromosome structural change which are probably lethal has been made in Chapters V and VI. The principal types likely to have a lethal effect are simple breaks (Figs. 30 A and 31 A, B), and asymmetrical exchanges (Figs. 30 D, E, G and 31 D, F). All these types result in the production of acentric fragments, which will be lost at division sooner or later owing to their lacking the centromere which ordinarily ensures their inclusion in a daughter nucleus. Different organisms are tolerant of the resulting genetic deficiencies in different degree. In *Drosophila melanogaster*, the loss of even 5 % of the X chromosome has a dominant lethal effect, while in maize the loss of even a whole chromosome may be viable.

Most of the types of structural change mentioned above result also in the production of dicentric chromosomes and chromatids. Dicentrics probably often have a lethal effect apart from their association with acentric fragments. This may be explained partly by the genetic unbalance resulting from unequal breakage of dicentric chromosomes which form bridges at anaphase, and partly by the mechanical difficulties experienced by a dividing cell in which an anaphase bridge forms (cp. Chapter V, p. 163). Bridge formation is however not lethal in all materials, since the bridge may break, and the daughter cells may then survive if the genetic unbalance is not too severe.[1]

Simple breaks which do not restitute, and asymmetrical exchanges (other than minute interstitial deletions which behave as recessive lethals rather than dominant lethals) are certainly lethal in *Drosophila* since evidence of them is never found in the salivary chromosomes. In plant material, which is more tolerant of genetic unbalance, a certain number of these aberrations may be viable though probably most are not. Symmetrical exchanges are viable in all organisms investigated, but they result in partial sterility (see p. 338).

Lethal types of chromosome structural change will lead to the

[1] If the triploid zygote from which the endosperm tissue of maize develops contains a broken chromosome, sister-union of chromatids occurs at the breakage point and an anaphase bridge results. This bridge breaks, and in each daughter cell sister-union occurs at the breakage ends resulting in bridges at the next division. The process of bridge formation, breakage, and sister-union thus continues in successive cell generations in the endosperm (McClintock, B. 1941*a*).

death of the cells concerned at or following division. It has already been pointed out that it is at this time that cells die which are killed by moderate doses of radiation. It is evident therefore that lethal chromosome structural changes must account for some of the cells which die at or following division, but it is not yet possible to say in general whether the majority, or only a minority, are to be accounted for in this way. One can attempt to answer the question in the case of the few materials, particularly *Drosophila* sperm and *Tradescantia* microspores, in which chromosome structural changes have been sufficiently investigated.

Drosophila melanogaster

Changes in *Drosophila* sperm which result in the egg fertilized by an irradiated sperm failing to develop to the adult stage have been discussed in detail in Chapter v under the heading Dominant Lethals (p. 161). The conclusion was reached that the whole lethal effect could be explained as being due to lethal types of chromosome structural change. The demonstration, though strongly suggestive, was not completely convincing since in *Drosophila* only viable types of structural change have been studied. In particular, the proportion of breakage ends which undergo sister-union in preference to joining with other breakage ends has to be determined from experiments on dominant lethals and sex-ratio distortion, since it cannot be determined from a study of salivary chromosomes. It is probable that the whole dominant lethal effect obtained by irradiating *Drosophila* sperm can be explained in terms of chromosome structural change, but the evidence is at present circumstantial.

When *Drosophila* eggs are irradiated before fertilization, a higher proportion of the zygotes fail to develop into adult flies than when it is the sperm which are irradiated by the same dose.[1] Yet many fewer chromosome interchanges are produced by irradiating unfertilized eggs than by irradiating sperm[2] with the same dose.

It has been remarked[3] that this observation shows that while it may be correct to ascribe the whole dominant lethal effect obtained by irradiating *Drosophila* sperm to chromosome structural change, some of the dominant lethal effect produced in

[1] See, e.g., Sonnenblick, B.P. (1940).
[2] See, e.g., Glass, H.B. (1940).   [3] Muller, H.J. (1938).

unfertilized eggs must be due to some other cause (e.g. an effect on the cytoplasm). This conclusion, though superficially the obvious deduction from the data, does not follow if chromosome structural changes and dominant lethals are interpreted as in Chapter v (p. 166). Two constants α and q enter into the expression for the yields of chromosome structural change and dominant lethals. α is the number of breaks primarily produced per nucleus per 1000 r. It is probably determined mainly by the dimensions of the chromosome thread and is therefore probably the same in the egg and the sperm. q is the probability that a breakage end primarily produced shall join with another breakage end in preference to remaining unjoined until the time of chromosome split, and then undergoing sister-union. q will be determined by such factors as the time of split and the freedom of movement of the chromosomes, and may very well be different in the egg and sperm nuclei.

Taking α to be the same for egg and sperm nuclei, the experimental fact that many fewer viable chromosome structural changes are produced in the egg than in the sperm by a given dose D means that q is much smaller in the egg than in the sperm, in which it has the value 0·76.[1]

The expression for the proportion of the fertilized eggs which develop to the adult stage when one of the gametes has been given a dose D is given by equation V-2 (p. 167) as $e^{-\alpha D} S_2$. $e^{-\alpha D}$ depends on α but not on q, and is less than unity. S_2 depends on q as well as α, takes the value unity when $q=0$, and is greater than unity when q is greater than 0. Experimentally the proportion of fertilized eggs which hatch when the egg has been irradiated diminishes approximately exponentially with the dose.[2] Evidently q for the egg, which we have already deduced is smaller than q for the sperm, is sufficiently nearly zero for S_2 to be practically unity.[3]

[1] More precisely, in terms of the theory of Chapter v, the proportion of viable gametes having structural change is $1-S_1/S_2$ (p. 167), and as is seen in Table 42, this quantity increases with increase of $\alpha q D$. α and D being the same in egg and sperm, q must therefore be smaller in the egg.

[2] Sonnenblick, B.P. (1940).

[3] The assumption that q is nearly zero is equivalent to assuming that nearly all the breaks primarily produced undergo sister-union, and therefore behave as lethals, in preference to restituting or exchanging with other breaks.

Thus when it is the unfertilized egg which is irradiated, the proportion of fertilized eggs which develop to the adult stage is $e^{-\alpha D}$. This is smaller than when it is the sperm which is irradiated on account of the omission of the term S_2.

Making this argument quantitative, it can readily be shown that a dose which results in 50 % of dominant lethals when given to the sperm will result in 86 % of dominant lethals when given to the fertilized egg.[1] This deduction may be compared with the experimental results of Sonnenblick, who found that a dose which gave 40 % of dominant lethals when delivered to the sperm gave 82 % of dominant lethals when delivered to the unfertilized egg, while a dose which gave 56 % of dominant lethals when delivered to the sperm gave 94 % of dominant lethals when delivered to the egg.

It is evident that dominant lethals in unfertilized eggs, as well as in sperm, can be explained by lethal types of chromosome structural change.

Several authors[2] have studied the lethal action of X-rays on fertilized eggs of *Drosophila*. The general method employed is to provide the flies with food on which to deposit their eggs, to collect a batch of eggs, and to irradiate them at a time after laying which varies in different experiments from a few minutes to a few hours. The proportion of eggs which hatch is plotted as a function of the dose, sigmoid survival curves being obtained, and the dose required to reduce the proportion hatching to 50 % is determined. The sensitivity of the eggs to radiation varies with the stage of development, the 50 % dose having a minimum value of 160 r. 90–120 min. after laying.[3] The 50 % dose for a batch of eggs irradiated within 30 min. of laying when usually not more than one or two nuclear divisions have occurred, is 290 r.

Direct cytological observations have not been made to determine whether chromosome structural changes account for the

[1] Taking $q = 0.76$ for sperm, 50 % survival is obtained at a dose which makes $\alpha D = 1.95$, whence $\alpha q D = 1.48$. With these values of αD and $\alpha q D$, $e^{-\alpha D} = 0.142$ and $S_2 = 3.5$ (interpolating in Table 42 for S_2). For unfertilized eggs irradiated by the same dose we take $\alpha D = 1.95$ as for sperm. We take $q \sim 0$ so that $S_2 = 1$. Thus the survival is now $e^{-\alpha D} = 0.142$, which means 86 % of dominant lethals.

[2] Packard, C. (1926) and later papers; Henshaw, P.S. & Henshaw, C.T. (1933) and later papers.

[3] Packard, C. (1935).

lethal effect. The best one can do therefore at present is to attempt to make an estimate of the proportion of nuclei likely to suffer lethal chromosome structural change on the basis of the data available for sperm and unfertilized eggs. Presumably, for a diploid nucleus, α, the number of breaks primarily produced per 1000 r. will have a value twice as great as that for a haploid gamete, i.e. $\alpha = 1.5$ for a diploid nucleus. No data are available from which to estimate the value of q. If we assume it to be small, as appears to be the case in the unfertilized egg, the proportion of irradiated diploid nuclei which survive a dose D will be $e^{-\alpha D}$. With a dose of 290 r., $\alpha D = 0.29 \times 1.5 = 0.435$ and $e^{-\alpha D} = 0.65$. Thus the dose of 290 r., which is observed experimentally to kill half of the eggs irradiated at an early stage of development, is sufficient to produce lethal chromosome structural changes in about one-third of the nuclei in an egg. This calculation suggests that lethal chromosome structural changes may play an important part in accounting for the killing of *Drosophila* eggs in an early stage of development.

To sum up our discussion of *Drosophila*, we may say that there are strong indications that lethal chromosome structural changes play a major part in accounting for the failure of eggs to hatch after the irradiation of the sperm or egg prior to fertilization, or of the egg soon after fertilization. The evidence is, however, at present circumstantial.

Tradescantia

Chromosome structural changes visible at the metaphase of the first haploid mitosis (pollen grain division) in *Tradescantia* following irradiation of the microspore prior to this division have been extensively investigated.[1] The subsequent behaviour of the pollen grain has also been studied.[2]

At the pollen grain mitosis (in unirradiated material) nuclear division but not cell division occurs, so that the two daughter nuclei remain in the same cell. Of the two nuclei one, the *generative nucleus*, will divide again after the germination of the pollen grain and give rise to the *sperm nuclei*, which take part in fertilization. The other nucleus, the *vegetative nucleus*, plays no role after germination has occurred.

[1] See Chapters VI and VII.   [2] Koller, P.C. (1943).

During the week which follows the pollen grain mitosis, the two nuclei become distinguishable, the generative nucleus developing into a long crescent-shaped body.

When at the end of this time the mature pollen grain falls onto the stigma of a plant or a smear of artificial medium on a microscope slide, germination occurs. A *pollen tube* grows out of the pollen grain, the elongated generative nucleus enters the pollen tube, and undergoes division (the second haploid mitosis) there.

If the microspore has been irradiated prior to the pollen grain mitosis, after this division small nuclei (*micronuclei*) additional to the generative and vegetative nuclei may be seen. These are acentric chromosome fragments which have failed to be included in either daughter nucleus at the pollen grain mitosis.

TABLE 78. Correlation of the presence of micronuclei with failure to develop after irradiation [1]

		Micronuclei present	Micronuclei absent
Pollen-grain differentiation:	Normal	60	165
	Suppressed	217	15
Pollen tube:	Normal	34	92
	Absent	330	100

A microspore which has received a dose of a few hundred roentgens during early prophase of the pollen grain division, or the interphase preceding, completes this division. In some cells subsequent development proceeds normally, in others the generative nucleus may fail to differentiate and the pollen grain fail to germinate, and so die without having been able to carry out its function of fertilization.

It has been found that there is a marked correlation between failure to develop normally and the existence of micronuclei, i.e. of acentric chromosome fragments. This is illustrated in Table 78, which shows that of the pollen grains which fail to develop pollen tubes, or in which the generative nucleus fails to differentiate, the majority have micronuclei, while of those which develop normally the majority are without micronuclei.

These observations on *Tradescantia* give strong support for the belief that chromosome structural changes resulting in acentric fragments which fail to be included in the daughter nuclei are often lethal.

[1] Koller, P.C. (1943). A dose of 360 r. was given 8 days before dehiscing.

A further demonstration of the relation between chromosome structural changes and lethal effect is provided by Fig. 58. The curve shows the proportion of pollen grain metaphases showing lethal types of chromosome structural change (i.e. chromatid

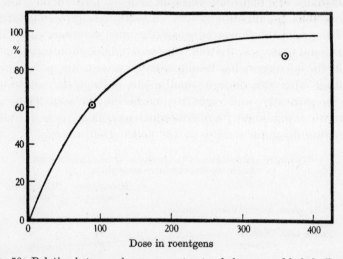

FIG. 58. Relation between chromosome structural changes and lethal effect in *Tradescantia* pollen grains. Curve shows percentage of pollen grain metaphases showing lethal types of chromosome structural change. Experimental points (Koller) show percentage of pollen grains failing to germinate.

breaks, isochromatid breaks, and asymmetrical exchanges) as a function of the dose given 24 hr. before pollen grain metaphase.[1] The points show the proportions of pollen grains which fail to germinate as a result of doses of 90 and 360 r. given at about this same stage.[2]

[1] Authors have not scored aberrations in this particular way. The curve has been constructed in the following fashion, making use of the coefficients of aberration production listed in Table 64, p. 241. The yields of chromatid and isochromatid breaks per cell produced by a dose D roentgens are $0{\cdot}725 \times 10^{-2} D$ and $0{\cdot}271 \times 10^{-2} D$ respectively. The yield of chromatid exchanges produced by dose D is $1{\cdot}81 \times 10^{-5} GD^2$ (for the factor G see equation (VII-8), p. 264), and of these half are asymmetrical. Thus the mean number of lethal aberrations induced per cell by dose D is $m = 0{\cdot}996 \times 10^{-2} D + 0{\cdot}905 \times 10^{-5} D^2 G$. The proportion of cells having one or more lethal aberrations is $1 - e^{-m}$ with this value of m.

[2] Koller, P.C. (1943). Koller's figures for irradiation made 9 days before dehiscing have been used. The proportions of pollen grains failing to germinate are somewhat higher if the irradiation is made 10 days before dehiscing, somewhat lower if made 8 days before dehiscing.

There is fair agreement between the points and the curve, showing that chromosome structural changes are able to account for the failure of pollen grains to germinate. The fact that at high doses the proportion of pollen grains failing to germinate is not so high as the proportion of pollen grain metaphases having lethal types of chromosome structural changes suggests that some of the 'lethal' types of chromosome structural change do not prevent pollen grain germination. These may be the smaller deficiencies; or it may be that only half of the chromatid breaks are lethal, namely, those which cause deficiencies in the vegetative nucleus.[1]

The bean root

The killing of bean roots by the irradiation of the region near the root-tip in which the cells are rapidly dividing has been extensively studied.[2]

In the unirradiated root-tip, half of the daughter cells formed in the dividing region remain there and undergo further division. The remainder do not divide further but elongate. It is the addition of these elongated cells to the root behind the dividing region which accounts for the increase of length of the root.

If the root-tip is irradiated, the rate of increase of length of the root diminishes. This is due to some of the cells in the dividing region being killed. As in other rapidly dividing tissues irradiated by moderate doses, the cells break down at division. After small doses the rate of increase of length of the root eventually returns to normal, indicating that a sufficient number of cells in the dividing region have survived to restore (by division) the normal cell population in this region. With larger doses the growth of the root is not resumed, and it dies, presumably indicating that nearly all the cells in the dividing region have been killed.

There are a number of indirect lines of evidence suggesting that the killing of bean roots by radiation is largely to be ascribed to lethal types of chromosome structural change. The doses of γ-rays, X-rays, neutrons, and α-rays needed to kill 50 %

[1] Making the assumption suggested by Koller that it is the vegetative nucleus and not the generative nucleus which conditions the response between the stigma or culture medium and pollen grain, and initiates germination.

[2] Mottram, J.C. (1933 a,b, 1935 a,b); Gray, L.H. & Read, J. (1942 a, b, c, d); Gray, L.H., Read, J. & Poynter, M. (1943).

of the bean roots irradiated have been found to be 651, 435, 75 and 72 energy-units respectively, the ratios of efficiencies of the four radiations for equal ionizations in the tissue thus being $1\cdot0 : 1\cdot5 : 8\cdot7 : 9\cdot0$.[1] This is the same sequence of efficiencies as that found for the production of lethal types of chromosome structural change in *Tradescantia* microspores, in which the relative efficiencies are $1\cdot0 : 1\cdot1 : 2\cdot8 : 3\cdot8$.[2] The fact that the sequence of efficiencies is the same suggests that the mechanism involved may be the same in the two cases, while numerical differences may reasonably be expected for chromosomes of different size.

It has also been found that lowering the temperature increases the proportion of bean roots killed by a given dose.[3] It is known that in *Tradescantia* the number of cells showing chromosome structural changes is increased by lowering the temperature (cp. Fig. 32, p. 221). It has also been pointed out that the effect of a given dose on the bean root is reduced by splitting the dose into fractions, with rest intervals between, in a manner which is consistent with the reduction of effect being due to there being a smaller number of chromosome exchanges produced in experiments in which there are long rest periods, owing to breakage ends formed by the first fractional dose having all restituted by the time the second fractional dose is given.[4]

Marshak[5] has determined the proportion of anaphase figures in the bean root-tip which show chromosome fragments or bridges at various times ranging from 3 to 24 hr. after irradiation. The abnormal anaphases at 3 hr. are probably mainly the result of 'physiological' changes in the chromosomes, the abnormal anaphases at the later times are probably mainly chromosome structural changes.[6] The ratios of efficiencies he

[1] Gray, L.H., Read, J. & Poynter, M. (1943).

[2] As before the yield of lethal types of chromosome structural change is taken to be the sum of chromatid breaks, isochromatid breaks, and half the chromatid interchanges at 100 r. The yields for X-rays, neutrons, and α-rays are taken from Table 64, p. 241, and the relative yields for X-rays and γ-rays are taken from Table 63, p. 239.

[3] Mottram, J.C. (1935b).

[4] Gray, L.H. (1944b).

[5] Marshak, A. (1942a). In interpreting his results with neutrons, we have assumed, as usual, that 1 n-unit $= 2\cdot5$ v-units.

[6] The terms 'physiological' change and 'structural' change are used in the sense of Chapter VI, p. 192.

finds for X-rays and neutrons at these later times range from 1 : 3·2 to 1 : 6·7, which are consistent with the ratio 1 : 5·8 found for the lethal effect on bean roots.

Marshak finds that the proportion of anaphase figures which show fragments or bridges can be represented by the formula $1 - e^{-kD}$ where D is the dose of X-rays in roentgens and k takes values ranging from 0·011 at 3 hr. after irradiation to 0·0022 24 hr. after irradiation. With a dose of 435 r., which is the dose killing 50 % of the bean roots in Gray and Read's experiments, the proportion of anaphases which have bridges or fragments is deduced from this formula to range from 99 % 3 hr. after irradiation to 62 % 24 hr. later. The proportion of anaphases with bridges or fragments is thus high enough to account for the death of the root, presuming that the daughter cells from these abnormal anaphases usually die.

Lethal physiological changes in chromosomes

As was mentioned in Chapter VI (pp. 192–7), irradiation of cells already sufficiently advanced in division at the time of irradiation not to suffer the delay in division experienced by cells not so far advanced may result in appearances at metaphase and anaphase suggestive of a surface stickiness of the chromosomes. At metaphase the chromosomes clump together, and at anaphase the separation of sister chromatids is hindered.

Division figures of this sort obtained in chick tissue,[1] in grasshopper neuroblasts,[2] and in onion root-tips,[3] are illustrated in Plate IV E–K. Similar appearances have been described in other tissues.[4]

In severe cases the bridge formed at anaphase may be too strong to break, and division will not then be completed. In other cases where division is completed, the daughter cells may

[1] Lasnitzki, I. (1943 b). Chick tissue in culture was irradiated for 25 min. at 100 r./min. and fixed immediately after the exposure. Cp. also Juul, J. & Kemp, T. (1933).

[2] Carlson, J.G. (1941 b). *Chortophaga* neuroblasts were irradiated at 200 r./min. for 40 min. and fixed 15 min. after exposure.

[3] Marshak, A. & Hudson, J.C. (1937). Onion root-tips were given 50 r. and fixed 3 hr. later.

[4] E.g. Amphibia (Alberti, W. & Politzer, G. 1924); grasshopper spermatocytes (White, M.J.D. 1937); plant microspores (Marquardt, H. 1938; Koller, P.C. 1943); plant root-tips (Pekarek, J. 1927; Marquardt, H. 1938; Sax, K. 1941 a).

be genetically unbalanced owing to the breakage of anaphase bridges. Thus the physiological effect is probably often lethal.

The metaphase and anaphase figures suggesting a surface stickiness of the chromosomes are typically seen soon after irradiation in cells too far advanced in division at the time of irradiation to suffer the temporary inhibition of division. Abnormal anaphases are also seen among the cells which divide some hours later, after the end of the period of inhibited division. These abnormal anaphases are supposed to be due to structural changes rather than to physiological changes in the chromosomes,[1] though the distinction between the two types of abnormal anaphase can only be made in materials favourable for cytological study. In view of the evidence presented in the preceding sections it is likely that cells showing abnormal anaphase configurations will eventually die and be included among the count of degenerate cells.

There is evidence[2] that in animal tissues many of the cells which die while attempting division after the cessation of the period of temporary inhibition do so at an early stage of division, and are recorded as degenerating cells without ever being included in the metaphase and anaphase counts. Such casualties cannot plausibly be accounted for by chromosome structural changes since the genetic unbalance or mechanical difficulties at anaphase presumed to account for the lethal effect of chromosome structural changes cannot take effect until the chromosomes divide. They are not at present understood.

Hereditary partial-sterility

Chromosome structural changes which do not lead to the production of acentric fragments or dicentric chromosomes do not have a lethal effect on cells carrying them. There are neither lagging fragments nor bridges at division, and each cell has a complete complement of genes.

If, as the result of irradiating an adult organism, a symmetrical interchange between two chromosomes is induced in one of the

[1] According to Marquardt, H. (1938), using *Bellevalia romana*; Carlson, J.G. (1941*b*), using grasshopper neuroblasts; Sax, K. (1941*a*), using onion root-tips.

[2] According to the observations of Tansley, K., Spear, F.G. & Glücksmann, A. (1937); Spear, F.G. & Glücksmann, A. (1938); Lasnitzki, I. (1940).

gametes, the F_1 individual resulting from the fusion of this gamete with a normal egg or sperm, though phenotypically normal, is usually partially sterile. The explanation of this phenomenon is made clear by Fig. 59. In this figure the complete

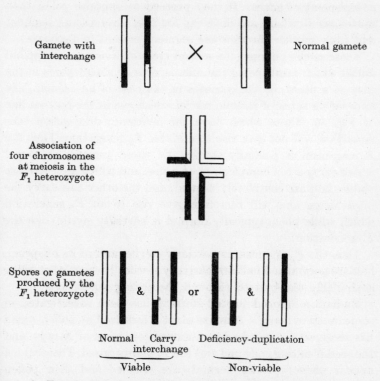

FIG. 59. Hereditary partial-sterility caused by symmetrical chromosome interchange.

chromosome complements of the various gametes and zygotes are not shown, but only the two chromosomes which take part in the interchange.

The haploid gametes, one carrying the interchange and one normal, are represented in the top line. The second line represents the constitution of a (diploid) cell of the F_1 individual. At meiotic prophase pairing normally occurs in twos between the homologous chromosomes derived from male and female parent. When an interchange is present pairing of homologous chromosome segments requires the association of four chromosomes instead of two, as represented in the second line of Fig. 59. At meiotic first anaphase the four chromosomes separate, two going to each pole. If the chromosomes which in the second line of Fig. 59 are drawn diagonally opposite to one another proceed to the same pole, then two nuclei result which each have a full complement of genes. If they proceed to opposite poles, then nuclei result which are deficient for one chromosome segment and have another chromosome segment present in duplicate.

These nuclei undergo the second meiotic division, the haploid nuclei which result being the microspores or macrospores in the case of a plant, or the gametes in the case of an animal. The alternative types of haploid nuclei are shown in the bottom line of Fig. 59. Those which have the deficiency-duplication configuration will not give rise to a viable F_2 generation. Thus the F_1 organism is partially sterile. Of those gametes or spores which carry a full complement of genes, and which are therefore viable, half are completely normal, and the other half carry the interchange and will therefore give rise to an F_2 generation which while phenotypically normal is partially sterile, like the F_1 generation.

Thus the F_1 organism is partially sterile, and of its offspring, half are normal and half are partially sterile. The partial-sterility is similarly transmitted through succeeding generations.

Radiation-induced partial-sterility has been investigated in some detail in maize.[1] A maize plant is fertilized by pollen which has received a dose of a few hundred roentgens of X-rays, and the seed resulting collected and the F_1 plants raised. Those which appear phenotypically normal[2] are selected and their pollen examined. Some of the pollen grains are visibly abnormal, fail to germinate, and are thus incapable of fertilizing. The genetic unbalance has resulted in breakdown, presumably at the microspore division.

In the case of animals where no haploid mitosis takes place

[1] Stadler, L.J. (1930b, 1931). That the explanation of the partial-sterility is chromosome interchange is confirmed by the observation of rings of four chromosomes at diakinesis in the pollen mother cells.

[2] Some plants of stunted growth are obtained in an experiment of this sort owing to simple breaks having been induced in some of the irradiated pollen. The F_1 plants from this pollen are heterozygous for a deficiency. If they flower they will be partially sterile owing to deficiencies being present in half of the spores. This partial-sterility is not transmitted to later generations.

after meiosis, but the products of meiosis are themselves the gametes, a deficiency-duplication does not render a sperm incapable of fertilizing. The zygote however does not give rise to a normal embryo, but death occurs at some stage presumably dependent on the extent of the genetic unbalance.

Partial-sterility has been investigated in mice.[1] A dose of a few hundred roentgens is given to either a male or a female parent and the F_1 offspring are mated to normal animals. Some of these F_1 females, mated to normal males, are found consistently to produce small litters. In the same way, some of the F_1 males mated to normal females are found consistently to father small litters. Investigation shows that the small litters are due to the death of some of the embryos in the uterus. Half of the living mice belonging to these small litters are completely normal, half, while apparently normal, in turn give rise to small litters when mated to normal mice.

Recapitulation and application to rapidly dividing animal tissues

It is probably true to say that a sufficiently large dose of radiation given to any cell at any stage will cause its immediate death. Such effects, produced by doses of radiation which cause serious amounts of chemical change in all parts of the cell, are of no great theoretical interest or practical importance. Much more significant is the fact that it has been found in a great variety of cells that moderate doses of radiation, of the order of a few hundred or a few thousand roentgens, which must produce a rather slight amount of chemical change in the cell as a whole, are able to cause the death of a cell, not immediately, but only at or following nuclear division.

This behaviour suggests that the killing of cells by moderate doses of radiation may be due to an effect on the genes or chromosomes.

In the case of the vegetative bacteria and a large virus (vaccinia) the proportion of the organisms which are killed

---

[1] By Snell, G.D. (1935, 1939, 1941); Snell, G.D. & Ames, F.B. (1939). That the cause of the partial-sterility is a chromosome interchange has been confirmed cytologically by Koller, P.C. & Auerbach, C.A. (1941), who observed the association of a ring or chain of four chromosomes at spermatogenesis.

depends upon the conditions of the irradiation (dose, intensity, temperature, and ion-density) in precisely the same manner as does the proportion of irradiated *Drosophila* sperm in which gene mutation is induced. We have therefore interpreted the killing of bacteria and of large viruses as lethal gene mutation. This interpretation is provisional since at present practically nothing is known about the genetics of bacteria and viruses. It seems, however, to be a plausible interpretation since not only can the radiation results be understood on this basis, but also the number of genes per cell inferred from the radiation experiments is smaller for a bacterium than for *Drosophila*, and smaller for vaccinia virus than for a bacterium. If the interpretation of lethal action as lethal mutation is extended also to the small crystallizable viruses, the radiation results indicate that these viruses are single naked genes. Thus the gene number per cell needed to explain the lethal action of radiation as lethal mutation increases from viruses to bacteria, and from bacteria to *Drosophila*, as would be expected on general grounds.

There is no evidence to suggest that gene mutation plays a large part in the killing of organisms other than bacteria or viruses. In *Drosophila*, recessive lethal gene mutations are produced by irradiation in the eggs and sperm and presumably also in diploid cells. But recessive lethal mutation in a diploid cell will not be lethal unless it is in the X chromosome of the male, owing to the presence of a normal allelomorph in the same cell.

In those organisms in which chromosome structural changes, as well as lethal actions, have been investigated, namely *Drosophila* sperm and eggs, *Tradescantia* pollen, and bean root-tips, fairly strong evidence, though at present circumstantial evidence, has been presented for the view that the main cause of the lethal effect is the production of types of chromosome structural change which lead to bridges at division or genetic unbalance after division.

The killing of the cells of rapidly dividing animal tissues by radiation is of practical importance in view of the use of radiations in the treatment of cancer. Unfortunately the cells of human tissues and of the tissues of the usual laboratory animals have many and small chromosomes, and it is not possible with these cells to make a detailed analysis of the changes suffered by chromosomes as a result of irradiation, such as is possible in the most favourable cytological materials. In view of the remarkable similarity of behaviour of the chromosomes of widely different species under normal conditions, it is reasonable to expect some similarity in their behaviour after irradiation. It seems reasonable therefore to take the results obtained with *Drosophila* and *Tradescantia* as a guide in interpreting experimental observations with rapidly dividing animal tissues.

In the first place there is ample evidence (detailed earlier in this chapter) that the cells which degenerate after the irradiation of these tissues with moderate doses commence to do so only at division. This result is consistent with the idea that the cause of degeneration is some change in the chromosomes. Two types of change in the chromosomes have been discussed, structural changes and physiological changes.[1] Structural changes in these materials are likely to be recognizable, if at all, only at anaphase (by the production of bridges and fragments)[2] and after division (by the presence of micronuclei).

Chromosome structural changes may be produced at any stage in the life cycle of a cell, and will only lead to degeneration after division. Thus when a rapidly dividing tissue is irradiated, degenerate cells may be expected to commence to appear when mitotic activity is resumed after the temporary inhibition. They will continue to appear for a time at least equal to the intermitotic period, since each batch of cells entering division will provide its quota of cells with lethal types of structural change. This in fact is what is observed (cp. Fig. 54, p. 309).

The second type of change suffered by the chromosomes as a result of irradiation (the 'physiological' effect) is a surface stickiness, possibly accounted for by the nucleic acid being deposited on the chromosomes in a fluid unpolymerized state.[3] The physiological effect is typically exhibited by cells already in division at the time of irradiation, and results in clumped metaphases and bridges at anaphase, such as are illustrated in varying degrees of severity in Plate IV E–K.[4] The physiological effect presumably accounts for the abnormal division figures observed

---

[1] These terms are used throughout in the sense of Chapter VI, p. 192.

[2] Thus Marshak, A. (1942) found that during the 24 hr. following the irradiation of a mouse lymphoma by 600 roentgens of X-rays, at least 50 % of the cells observed at anaphase showed bridges or fragments.

[3] Darlington, C.D. (1942).   [4] See footnotes 1, 2, 3, on p. 337.

in rapidly dividing animal tissues in the hours immediately following irradiation, *before* the resumption of mitotic activity.[1]

If, as is said to be the case,[2] recovery is complete from the physiological effect by the time mitotic activity is resumed, then degeneration of cells which takes place after the resumption of mitotic activity cannot be ascribed to the physiological effect. In this event the physiological effect can play only a small part in causing the death of a rapidly dividing tissue (at any rate as a result of a single irradiation) since, even if all the cells in division at the time of irradiation are affected, these constitute only a small fraction of the total number of cells in the tissue. However, it must be remembered that the conclusion, that the physiological effect on the chromosomes is not exhibited by cells which enter division after the resumption of mitotic activity, was reached by studying cells with a longer time scale and may not apply to rapidly dividing animal tissues.

In view of the difficulties of direct cytological observation, one may hope to obtain evidence of whether chromosome structural changes play a large or a small part in accounting for the killing of cells in rapidly dividing animal tissues by studying the manner in which the proportion of cells affected varies with the dose, with the intensity, and with the type of radiation, and comparing with the known behaviour of chromosome structural changes in this respect.

[1] E.g. by Lasnitzki, I. (1943*a*).   [2] Cp. Chapter VI, p. 196.

Appendix I

SUPPLEMENTARY CALCULATIONS[1]

The relation between dosage in air and energy dissipation in tissue [2]

X- and γ-rays. The roentgen is a unit based on the production of ionization in air. Dosage expressed in roentgens can readily be converted to terms of energy dissipation per unit volume of air. To convert a dose in roentgens into terms of energy dissipation in tissue, it therefore suffices to know the ratio of the energy dissipations per unit volume in tissue and in air for the particular radiation.

For X- or γ-rays this is the ratio of the values of $\tau + \sigma_a$, where τ is the photoelectric coefficient, and σ_a is the part of the Compton scattering coefficient σ which corresponds to the transference of energy to a recoil electron. σ_s, the part of the scattering coefficient which corresponds to the energy of the scattered radiation is not relevant here, since the radiation scattered in a small volume of irradiated air or tissue will not usually contribute to the ionization in that small volume. Where a large volume of tissue is being irradiated, the scattered radiation from the surrounding tissue does of course add to the ionization in a given cell, but since the scattered radiation is included in the dose in roentgens, providing the ionization measurement has been made in a cavity in the tissue or in a suitable phantom, it is still correct to use $\tau + \sigma_a$ and not $\tau + \sigma_a + \sigma_s$.

On the other hand, τ is the total photoelectric absorption coefficient without subtraction of that part of it corresponding to energy of the excited characteristic radiation of the absorbing atom, since such characteristic radiation from light elements has a short range and will in fact often be absorbed in the same cell in which it is produced.

$_e\sigma_a$, the scattering-absorption coefficient per electron, is

---

[1] This Appendix comprises a number of calculations, the results of which have been required in the text, but the details of which are only of technical interest.

[2] Table 2 (p. 8) and Table 3 (p. 12) are based on this section.

independent of the composition of the scatterer, and is given by the well-known formula of Klein and Nishina:[1]

$$_e\sigma_a = \frac{2\pi e^4}{m^2 c^4}\left[\frac{2(1+\alpha)^2}{\alpha^2(1+2\alpha)} - \frac{1+3\alpha}{(1+2\alpha)^2} + \frac{(1+\alpha)(1+2\alpha-2\alpha^2)}{\alpha^2(1+2\alpha)^2}\right.$$

$$\left. - \frac{4\alpha^2}{3(1+2\alpha)^3} - \left(\frac{1+\alpha}{\alpha^3} - \frac{1}{2\alpha} + \frac{1}{2\alpha^3}\right)\log(1+2\alpha)\right], \quad \text{(Ap.-1)}$$

where $\alpha = h\nu/mc^2$, $e = 4{\cdot}8025 \times 10^{-10}$, $m = 9{\cdot}1066 \times 10^{-28}$,

$$c = 2{\cdot}99776 \times 10^{10}, \quad h = 6{\cdot}624 \times 10^{-27}.$$

For small values of α this may be expanded as

$$_e\sigma_a = \frac{8\pi e^4}{3m^2 c^4}[\alpha - 4{\cdot}2\alpha^2 + 14{\cdot}7\alpha^3 - 46\tfrac{6}{35}\alpha^4 + \ldots].$$

$_e\sigma_a$ is tabulated in Table 80 for various wave-lengths of radiation. By multiplying by the number of electrons per gram in any given tissue, the value of σ_a/ρ, the scattering-absorption coefficient per gram for the tissue, may be calculated. The number of electrons per gram is readily calculated from the elementary analysis of the tissue with the aid of the figures given in Table 79 of the number of electrons per gram in various elements likely to occur in tissue.

τ, the photoelectric coefficient, has to be obtained from experimental values of the absorption coefficient $\mu = \tau + \sigma$ of the elements, by subtracting from these experimental values the theoretical (Klein-Nishina) values of σ, the scattering coefficient.[2] The scattering coefficient per electron is

$$_e\sigma = \frac{2\pi e^4}{m^2 c^4}\left[\frac{2(1+\alpha)^2}{\alpha^2(1+2\alpha)} - \frac{1+\alpha}{\alpha^3}\log(1+2\alpha)\right.$$

$$\left. + \frac{1}{2\alpha}\log(1+2\alpha) - \frac{1+3\alpha}{(1+2\alpha)^2}\right], \quad \text{(Ap.-2)}$$

[1] Klein, O. & Nishina, Y. (1929). Note that the formulae of Klein and Nishina are incorrectly quoted in several text-books, e.g. Rutherford, E., Chadwick, J. & Ellis, C.D. (1930); Kirchner, F. (1930); Rasetti, F. (1937).

[2] For wave-lengths exceeding 0·2A., the scattering experimentally observed exceeds the Klein-Nishina scattering, owing to the coherent scattering. However, for wave-lengths long enough for the difference to be significant, the scattering coefficient is so much less than the photoelectric coefficient that little error is made in the estimation of the latter by the use of the Klein-Nishina formula.

or for small values of α,

$$_e\sigma = \frac{8\pi e^4}{3m^2c^4}[1 - 2\alpha + 5\cdot2\alpha^2 - 13\cdot3\alpha^3 + 32\tfrac{24}{35}\alpha^4 \ldots].$$

TABLE 79. Photoelectric absorption coefficients (τ/ρ)

Element	Z	A	Electrons per g. $\times 10^{-23}$	Photoelectric coefficient $0\cdot0089 Z^{4\cdot1}/A$	Power of λ in photoelectric law
H	1	1·008	5·975	0·009	
C	6	12·01	3·009	1·150	3·05
N	7	14·01	3·010	1·854	
O	8	16·00	3·012	2·809	
Na	11	23·00	2·881	7·200	
Mg	12	24·32	2·972	9·74	
Al	13	26·97	2·903	12·20	
Si	14	28·06	3·005	15·88	
P	15	31·02	2·912	19·06	
S	16	32·06	3·006	24·04	2·85
Cl	17	35·46	2·888	27·86	
A	18	39·94	2·714	31·29	
K	19	39·10	2·927	39·8	
Ca	20	40·08	3·005	48·1	
Fe	26	55·84	2·804	101·0	
Air			3·007	$\tau/\rho = 2\cdot05\lambda^{3\cdot05} + 0\cdot41\lambda^{2\cdot85}$	
H$_2$O			3·343	$\tau/\rho = 2\cdot50\lambda^{3\cdot05}$	
Virus protein (p. 7)			3·216	$\tau/\rho = 1\cdot56\lambda^{3\cdot05} + 0\cdot67\lambda^{2\cdot85}$	
Wet tissue (p. 7)			3·307	$\tau/\rho = 2\cdot26\lambda^{3\cdot05} + 0\cdot27\lambda^{2\cdot85}$	

$_e\sigma$ is tabulated in Table 80, while a convenient compilation of experimental values of μ/ρ for various wave-lengths has been made by S.J.M. Allen.[1] It is found that when the theoretical values of σ/ρ are subtracted from Allen's μ/ρ values, and the τ/ρ values so obtained converted into values of τ_{at}, the coefficient per atom, they may be satisfactorily fitted by a formula of the usual type $\tau_{at} \propto Z^m \lambda^n$, the values obtained for the coefficients m and n being indicated in the equations

$$\tau_{at} = 1\cdot48 \times 10^{-26} Z^{4\cdot1} \lambda^n, \quad \text{with } n = 3\cdot05 \text{ for elements C, N, O,}$$
$$n = 2\cdot85 \text{ for elements Na–Fe.}$$

Thus the photoelectric coefficient per gram is

$$\frac{\tau}{\rho} = 0\cdot0089 \left(\frac{Z^{4\cdot1}}{A}\right) \lambda^n, \qquad \text{(Ap.-3)}$$

Z and A being the atomic number and atomic weight and n

[1] In the appendix to Compton, A.H. & Allison, S.K. (1935), *X-rays in Theory and Experiment*.

taking the values given above.[1] λ is the wave-length in Angstroms.

The values of $0.0089 Z^{4.1}/A$ are tabulated for light elements in Table 79. With the aid of this table a formula for the photoelectric

TABLE 80. Klein-Nishina coefficients

Radiation			Coefficients[1] $\times 10^{25}$			Photoelectric absorption in water
α	λ (A.)	$h\nu$ (ekV.)	$_e\sigma$	$_e\sigma_a$	$_e\sigma_s$	τ/ρ
0.01	2.4265	5.108	6.524	0.0638	6.460	37.33
0.02	1.2132	10.22	6.401	0.1226	6.278	4.508
0.03	0.8088	15.32	6.284	0.1769	6.107	1.309
0.04	0.6066	20.43	6.172	0.2269	5.945	0.5442
0.05	0.4853	25.54	6.066	0.2731	5.793	0.2755
0.06	0.4044	30.65	5.962	0.3156	5.647	0.1581
0.07	0.3466	35.76	5.862	0.3517	5.511	0.09874
0.08	0.3033	40.87	5.773	0.3936	5.380	0.06570
0.09	0.2696	45.98	5.684	0.4276	5.256	0.04590
0.10	0.2426	51.08	5.599	0.4605	5.138	0.03328
0.11	0.2206	56.19	5.516	0.4896	5.026	0.02488
0.12	0.2022	61.30	5.437	0.5176	4.919	0.01908
0.15	0.1618	76.62	5.218	0.5914	4.627	0.00966
0.18	0.1348	91.95	5.023	0.6497	4.373	0.00553
0.21	0.1155	107.3	4.847	0.7031	4.144	0.00346
0.24	0.1011	122.6	4.687	0.7455	3.942	0.00230
0.27	0.0899	137.9	4.542	0.7812	3.761	0.00161
0.30	0.0809	153.2	4.409	0.8118	3.598	0.00117
0.33	0.0735	168.6	4.287	0.8375	3.450	0.00087
0.36	0.0674	183.9	4.175	0.8601	3.315	0.00067
0.39	0.0622	199.2	4.070	0.8793	3.191	0.00052
0.42	0.0578	214.6	3.973	0.8953	3.078	0.00042
0.45	0.0539	229.9	3.883	0.9091	2.974	0.00034
0.48	0.0506	245.2	3.798	0.9217	2.877	0.00028
0.6	0.04044	306.5	3.507	0.9562	2.551	0.00014
0.8	0.03033	408.7	3.140	0.9820	2.158	0.00006
1.0	0.02426	510.8	2.866	0.9873	1.879	0.00003
1.2	0.02022	613.0	2.650	0.9823	1.668	0.00002
1.4	0.01733	715.2	2.474	0.9711	1.503	0.00001
1.6	0.01517	817.3	2.326	0.9580	1.368	0.00001
1.8	0.01348	919.5	2.200	0.9424	1.258	0.00000
2.0	0.01213	1022	2.090	0.9260	1.164	0.00000

[1] Multiply these coefficients by 3.343×10^{23} to obtain coefficients per gram in water, i.e. multiply the actual numbers listed by 0.03343.

absorption coefficient of a tissue can be calculated if its elementary analysis is available. Formulae of this sort for air, water, virus protein and a wet tissue (for the composition assumed for the two last-named see p. 7) are included in Table 79.

[1] A more complicated empirical formula, covering all elements, has been given by Victoreen, J.A. (1943).

The formulae may be compared with the empirical formulae given by Küstner[1] and Müller,[2] namely,

$$\text{Air } \tau/\rho = 2\cdot 33\lambda^{3\cdot 13}, \qquad \text{Water } \tau/\rho = 2\cdot 54\lambda^{3\cdot 22}.$$

These formulae are based on actual absorption measurements on the two substances, the allowance for scattering being made in a semi-empirical fashion which seems less satisfactory than the use of the Klein-Nishina formula, especially in the present circumstances when the Klein-Nishina formula has in any case to be used to deduce $\tau + \sigma_a$ from τ. Küstner and Trubestein[1] give empirical formulae for various tissues based on actual absorption measurements with these tissues. We have preferred however to calculate $\tau + \sigma_a$ from the elementary analysis combined with the τ coefficients for the separate elements, since in this way a more consistent set of figures less affected by experimental error in any one experiment is likely to be achieved. The assumption implicit in this procedure that the photoelectric absorption of an atom is independent of its chemical combination is unlikely to be wrong since photoelectric absorption is mainly by the innermost electron shells.

Formula (3) is only valid on the short wave side of the K absorption edge. For long waves, e.g. $\lambda > 1\text{A.}$, it has been thought best not to employ an empirical formula, but instead to use Allen's values of μ/ρ and to subtract from them σ_s/ρ, so obtaining $\mu/\rho - \sigma_s/\rho = \tau/\rho + \sigma_a/\rho$.

The ratio of the absorption coefficients in tissue and air of a particular wave-length depends on the elementary analysis of the tissue, and the data in Tables 79 and 80 suffice to calculate the ratio if the elementary analysis is known. The calculation has been carried out for two compositions of tissue and water, at selected wave-lengths, and the results employed in Table 2 (p. 8).

α-rays, protons and electrons. If the experimental material is irradiated in a thin layer by a beam of particles, e.g. α-rays, protons or electrons, then the ratio of the energy dissipation in tissue and air is simply the ratio of the stopping powers of the two substances. These are calculated by the Bethe formula (cp. pp. 350–2).

[1] Küstner, H. & Trubestein, H. (1937).
[2] Müller, I. (1938).

Neutrons. The unit of neutron dose used in Table 2 is the 'v' (cp. p. 20), which is based upon the ionization produced in a small air cavity in irradiated water. Ionization in the air cavity may be converted into energy dissipation in air by assuming 35 eV. per ion pair. Nearly all the ionization in the air, and energy dissipation in the water, being produced by the protons which are set into motion by the neutrons, the energy dissipation in the water can be obtained from the energy dissipation in the air by multiplying by the stopping power ratio of the two substances for protons of the appropriate energy. We thus readily obtain the figures given in Table 2 for the energy dissipation in water corresponding to $1v$ of neutrons.

To calculate now the ratio of the energy dissipation in a tissue of known composition to that of water we note that the energy of recoil of an atom of atomic weight A for a given neutron energy is proportional to $A/(1+A)^2$, while the number of recoil atoms per unit volume with a given neutron dose is proportional to $p\sigma/A$, where p is the fractional content, by weight, of the particular element in the tissue, and σ is its atomic cross-section for neutron scattering. We proceed therefore by evaluating $\Sigma \dfrac{p\sigma}{(1+A)^2}$ for all the elements in the tissue, and compare the figure obtained with the corresponding figure for H_2O. In the case of nitrogen, disintegration resulting in the emission of an α-particle or a proton occurs as well as nuclear recoil, and has to be taken into account in the calculations.[1] The values of Kikuchi and Aoki[2] have been used for the scattering cross-sections, and of Baldinger and Huber[3] for the disintegration cross-sections.

Spatial distribution of ionization in irradiated tissue[4]

Theoretical formulae exist for computing data of the sort given in Tables 10–18, and have been conveniently summarized by Bethe.[5] A certain amount of experimental data also exists, usually for air or other gases. Since it would in any case be

[1] As pointed out by Gray, L.H. & Read, J. (1939).
[2] Kikuchi, S. & Aoki, H. (1939).
[3] Baldinger, E. & Huber, P. (1939 *a,b*).
[4] This section describes the manner in which Tables 10–18 of Chapter I have been obtained.
[5] Bethe, H.A. (1933).

necessary to appeal to theory to fill in the gaps in the experimental data, a more consistent set of figures can probably be obtained by relying entirely on the theory, which has therefore been done in this book. Calculated values of energy dissipation and primary ionization by electrons in gases agree with experiment within the accuracy of the experiments, which is not very great. For α-rays, energy dissipation is known experimentally rather accurately, and experimental values are here more reliable than theory.[1] As regards δ-ray production, experimental data are rather meagre. Experiments by Hornbeck and Howell[2] and Shearin and Pardue[3] indicate that the theory is correct for the production of fast δ-rays by fast primary electrons. Experiments by Alper[4] agree fairly well with theory for fast α-rays, but give fewer δ-rays than theory for slow α-rays. The experimental difficulties of the measurement of ranges of δ-rays of a few hundred volts must however be very great, and we have preferred to rely on the theory.

The principal uncertainty in the application of the Bethe formulae is the modification of formulae derived for hydrogen-like atoms to more complicated atoms and molecules. A value has to be assigned to $\bar{E}$, the effective mean ionization potential of the atom or molecule. The value of $\bar{E}$ is usually chosen for each element to give agreement between the theoretical stopping power formula and the experimental stopping power of the element for α-rays. Mano[5] has given a table of values of $\bar{E}$ which includes, hydrogen 16, air 86, and oxygen 100 eV., based on α-particle range measurements in these gases, and suggests $\bar{E} = 45$ eV. for H_2O, interpolated between the values for hydrogen and oxygen. This value of $\bar{E}$ we first used in calculating energy dissipation in tissue. However, it appears that Mano obtained $\bar{E} = 45$ eV. for H_2O by considering $Z = (2+8)/3 = 3 \cdot 33$ as a 'weighted mean atomic number' of the constituent atoms, and this procedure is probably unsound. Gray[6] in a recent review has given evidence indicating that the Bragg additive law relating the α-ray stopping powers of molecules to those of their con-

[1] Cp. Livingston, M.S. & Bethe, H.A. (1937).
[2] Hornbeck, G. & Howell, I. (1941).
[3] Shearin, P.E. & Pardue, T.E. (1942).
[4] Alper, T. (1932).
[5] Mano, G. (1934).
[6] Gray, L.H. (1944 a).

stituent atoms is obeyed to one or two per cent. If we accept the additive law, then it is evident from its position in the formula that the effective value of $\bar{E}$ should be calculated from the values 16 and 100 eV. of the hydrogen and oxygen atoms as a weighted geometric mean, viz. $(16^2 \times 100^8)^{0 \cdot 1} = 69$ eV. It is unfortunate that the only direct measurements of α-particle ranges in water[1] are the principal exceptions to the additive rule. These experiments give 32μ for the range of a polonium α-particle, and 60μ for a Ra C' α-particle, considerably lower than the values given in Table 12, which are derived from the accurately known (mean) ranges in air with the aid of the theoretical stopping power ratio of water and air, the latter calculated with $\bar{E} = 69$ eV. for water. These experimental ranges would be more nearly in agreement with $\bar{E} = 45$ eV. for water. However, we have accepted Gray's conclusion that the weight of the evidence is in favour of the validity of the additive law. We therefore recalculated the stopping powers on the basis of $\bar{E} = 69$ eV., and all stopping powers and ranges for α-particles, protons and electrons in Chapter I are calculated on this basis. We did not think it necessary, however, to recalculate some of the data in subsequent tables, and Table 15 of Chapter I and Tables 81, 82 and 83 of the Appendix are based on an energy-range relation for electrons which has been computed with $\bar{E} = 45$ eV. instead of $\bar{E} = 69$ eV. which we now prefer. The difference is small, since $\bar{E}$ has less influence on electron stopping power than on α-particle stopping power, and in any event there is no certainty that $\bar{E}$ values deduced from α-particle stopping power experiments are the best values to use in computing electron stopping powers.

In calculating stopping powers all the extranuclear electrons, 10 per molecule for H_2O, are considered effective. In calculating primary specific ionization the contribution of the two innermost electrons is ignored, and 18 eV. taken as the effective ionization potential of the remaining 8 electrons. Similarly, in calculating δ-ray energy distributions, only the 8 outermost electrons are considered. δ-rays of energy less than 100 eV. are not considered; the calculation of their number is more difficult than for more energetic δ-rays, and in any event such short δ-rays are from our point of view better regarded as ion-clusters.

[1] Michl, W. (1914); Philipp, K. (1923).

TARGET THEORY CALCULATIONS[1]

The overlapping factor F

An ionizing particle produces ion-clusters at a mean separation of L along its path. An effective hit is supposed scored when in its passage through the target, taken to be a sphere of radius r, one or more ion-clusters (i.e. primary ionizations) are produced in the target. It is required to calculate the average number of effective hits per target when a dose of radiation corresponding to the production of n ion-clusters per unit volume is delivered. The dose given corresponds to the production of an average of $n\tfrac{4}{3}\pi r^3$ ion-clusters in a sphere of radius r. Since the mean path of an ionizing particle in a sphere is $4r/3$, and the separation of consecutive ion-clusters averages L, an average of $\tfrac{4}{3}r/L$ ion-clusters are produced by each ionizing particle passing through the sphere. Hence

$$n\tfrac{4}{3}\pi r^3/(\tfrac{4}{3}r/L) = nL\,\pi r^2 \qquad \text{(Ap.-4)}$$

ionizing particles cross the sphere. $nL\,2\pi y\,dy$ will cross the sphere at distances between y and $y+dy$ from its centre, and such particles will have a path length of $2x = 2(r^2 - y^2)^{\frac{1}{2}}$ in the sphere (see Fig. 60). In this path length the mean number of ion-clusters produced is $2x/L$, and since clusters are produced not at equal intervals but according to the mean free path law, $(1-\mathrm{e}^{-2x/L})$ is the probability that at least one cluster will be produced. Hence the mean number of hits per target is

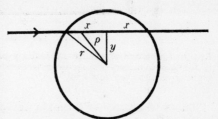

FIG. 60. Illustrating target calculation.

$$nL\int_0^r (1-\mathrm{e}^{-2x/L})\,2\pi y\,dy = 2\pi nL \int_0^r (1-\mathrm{e}^{-2x/L})\,x\,dx,$$

since $x^2 + y^2 = r^2$,

$$= \pi r^2\, nL\,\{1 - 2(1-\mathrm{e}^{-\xi})/\xi^2 + 2\mathrm{e}^{-\xi}/\xi\}, \quad \text{where } \xi = 2r/L,$$

$$= \tfrac{4}{3}\pi r^3\, n/F, \quad \text{where } F = (2\xi/3)/\{1 - 2(1-\mathrm{e}^{-\xi})/\xi^2 + 2\mathrm{e}^{-\xi}/\xi\},$$

$$\text{(Ap.-5)}$$

[1] This section supplements Chapter III.

If the dose corresponds to the production of an average of one ion-cluster per volume $\frac{4}{3}\pi r^3$, then $1/F$ is the mean number of hits per target. As shown in curve A of Fig. 61, this is unity for a radiation of low ion-density ($L \gg 2r$), and diminishes as the number of ionizations per micron path increases.

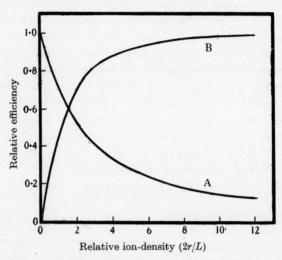

FIG. 61. Relative efficiencies of ionizing particles of different ion-densities: A, for equal numbers of ion-clusters per unit volume; B, for equal numbers of ionizing particles per unit area.

This is the proof of formula (III-2), p. 85. F is tabulated as a function of ξ in Table 26, p. 86.

If the dose corresponds to the passage of an average of one ionizing particle per target, then the mean number of hits per target is easily shown to be

$$1 - 2(1 - e^{-\xi})/\xi^2 + 2e^{-\xi}/\xi = 2\xi/3F. \qquad \text{(Ap.-6)}$$

This tends to the value unity for a densely ionizing particle ($L \ll 2r$), since such a particle passing through the target is certain to produce ionization in it, and is less than unity for less densely ionizing particles. It is plotted as curve B of Fig. 61.

Fig. 61 shows that for a given number of ionizing particles per unit area, densely ionizing particles are more effective than less densely ionizing particles, but that *per ionization*, densely ionizing particles are less effective in this type of action.

Target without sharp boundary

Suppose that an ion cluster produced at a distance ρ from the centre of the target has a probability $e^{-\rho^2/b^2}$ of being an effective 'hit'. Then an ionizing particle which passes through the target at a distance y from the centre (cp. Fig. 60) will produce a number dx/L of ion-clusters in any path length dx, and hence altogether will produce a mean number $\int_{-\infty}^{+\infty} e^{-\rho^2/b^2} dx/L$ of effective hits in the target. This integral evaluates to

$$(e^{-y^2/b^2}/L) \int_{-\infty}^{+\infty} e^{-x^2/b^2} dx = (b\sqrt{\pi}/L) e^{-y^2/b^2} = z, \text{ say.} \quad \text{(Ap.-7)}$$

This being the *mean* number of effective hits, the probability of there being at least one hit will be $(1-e^{-z})$. For a dose of radiation corresponding to the production of n ion-clusters per unit volume the number of ionizing particles passing at distances between y and $y+dy$ from the centre of the target is $nL\,2\pi y\,dy$ (as before), and hence the mean number of hits per target for this dose is $nL \int_0^\infty (1-e^{-z})\, 2\pi y\, dy$, integrating to

$$nL\,\pi b^2 \{0 \cdot 5772 + \log z_0 + Ei(z_0)\}, \quad \text{(Ap.-8)}$$

where $z_0 = b\sqrt{\pi}/L$ and $Ei(z_0) \equiv \int_{z_0}^\infty e^{-z} dz/z$ is the exponential integral.

For an ionizing particle of low ion-density ($L \gg b$), equation (8) tends to the value $nL\pi b^2 z_0 = n\pi^{\frac{3}{2}} b^3$, which can be written in the form $n\frac{4}{3}\pi(1\cdot 10 b)^3$. Comparing with the ordinary formula $n\frac{4}{3}\pi r^3$ for the mean number of ion-clusters produced in a target of definite radius r by a radiation (e.g. γ-rays) in which the ion-clusters are widely separated, it is evident that for such a radiation our target without sharp boundary will behave as a target of apparent radius $r = 1\cdot 10 b$.

If we now use a densely ionizing radiation with which we should expect, with a target of definite boundary, every ionizing particle passing through the target to give an effective hit, and deduce the apparent target radius R on this basis, then from equation (8) we have

$$\pi R^2 = \pi b^2 \{0 \cdot 5772 + \log z_0 + Ei(z_0)\},$$

whence

$$R = \frac{r}{1\cdot 1} \{0\cdot 5772 + \log(0\cdot 8 \times 2r/L) + Ei(0\cdot 8 \times 2r/L)\}^{\frac{1}{2}}, \quad \text{(Ap.-9)}$$

because

$$b = r/1\cdot 1 \quad \text{and} \quad z_0 = b\sqrt{\pi/L} = r\sqrt{\pi/(1\cdot 1 L)} = 0\cdot 8 \times 2r/L.$$

It is by substituting $2r/L = 5$ and 25 respectively in equation (9) that the radii of circles D and E of Fig. 12 (p. 97) have been calculated.

The associated volume calculation[1]

Secondary electrons of less than 100 eV. energy are assumed to produce clusters sufficiently compact to behave as units, and have an associated volume no bigger than that of an isolated primary ionization. Secondary electrons of more than 100 eV. energy are regarded as δ-rays. Spheres are described around each of the n ionizations in the δ-ray track as centres, as in Fig. 7D (p. 84), and the associated volume deduced, being $n\frac{4}{3}\pi r^3/F$, where F is the usual overlapping factor (Table 26, p. 86) calculated from the ratio (ξ) of the target diameter $2r$ to the mean separation L of primary ionizations in the path of the δ-ray. This is the associated volume of the δ-ray *additional* to the associated volume of the primary ionization at which the δ-ray originated. This last associated volume is not included with the δ-ray since it is already included in the calculation of the associated volume of the primary ionizing particle. The quantity $n\frac{4}{3}\pi r^3/F$ is in fact the term omitted from the cruder calculation of method III (p. 83) which ignores the distinction between a δ-ray and a cluster.

Calculation of associated volume for any energy of electron. The procedure will be made clearer by quoting a few figures. Let us consider how we calculate the associated volume for a primary electron of any desired energy completely absorbed in the tissue, for a target of diameter say 10 mμ, of the same density as water. We start by considering the last 1 ekV. of the electron track. Splitting the 1 ekV. into intervals 0–0.1, 0.1–0.2, 0.2–0.3 ekV., etc., we obtain by differencing Table 10 (p. 24) the ranges corresponding to these energy intervals, and hence, using Table 11

[1] Figs. 8 and 9 are based on the calculations described in this section.

(p. 25), the number of primary ionizations produced in each interval. Between 0·4 and 0·3 ekV., for example, the primary electron travels $0·0045\mu$ producing primary ionizations at the rate 584 per μ. Thus it produces $0·0045 \times 584 = 2·63$ primary ions, having a ξ value of $0·01 \times 584 = 5·84$, since $2r = 0·01\mu$ and $1/L = 584$ per μ. From Table 26 (p. 86) we read off the corresponding value $F = 4·131$. The associated volume between 0·4 and 0·3 ekV. is thus $\frac{4}{3}\pi r^3 \left(\frac{2·63}{4·131}\right)$. Working throughout in units of volume of $\frac{4}{3}\pi r^3$ we write this simply as 0·637. Adding up nine similarly calculated contributions for the other intervals we obtain 5·851 as the associated volume for the primary electron between 1 and 0 ekV., and similarly obtain associated volumes for the intervals 0·9, 0·8, 0·7, ..., 0·1 ekV. to 0.

We have now to allow for the δ-rays. Instead of calculating separately for each 0·1 ekV. interval, we consider it sufficient to calculate the δ-ray production per unit length of path by a 0·5 ekV. primary electron, and multiply it by $0·05344\mu$, the range of the primary electron between 1 and 0 ekV. From Table 16 (p. 28) we find that an 0·5 ekV. primary electron produces δ-rays of energy between 0·1 and 0·15 ekV. at the rate 9·12 per μ path, and of energy between 0·15 and 0·25 ekV. at the rate 7·38 per μ path, and none more energetic than this. The actual number of δ-rays is thus 0·488 between 0·1 and 0·15 ekV. (mean energy 0·125 ekV.), and 0·394 between 0·15 and 0·25 ekV. (mean 0·2 ekV.). The associated volumes for an electron of 0·125 and 0·2 ekV. we can interpolate in the table of associated volumes we have just calculated; they are 0·453 and 0·738. Thus δ-rays add $0·453 \times 0·488 + 0·738 \times 0·394 = 0·512$ to the associated volume 5·851. This represents an addition of 8·74 % for the δ-ray contribution to the associated volume of a 1 ekV. electron. Taking this to be sufficiently nearly constant, we correct by the same percentage the table of associated volumes for smaller electron energies.

The next stage is to consider the range interval 2 to 1 ekV. of the primary electron. This we divide up into subranges 1·0–1·2, 1·2–1·4, ..., 1·8–2·0 ekV., and calculate the associated volume for the primary ions in each subrange as before. We then read off the δ-ray production by a primary electron of 1·5 ekV. from Table 16, tabulating the numbers of δ-rays of energies 0·1–0·15, 0·15–0·25,

0·25–0·35 ekV., etc., up to the maximum δ-ray energy of 0·75 ekV. The centres of these ranges are 0·125, 0·2, 0·3, ..., 0·7 ekV., and we now know the associated volume for an electron of any energy up to 1 ekV. Thus we can deduce the associated volume for all the δ-rays. The δ-rays are found to add 9·30 % to the associated volume of the primary electron between 2 and 1 ekV., and it is assumed that the same correction percentage is valid for 1·8–1, 1·6–1, etc. Thus we deduce the associated volumes, δ-ray contributions included, for any electron energy up to 2 ekV.

Building up in further steps of 2–4, 4–8, 8–16 ekV., etc., we finally obtain tables of the associated volume for any electron energy. The computation has been carried up to 480 ekV. for target diameters of 4, 10, 20, 40 and 80 mμ, and Table 81 gives these figures.

By this stepwise method of calculation it is seen that we automatically allow (except in the first step) for 'δ-rays produced by δ-rays'.

α-rays and protons. Table 81 is used when the primary particles are electrons, i.e. in X-ray, β-ray or γ-ray irradiations. The early part of it is also used to make allowance for the δ-rays when the primary particles are α-rays or protons. In this case we read off from Table 17 A or B (pp. 30, 31) the number of δ-rays having energies in the intervals 0·1–0·15, 0·15–0·25, 0·25–0·35 ekV., etc., multiply each by the associated volume, given in Table 81, of an electron of energy corresponding, and so obtain the contribution due to the δ-rays. This is to be added to the associated volume for the primary ionizations, which is simply $1/(LF)$ per micron path, where $1/L$ is the number of primary ionizations per micron and F the overlapping factor, calculated as usual. In this manner we have drawn up Tables 82 and 83 of associated volume per micron path or ekV. energy loss by α-rays and protons of the stated energies. For the sake of interest we have shown separately the contributions of primary particle and of δ-rays of ⩾ 100 eV. energy. The totals, also given, are the figures used in calculation of the 37 % dose.

If the target material is of density ρ higher than water, then the ionizations are correspondingly closer together and the overlapping for a target diameter $2r$ and density ρ is as great as it would be for a target diameter $2r\rho$ in tissue of unit density. The

TABLE 81. Associated volume per electron, for electrons of energies 0·1–480 ekV.

Electron energy ekV.	Target diameter, $2r\rho$, in mμ				
	4	10	20	40	80
0·1	0·956	0·3867	0·1933	0·0965	0·0482
0·2	1·934	0·7961	0·3988	0·1993	0·0997
0·3	2·994	1·277	0·6435	0·3222	0·1611
0·4	4·129	1·832	0·9315	0·4678	0·2340
0·5	5·318	2·464	1·261	0·6350	0·3180
0·6	6·546	3·142	1·634	0·8236	0·4111
0·7	7·804	3·871	2·040	1·033	0·5167
0·8	9·084	4·643	2·480	1·262	0·6325
0·9	10·38	5·472	2·954	1·511	0·7583
1·0	11·69	6·314	3·456	1·778	0·8939
1·2	14·46	8·104	4·543	2·361	1·191
1·4	17·25	9·996	5·728	3·012	1·525
1·6	20·06	11·95	7·000	3·726	1·894
1·8	22·88	13·97	8·348	4·499	2·298
2·0	25·71	16·04	9·762	5·342	2·734
2·5	32·96	21·42	13·55	7·622	3·952
3·0	40·22	26·99	17·62	10·18	5·355
3·5	47·47	32·69	21·92	12·99	6·932
4	54·70	38·48	26·39	16·00	8·672
5	69·48	50·42	35·81	22·56	12·57
6	84·17	62·53	45·64	29·69	16·99
7	98·77	74·74	55·76	37·27	21·88
8	113·3	87·01	66·10	45·22	27·17
9	128·0	99·53	76·71	53·50	32·81
10	143·0	112·3	87·67	62·15	38·82
11	157·6	124·9	98·46	70·81	44·95
12	172·2	137·5	109·4	79·70	51·38
13	186·8	150·1	120·4	88·75	58·02
14	201·2	162·6	131·4	97·87	64·81
15	215·5	175·1	142·5	107·1	71·78
20	288·0	238·6	199·0	155·1	108·9
25	359·6	301·7	255·9	204·6	148·8
30	430·5	364·6	313·0	255·0	190·5
40	573·8	492·2	429·6	359·0	278·6
50	716·5	619·8	547·1	465·4	370·7
60	855·9	744·9	662·7	570·7	463·6
70	998·1	872·6	781·0	678·8	559·3
80	1139	999·5	898·9	785·8	656·0
90	1280	1126	1016	894·0	753·3
100	1419	1252	1134	1002	851·1
110	1562	1380	1254	1113	951·9
120	1700	1505	1370	1221	1051
150	2122	1886	1725	1551	1352
180	2539	2263	2077	1880	1655
210	2953	2638	2427	2207	1957
240	3367	3012	2778	2535	2261
300	4201	3768	3486	3198	2876
360	5028	4518	4189	3857	3489
420	5849	5263	4887	4513	4101
480	6674	6012	5589	5173	4717

Note. For tissue densities, ρ, other than 1 g./cm.³, the figures headed 'target diameter' are to be regarded as values of $2r\rho$, not of $2r$. The associated volumes are in units of $\frac{4}{3}\pi r^3$.

same tables suffice for different density tissues providing the rules given in the footnotes to the tables are adhered to.

A complication arises in the case of α-rays not present for less densely ionizing radiations. With rather large targets it may

TABLE 82. Associated volumes for α-rays

(Per micron track, and per ekV. energy dissipation.)

α-ray energy eMV.		Target diameter, $2r\rho$, in mμ				
		4	10	20	40	80
1	Due to primary ions	373·3	149·9	74·99	37·50	18·76
	Due to δ-rays	682·2	169·7	53·24	15·08	3·98
	Total, per μ track	1055·5	319·6	128·2	52·58	22·73
	Total, per ekV.	3·999	1·211	0·4858	0·1992	0·0861
2	Due to primary ions	369·4	149·6	74·95	37·50	18·75
	Due to δ-rays	641·3	210·9	82·69	28·91	9·16
	Total, per μ track	1010·7	360·6	157·6	66·41	27·91
	Total, per ekV.	5·740	2·048	0·8952	0·3771	0·1585
3	Due to primary ions	363·6	149·3	74·91	37·49	18·75
	Due to δ-rays	512·5	211·9	90·44	32·61	11·32
	Total, per μ track	876·1	361·2	165·3	70·10	30·07
	Total, per ekV.	6·509	2·685	1·228	0·5208	0·2234
4	Due to primary ions	356·6	148·8	74·85	37·48	18·75
	Due to δ-rays	431·4	200·0	91·03	38·85	15·27
	Total, per μ track	788·0	348·8	165·9	76·33	34·02
	Total, per ekV.	7·156	3·168	1·506	0·6932	0·3089
5	Due to primary ions	348·3	148·2	74·78	37·47	18·75
	Due to δ-rays	375·0	187·8	88·64	39·83	16·52
	Total, per μ track	723·3	336·0	163·4	77·31	35·27
	Total, per ekV.	7·713	3·583	1·743	0·8244	0·3761
6	Due to primary ions	339·4	147·6	74·70	37·46	18·75
	Due to δ-rays	333·4	170·5	86·17	39·73	18·34
	Total, per μ track	672·8	318·1	160·9	77·19	37·09
	Total, per ekV.	8·203	3·878	1·961	0·9412	0·4522
7	Due to primary ions	330·1	146·8	74·61	37·46	18·74
	Due to δ-rays	301·1	156·7	83·18	39·64	17·51
	Total, per μ track	631·2	303·5	157·8	77·10	36·26
	Total, per ekV.	8·635	4·152	2·159	1·055	0·4960
8	Due to primary ions	320·7	145·9	74·48	37·43	18·75
	Due to δ-rays	275·4	145·6	81·80	38·82	17·71
	Total, per μ track	596·1	291·5	156·3	76·25	36·45
	Total, per ekV.	9·019	4·411	2·365	1·154	0·5516

Note. For tissue density ρ, other than $1\,\text{g./cm.}^3$, the figures headed 'target diameter' should be regarded as values of $2r\rho$ instead of $2r$. Further, the figures of associated volume per micron of α-particle track, i.e. the first three figures of each group of four are to be multiplied by ρ. The associated volumes are in units of $\frac{4}{3}\pi r^3$.

happen that the associated volumes constructed round two δ-ray tracks may merge, as in Fig. 7F, p. 84. It would be difficult to allow for this effect by a rigid calculation, but the following method, which has been adopted, should prevent any serious

ASSOCIATED VOLUME

error arising. Instead of taking into account all δ-rays of energy exceeding 100eV., account is taken only of δ-rays of energy exceeding energy W, where W is chosen so that the number of δ-rays of energy $\geqslant W$ is 1 per distance $2r$ along the α-ray track.

TABLE 83. Associated volumes for protons

(Per micron track, and per ekV. energy dissipation.)

Proton energy eMV.		Target diameter, $2r\rho$, in mμ				
		4	10	20	40	80
1	Due to primary ions	235·3	132·8	72·64	37·20	18·71
	Due to δ-rays	108·3	52·8	28·33	14·56	7·33
	Total, per μ track	343·6	185·6	101·0	51·76	26·04
	Total, per ekV.	12·41	6·704	3·646	1·869	0·9404
2	Due to primary ions	160·2	109·4	67·59	36·51	18·63
	Due to δ-rays	69·1	36·5	20·82	11·17	5·72
	Total, per μ track	229·4	145·9	88·41	47·67	24·35
	Total, per ekV.	13·78	8·761	5·310	2·863	1·463
3	Due to primary ions	122·2	91·65	61·88	35·50	18·50
	Due to δ-rays	52·1	28·77	17·06	9·47	4·96
	Total per μ track	174·3	120·4	78·94	44·98	23·46
	Total, per ekV.	14·29	9·870	6·470	3·686	1·922
4	Due to primary ions	99·29	78·73	56·55	34·29	18·33
	Due to δ-rays	42·39	24·08	14·68	8·38	4·48
	Total, per μ track	141·7	102·8	71·23	42·67	22·81
	Total, per ekV.	14·54	10·55	7·312	4·380	2·342
6	Due to primary ions	72·87	61·56	47·83	31·67	17·89
	Due to δ-rays	31·36	18·50	11·73	6·97	3·86
	Total, per μ track	104·2	80·06	59·56	38·64	21·75
	Total, per ekV.	14·77	11·34	8·439	5·474	3·081
8	Due to primary ions	57·95	50·73	41·34	29·15	17·36
	Due to δ-rays	25·18	15·22	9·88	6·05	3·45
	Total, per μ track	83·13	65·95	51·22	35·20	20·80
	Total, per ekV.	14·85	11·78	9·148	6·286	3·716
10	Due to primary ions	48·33	43·26	36·41	26·88	16·77
	Due to δ-rays	21·18	13·03	8·61	5·52	3·15
	Total, per μ track	69·51	56·30	45·02	32·40	19·92
	Total, per ekV.	14·88	12·05	9·637	6·936	4·263

Note. For tissue densities ρ other than 1 g./cm.^{3} the figures headed 'target diameter' should be regarded as values of $2r\rho$ instead of $2r$. Further, the figures of associated volume per micron of proton track, i.e. the first three figures in each group of four, are to be multiplied by ρ. The associated volumes are in units of $\frac{4}{3}\pi r^3$.

The correction is small except for the larger sizes of target. The correction is of course only applied when, on working out the value of W, it is found to exceed 100eV.

Tables 82 and 83 refer to definite energies of α-particles and protons, not to complete absorption in the tissue. They are thus directly suitable for use when thin films are irradiated by α-rays or protons which do not suffer complete absorption in the film. When complete absorption occurs in the tissue, as in neutron experiments or α-ray experiments using dissolved radon, in the absence of a calculation made specifically for this case an average particle energy should be used.

Table 81 refers to electrons completely absorbed in the tissue and is thus suitable for use with X-rays and γ-rays. Unity should be added to the figures of associated volume per electron to allow for the two ends of the electron track. This has not been done already in this table since it is required to be useable also for δ-ray calculations for which the addition would be incorrect. (The figures in Table 81 give the associated volumes represented in Fig. 7 E, p. 84; the addition of unity to the associated volume per electron is required to give the associated volumes represented in Fig. 7 C.)

If an experiment is made using β-rays irradiating thin films of material, in which the β-rays are not completely absorbed, the table may be used by differencing.

Conversion of associated volumes to 37 % *dose*. The calculation of the 37 % dose for any given size of target, i.e. that dose which corresponds to an average of one hit per target, is now straightforward. A dose corresponding to 1 ekV. energy dissipation per μ^3 gives an associated volume of $V \times \frac{4}{3}\pi r^3$ per μ^3, where V is the figure for associated volume per ekV. read off from Tables 82 or 83 or deduced from Table 81 for the appropriate radiation and target size. $1 - \exp(-V \times \frac{4}{3}\pi r^3)$ is thus the probability of securing a hit with a dose of 1 ekV. per μ^3. The 37 % dose, i.e. the dose giving an average of one hit per target, is thus $1/(V\frac{4}{3}\pi r^3)$ ekV. per μ^3. We can now deduce the dose in roentgens by reference to Table 2 (p. 8). If the density is not unity but ρ g./cm.^{3}, the inactivation dose is $1/(V\rho\frac{4}{3}\pi r^3)$ ekV. per 10^{-12} g.

As an example let us calculate the inactivation dose to be expected for a virus having target diameter $2r = 14$ mμ irradiated by α-rays of 5 eMV., the density being $\rho = 1\cdot4$ g./cm.^{3} We have $2r\rho = 14 \times 1\cdot4 = 19\cdot6$ mμ, and interpolating in Table 82 between the columns $2r\rho = 20$ and $2r\rho = 10$ we find that for a 5 eMV. α-ray the associated volume is $V = 1\cdot78$ (units of $\frac{4}{3}\pi r^3$) per ekV. Thus the inactivation dose is

$$\frac{1}{1\cdot78 \times 1\cdot4 \times \frac{4}{3}\pi(0\cdot007)^3} = 2\cdot79 \times 10^5 \text{ekV. per } 10^{-12} \text{ g.}$$

In Table 2 (p. 8) we read that 1 r. is 66·65 eV. per 10^{-12} g. in virus protein. Thus the inactivation dose is

$$2{\cdot}79 \times 10^8/66{\cdot}65 = 4{\cdot}19 \times 10^6 \text{ r.}$$

In this manner the curves of Figs. 8 and 9 and thence of Figs. 10 and 11 of Chapter III have been calculated.

Appendix II

TEXTUAL REVISIONS AND ADDITIONS[1]

MUTATIONS

[p. 154] **The influence of temperature**

Stadler[2] has shown that the yield of visible mutations induced by irradiation of barley seed is the same, per unit dose, at 10, 20, 30, 40 and 50° C. Experiments by different workers recording sex-linked recessive lethals in *Drosophila* have given conflicting results. The experiments summarized in Table 40[3] show no change in the yield of mutations with a 30° change of temperature. More recently, King[4] determined the yield of sex-linked lethals produced by doses of 600–3600 r., given either at room temperature ($\sim 25°$ C.) or at $0.5°$ C. His results can be satisfactorily fitted by the usual formula $1 - e^{-mD}$ (p. 146), the values of m being:

at $0.5°$: $\quad 7.40 \pm 0.70$ lethals per 100 sperm per 1000 r.,

at $\sim 25°$: $\quad 3.43 \pm 0.29$ lethals per 100 sperm per 1000 r.,

showing a quite significant increase of mutation rate at low temperature.

Medvedev[5] also obtained a significantly higher yield at 0° C. than at $\sim 20°$ C., and at 19° C. than at 37° C.

[p. 154] **Mutation yields in different cells**

It is of considerable interest to know whether the probability of a given mutational step being induced by a given dose is a constant or is different for the same mutational step in different

---

1 Mainly in the form noted by the author.

2 Stadler, L.J. (1930 *a*).

3 Timoféeff-Ressovsky, N.W. & Zimmer, K.G. (1939). Makhijani, J.K. (1944), cp. Muller, H.J. (1940), also obtained no difference in the yields at ~ 5 and $\sim 37°$ C.

4 King, E.D. (1947). The author states that the presence of ice surrounding the vial containing the flies in the lower temperature experiment increased the dose by 5·6 %, which has not been allowed for in calculating the yields.

5 Medvedev, N.N. (1935, 1938).

cells. There is evidence[1] that the mutation rate is less in unripe sperm (presumably spermatogonia) than in ripe sperm, the ratios being 0·18:1 for sex-linked lethals, 0·67:1 for sex-linked visibles, and 0·58:1 for second-chromosome lethals.

That the difference between ripe and unripe sperm is so much more marked in the case of sex-linked lethals than in the case of second-chromosome lethals is presumably due to the fact that the spermatogonia are hemizygous for these lethals but heterozygous for the second-chromosome lethals, and strongly suggests that the apparently low yield of sex-linked lethals in unripe sperm is mainly due to germinal selection, i.e. the death of spermatogonia containing the lethals.[2]

The yield of visible mutations in barley produced by a given dose is four times as great when germinating seeds are irradiated than when dormant seeds are irradiated.[3]

DOMINANT LETHALS [p. 163]

A most convincing demonstration that dominant lethals are indeed due to damage to the chromosomes and not to the cytoplasm is afforded by an experiment which has been made using the parasitic wasp *Habrobracon*.[4] The eggs of this wasp, if fertilized, develop into diploid females; if they are not fertilized they develop into haploid (gynogenetic) males. Sperm which do not enter an egg are, of course, incapable of development. An unfertilized egg which receives a few thousand roentgens does not hatch, whether fertilized or not, subsequent to irradiation. But an unfertilized egg which receives a much heavier dose (order 3×10^4 r.) and is then fertilized by an untreated sperm may develop into a haploid (androgenetic) male in which the chromosomes of the egg play no role. Evidently the dose of 3×10^4 r. leaves the cytoplasm still capable of supporting the development of the intact sperm nucleus. This experiment demonstrates that the dominant lethal effect obtained with smaller doses to the egg must therefore be due to damage to the nucleus of the egg.

[1] Reviewed by Timoféeff-Ressovsky, N.W. (1937b).
[2] It is known that most sex-linked lethals are lethal to individual cells (when homozygous) and not merely to the whole organism (Demerec, M., 1936).
[3] Stadler, L.J. (1930b).   [4] Whiting, A.R. (1946).

The experiment with *Habrobracon* also demonstrates that damaged chromosomes are positively lethal to a cell and not merely negatively ineffective, since a fertilized egg in which either egg or sperm has been damaged by a moderate dose of radiation prior to fertilization does not hatch, but an egg in which either egg or sperm chromosomes are absent, or else have been rendered completely ineffective by a very heavy dose of radiation, hatches as a haploid organism.

CHROMOSOME STRUCTURAL DAMAGE

[p. 216] **Location of breaks**

When the observed distribution of breaks was compared with the distribution to be expected on the hypothesis that the probability of a break being produced in a given segment is simply proportional to the number of bands in the segment, the experimental distribution differed significantly from the distribution expected on this hypothesis, the difference lying not in any pronounced gradient of breakability from one end of the chromosome to the other, but in the existence of a number of regions in which the mean number of breaks per band was three or four times the average. It is believed that these segments contain heterochromatin, since in salivary gland cells pairing occurs among these regions, and between them and the proximal heterochromatin.[1]

[p. 217] **Frequency relations**

Instead of studying separately the distribution of chromatid breaks, isochromatid breaks and interchanges as has been done in Table 56, one can determine the proportion of cells which contain r chromatid breaks, together with s isochromatid breaks, together with t interchanges. If the three types of aberrations are distributed independently in Poisson distributions with mean values m_1, m_2 and m_3 respectively, the proportion expected is

$$(e^{-m_1}.m_1^r/r!)\,(e^{-m_2}.m_2^s/s!)\,(e^{-m_3}.m_3^t/t!)$$

The results of tests of this sort[2] on *Tradescantia* microspores show agreement with the hypothesis that the three types of aberration are distributed independently in Poisson distributions.

[1] For additional evidence and discussion see Kaufmann, B.P. (1944); also Slizynski, B.M. (1945).
[2] Catcheside, D.G., Lea, D.E. & Thoday, J.M. (1946*a*).

Modifying factors [p. 223]

While exposure of male *Drosophila* to infra-red radiation alone does not cause chromosome structural changes, exposure to infra-red before X-raying results in the yield of chromosome structural changes being increased.[1] Thus 24 hr. exposure to infra-red rays of wave-length 8000–20,000 A. increased the percentage of sperm showing chromosome structural changes after 4000r. from $31 \pm 4\%$ to $47 \pm 4\%$, an increase equivalent to raising the X-ray dose to between 5000 and 6000r. No significant changes were detected in the yields of dominant lethals, or of sex-linked recessive lethals[2] when the X-irradiation was preceded by infra-red.

There is probably also an increase in the yield of chromosome structural changes when the infra-red treatment follows the X-raying.[3]

Experiments have also been made to study the effect of a prior or subsequent treatment with ultra-violet light upon the yield of chromosome structural changes produced by X-rays. Kaufmann and Hollaender[4] irradiated *Drosophila* males with 4000r. of X-rays, and subsequently exposed them to a (surface) dose of $1\cdot8 \times 10^5$ ergs/cm.² of ultra-violet light of wave-length 2536A. Salivary-gland chromosomes of F_1 larvae were examined for gross chromosome structural changes. Chromosome structural changes were produced in $23\cdot2 \pm 2\cdot5\%$ of the sperm, compared to 30% produced by 4000r. without ultra-violet treatment. It appears that ultra-violet light (which does not itself induce gross

[1] Kaufmann, B.P., Hollaender, A. & Gay, E.H. (1946). Breaks were counted in the salivary-gland chromosomes of F_1 larvae. The infra-red radiation raised the body temperature of the flies by about 7°.

[2] Kaufmann, B.P. & Gay, E.H. (1947).

[3] Experiments to determine whether *subsequent* infra-red irradiation modifies the yield of chromosome structural changes produced by X-rays are complicated by the fact that infra-red accelerates the ripening of immature sperm. Since immature sperm are less sensitive to X-rays than mature, the result of prolonged infra-red treatment of male flies after X-raying and before mating is to diminish the yield of chromosome structural changes. However, if the females which have been mated to X-ray males are then exposed to infra-red radiation during the period of oviposition, a higher proportion of the F_1 larvae have chromosome structural changes than in the absence of infra-red treatment ($37\cdot5 \pm 3\%$ against $29\cdot5 \pm 2\cdot5\%$ after 4000r., according to Kaufmann, B.P. (1946b)).

[4] Kaufmann, B.P. & Hollaender, A. (1946).

chromosome structural changes) reduces the proportion of X-ray-induced breaks which take part in viable types of rearrangement. There appeared to be no corresponding changes in the proportion of X-ray-induced breaks which behaved as dominant lethals.

Swanson[1] similarly found that the yield of chromatid interchanges induced by X-rays in *Tradescantia* pollen tubes was reduced by ultra-violet light (which does not itself induce interchanges). Table 84 shows the effect of a 30 sec. exposure to ultra-violet light of wave-length 2536 A. upon the yield of chromatid interchanges produced by 247 r. of X-rays. The X-rays were given in every case 2 hr. after the germination of the pollen.

TABLE 84. Modification by ultra-violet light of the yield of chromatid interchanges produced by 247 r. of X-rays in *Tradescantia* pollen tubes. (Swanson (1944))

Time of ultra-violet exposure in relation to X-ray exposure	Yield of chromatid interchanges per 100 cells
No ultra-violet	0·94
Ultra-violet 1 hr. before X-rays	0·08
Ultra-violet immediately before X-rays	0·22
Ultra-violet immediately after X-rays	0·00
Ultra-violet ½ hr. after X-rays	0·72 ⎱ average 0·94
Ultra-violet 1 hr. after X-rays	1·16 ⎰

These results show that ultra-violet light given either before, or soon after, the X-rays is able drastically to reduce the proportion of X-ray-induced breaks which take part in chromatid interchanges. That no such effect is found when the ultra-violet light is given after ½ hr. or more has elapsed after the X-irradiation is evidence that interchanges in pollen tubes (as in microspores) are completed within this time. The yield of isochromatid breaks is not much affected by a prior or subsequent ultra-violet irradiation. The yield of chromatid breaks is reduced, despite the fact that ultra-violet light alone is able to produce chromatid breaks.

It appears therefore that the effect of ultra-violet light is to make restitution of an X-ray-induced break more probable, and to reduce the probability of two breaks exchanging, except in the case of breaks at the same loci in sister chromatids. Swanson considers that the effect of the ultra-violet light is to make the nucleic acid matrix of the chromosome firmer, so hindering

[1] Swanson, C.P. (1944).

interchange between breaks in different chromosomes, and also hindering separation of the two breakage ends to make a permanent break.

Sensitivity at different stages [p. 224]

If *Drosophila* males are irradiated, and then mated to successive batches of unirradiated females, the offspring from these matings are derived from sperms which, in the case of the first batch, are mature, and which in the later batches are at successively earlier stages of spermatogenesis at the time of irradiation. The yield of chromosome interchanges obtained in the successive batches progressively diminishes,[1] showing that the earlier stages of spermatogenesis are less sensitive to the production of interchanges than are the mature sperm.

Whiting[2] compared the sensitivity to X-rays of the oocytes [p. 225] of *Habrobracon* irradiated at first meiotic metaphase and at diplotene (first meiotic prophase) respectively. Lethal types of chromosome structural change led to death of the embryo in the morula stage, and could also be detected by the appearance of bridges and fragments at meiotic first and second anaphase and cleavage divisions.

Metaphase was many times more sensitive than diplotene, and the lethal effect was due to sister-union isochromatid breaks, which occurred only between the centromeres and the proximal chiasmata (deduced from the fact that bridges are seen at second meiotic anaphase, not at first anaphase). This part of the chromosome is in tension, since the chiasmata are resisting the separation of the centromeres, and the high sensitivity is explained by no restitution of primary breaks occurring.

Diplotene is considerably less sensitive, and sister-union isochromatid breaks occur between chiasmata, as well as between centromere and proximal chiasmata (deduced from the fact that bridges are seen at first anaphase as well as at second anaphase). Some interchanges between breakages in different chromosomes

[1] From 12 to 0% with 4000r. in the experiments of Catsch, A. & Radu, G. (1943). Irradiated males were mated to females homozygous for the second-chromosome gene *cinnibar* eye and the third chromosome gene *spineless*. The F_1 male offspring were tested for linkage between these genes.

[2] Whiting, A.R. (1945).

occur at the higher doses, and the yield of these interchanges is reduced by fractionation of dose, which affords evidence that a considerable proportion of primary breaks restitute. It seems probable that the lower sensitivity at diplotene is due to the lower tension in the chromatid threads permitting restitution of a large proportion of the breaks primarily produced.

It is noteworthy that the sensitivity of *Habrobracon* sperm[1] is comparable to that of the oocytes in first metaphase rather than diplotene, confirming the view[2] that restitution does not occur in sperm, and that the condition of the chromosomes in sperm is comparable to that in metaphase.

[p. 233] ## Dose relations

The yield of exchanges produced by neutrons is proportional to the dose and independent of intensity, while the yield of exchanges produced by X-rays is proportional to the square of the dose at high intensity and diminishes at lower intensities. This difference is believed on theoretical grounds (see p. 250) to be quantitative rather than qualitative; that is to say, it is believed that the yield of exchanges is, with both radiations, equal to the sum of two terms, a '1-hit' term proportional to the dose and independent of intensity, and a '2-hit' term proportional to the square of the dose and dependent upon intensity. It has been found possible[3] to detect the 1-hit component in X-ray experiments. The yield of interchanges per cell (y) produced by a given dose (x roentgens) was fitted by the formula $y = \alpha x + \beta x^2$. In these experiments four doses (approximately 25, 50, 100 and 150 r.) were given in a constant time, which was 251 min. in one series and 1·2 min. in another. The values of α and β obtained are given in Table 85. It is seen that the 1-hit component (α) is

TABLE 85. *Analysis of X-ray-induced interchanges into 1-hit and 2-hit components*

($y = \alpha x + \beta x^2$ is the number of c/c interchanges per nucleus produced by a dose of x roentgens.)

Duration of exposure (min.)	$\alpha \times 10^4$ per r.	$\beta \times 10^6$ per r.^{2}
251	2·66 ± 0·91	3·89 ± 0·92
1·2	1·85 ± 1·04	13·06 ± 1·19

1 Heidenthal, G. (1945).   2 Muller, H.J. (1941).
3 Catcheside, D.G., Lea, D.E. & Thoday, J.M. (1946 b).

not significantly changed by change of exposure time, while the 2-hit component diminishes with prolongation of exposure time.

An isochromatid break involves two breaks, and while it appears that, with all radiations, these are usually produced by the same ionizing particles which traverse both sister chromatids of a split chromosome, there is no reason to doubt that if sister chromatids were broken at about the same locus by *separate* ionizing particles an isochromatid break would result. Thus we should expect the yield of isochromatid breaks to be made up of two terms, a '1-hit' term proportional to the dose and independent of the intensity, and a '2-hit' term proportional to the square of the dose and diminishing with prolongation of exposure time at constant dose.

Of the data shown in Fig. 34B, the most complete are those of Sax, and while these may be fitted by a straight line the χ^2 test gives $P = 0.07$), the fit is considerably improved when they are fitted by the formula $y = \alpha x + \beta x^2$ (the χ^2 test gives $P = 0.6$, despite the loss of one degree of freedom).

Relative efficiencies of different types of radiation [p. 240]

Comparing the yields of exchanges, Kotval[1] found γ-rays less efficient than X-rays in the ratio 0.80 ± 0.07, while Catcheside *et al.*[2] found γ-rays less efficient than X-rays in the ratio 0.77 ± 0.05. Evidently, any difference that exists between γ-rays and X-rays is slight, and in the direction of γ-rays being the less efficient.

The following provisional explanation may be advanced for [p. 243] the difference in relative frequencies of the different types of chromosome aberration in X-ray and ultra-violet experiments. In the first place, the fact that isochromatid breaks are not produced by ultra-violet light presents no difficulty, since isochromatid breaks found in experiments with ionizing radiations are attributed to the passage of a single ionizing particle through both chromatids. Since ultra-violet quantal absorptions are not localized along tracts in this way, we should not expect to get isochromatid breaks in ultra-violet experiments.

The fact that exchanges between breaks in different chromosomes do not occur in ultra-violet experiments in which chromatid

[1] Kotval (unpublished).
[2] Catcheside, D.G., Lea, D.E. & Thoday, J.M. (1946*b*).

breaks are freely produced is more difficult to understand, and at first glance seems to demand the assumption that ultra-violet-induced breaks are different in nature from X-ray-induced breaks. However, we have seen (p. 368) that ultra-violet light preceding or following X-irradiation reduces the proportion of X-ray-induced breaks which take part in gross chromosome structural changes. Presumably ultra-violet light treatment simultaneous with the X-irradiation would have the same effect. Evidently, therefore, ultra-violet light has, besides the property of producing breaks, the additional property of modifying the chromosomes or the nuclear sap in some way which reduces the proportion of breaks which take part in exchange.[1] This would seem to be a sufficient explanation of the absence of exchanges in ultra-violet experiments on *Tradescantia* pollen tubes, and may apply also to *Drosophila* sperm and maize pollen.

The efficiency of ultra-violet light in producing breaks is extremely low. The yield of chromatid breaks obtained in Swanson's experiments, for example, corresponds to about 1 chromatid break observed per 10^{11} quanta incident on the chromatids.

[p. 259] ## Evaluation of breakage frequency

It has been assumed above that when two breaks are produced (simultaneously) at a separation less than $h = 1\mu$, exchange between them is certain. It would perhaps be more plausible to suppose, not that exchange is certain, but that union is random between the four breakage ends, which would make the probability of exchange recurring equal to two-thirds. If this modification is adopted, the values of ξ listed for Method III in Table 67 should be increased by about 20%, and the values of f correspondingly reduced.

It is arguable that we have no right to assume either that exchange is certain, or that exchange is random between the four breakage ends, since the arguments of the last section which led to the conclusion that most pairs of breaks which exchange are initially separated by a distance not exceeding 1μ did not give any estimate of the probability of exchange occurring under these circumstances. If this view is accepted, Method III cannot be used. In this event, however, it is permissible to

[1] Swanson, C.P. (1944), suggests that the action is on the matrix of the chromosomes.

reverse the argument and to use the values of ξ given by other methods to show that the probability of an exchange between two breaks produced at a separation not exceeding $h = 1\mu$ approaches unity.

Alternative derivation of G [1] [p. 264]

The experimental fact that the yield of interchanges produced by a given dose diminishes with increase of the exposure time must mean that two breaks at a given position produced at different times have a smaller probability of exchanging than the same two breaks at the same position would have if produced simultaneously. We may designate by $f(t)$ the factor by which the probability is reduced with increase of t; $f(t)$ diminishes from the value unity at $t = 0$.

Consider two breaks independently produced by an irradiation at uniform intensity which extends over time T. Each of them may occur with equal probability anywhere in the time T, and it is readily shown that the probability of the two breaks being separated by an interval between t and $t + dt$ is $2(1 - t/T)(dt/T)$. The ratio G of the number of exchanges produced by the given dose spread over time T to the number which would be produced by the same dose given in a very short time is evidently

$$G = \int_{t=0}^{T} 2(1 - t/T)(dt/T) f(t) = 2 \int_{x=0}^{1} f(Tx)(1-x) \, dx.$$

To evaluate G we must assume some plausible form for the function $f(t)$. The simplest function fulfilling the condition of being unity at $t = 0$ and diminishing gradually with increase of t is $f(t) = e^{-t/\tau}$. With this form of $f(t)$ we obtain

$$G = 2(\tau/T)^2 \{T/\tau - 1 + e^{-T/\tau}\}.$$

BACTERIA

Effect of intensity [p. 321]

Table 86 presents further data showing how the mean lethal dose is the same, within the error of the experiment, whether the irradiation is made at a low intensity and spread over a prolonged time, or at a high intensity and concentrated in a short time. This table is thus supplementary to Table 74 (p. 321).

[1] Catcheside, D.G., Lea, D.E. & Thoday, J.M. (1946*b*).

TABLE 86. Independence of mean lethal dose on intensity

Radiation	Organism	Intensity (r./min.)	M.L.D. (r.)	Reference
X-rays (0·71 A.)	Bacteriophage, C-16	57,500	$46·8 \times 10^3$	1
		1,730	47·3	
		114	46·9	
X-rays (0·71 A.)	B. dysenteriae, Y 6 R	40,000	$5·1 \times 10^3$	2
		2,180	5·3	
		83·4	5·2	
X-rays (1·54 A.)	Yeast, S. ellipsoideus	157,000	$10·1 \times 10^3$	2
		4,600	10·0	
		135	10·3	
Ultra-violet light 2536 A.		(ergs cm.$^{-2}$ sec.$^{-1}$)	(ergs cm.$^{-2}$)	
	B. dysenteriae, Y 6 R	45,000	245	2
		8,300	242	
		4,800	241	
		830	245	
		260	264	
		83	260	
		71	260	
		8	258	
	Yeast, S. ellipsoideus	140,000	752	2
		14,800	796	
		2,800	780	
		805	770	
		280	778	
		224	775	
		28	776	

1 Latarjet, R. (1942).   2 Latarjet, R. (1944).

[p. 325] Metabolic disturbances

A culture of bacteria rendered non-infective by a heavy dose of radiation will nevertheless continue to respire[1] and will support a limited multiplication of bacteriophage.[2] These properties are not retained by bacteria killed by heat or chemical disinfectant.

1 Bonét-Maury, P., Perault, R. & Erichsen, M.L. (1944).
2 Rouyer, M. & Latarjet, R. (1946).

BIBLIOGRAPHY

Titles given are paraphrases of the original titles in
many instances

AEBERSOLD, P.C. & LAWRENCE, J.H. (1942). Physiological effects of neutrons. *Ann. Rev. Physiol.* **4**, 25.

ALBERTI, W. & POLITZER, G. (1923). Influence of X-rays on cell division. Part I. *Arch. Mikr. Anat.* **100**, 83.

ALBERTI, W. & POLITZER, G. (1924). Influence of X-rays on cell division. Part II. *Arch. Mikr. Anat.* **103**, 284.

ALLSOPP, C.B. (1944). Radiochemistry, a review of recent progress. *Trans. Faraday Soc.* **40**, 79.

ALPER, T. (1932). Production of δ-rays by α-particles. *Z. Phys.* **76**, 172.

ALTENBURG, E. (1934). Production of mutations by ultra-violet light. *Amer. Nat.* **68**, 491.

ALTENBURG, E. (1936). Mutations by the polar cap method of treatment. *Biol. Zh.* **5**, 27.

ANDERSON, R.S. & HARRISON, B. (1943). Effect of X-rays on ascorbic acid. *J. Gen. Physiol.* **27**, 69.

ATWOOD, K. & ROLLEFSON, G.K. (1941). Efficiency of the primary photochemical process in solution. *J. Chem. Phys.* **9**, 506.

BÄCKSTRÖM, H.L.J. (1940). Filters for ultra-violet light. *Acta Radiol., Stockh.*, **11**, 327.

BADIAN, J. (1933). The chromatin and development cycle of bacteria. *Arch. Mikrobiol.* **4**, 409.

BAKER, S.L. (1935). Action of β-rays on bacteria and viruses. *Brit. J. Exp. Path.* **16**, 148.

BAKER, S.L. & NANAVUTTY, S.H. (1929). Action of ultra-violet light on bacteria and bacteriophage. *Brit. J. Exp. Path.* **10**, 45.

BALDINGER, E. & HUBER, P. (1939a). Disintegration of nitrogen by fast neutrons. *Nature, Lond.*, **143**, 894.

BALDINGER, E. & HUBER, P. (1939b). Disintegration of nitrogen by fast neutrons. *Helv. Phys. Acta*, **12**, 330.

BAUER, H. (1939a). Production by X-rays of chromosome mutations in *Drosophila*. *Chromosoma*, **1**, 343.

BAUER, H. (1939b). X-ray-induced chromosome mutations in ring-X chromosomes of *Drosophila*. *Naturwissenschaften*, **27**, 821.

BAUER, H., DEMEREC, M. & KAUFMANN, B.P. (1938). X-ray-induced chromosomal alterations in *Drosophila melanogaster*. *Genetics*, **23**, 610.

BAWDEN, F.C. (1943). *Plant Viruses and Virus Diseases*. Waltham, Massachusetts.

BAWDEN, F.C. & PIRIE, N.W. (1938). Chemical, physical, and serological properties of plant viruses. *Tabul. biol., Berl.*, **16**, 355.

BELGOVSKY, M.L. (1939). Frequency of X-ray-induced minute chromosome rearrangements in *Drosophila*. *Bull. Acad. Sci. U.R.S.S.*, p. 159.

BENFORD, F. (1936). Monochromator for ultra-violet light. *J. Opt. Soc. Amer.* **26**, 99.

BETHE, H. (1933). Quantum mechanics of one- and two-electron problems. *Handb. Phys.* **24** (i), 273.

BISHOP, D.W. (1942). Cytological demonstration of chromosome breaks soon after X-irradiation. *Genetics*, **27**, 132.

BONÉT-MAURY, P. & OLIVIER, H.R. (1939). Utilization of the biological effect of radiation in the study of immunity. *Act. Sci. Ind.* No. 725, p. 492.

BONNER, T.W. (1941). Neutron energy from D + D reaction. *Phys. Rev.* **59**, 237.

BONNER, T.W. & BRUBAKER, W.M. (1935). Neutron energy from Li + D disintegration. *Phys. Rev.* **48**, 742.

BONNER, T.W. & BRUBAKER, W.M. (1936). Neutron energy from disintegration of Be, B, and C by deuterons. *Phys. Rev.* **50**, 308.

BOWEN, E.J. (1942). *The Chemical Aspects of Light.* Oxford.

BOZEMAN, M.L. (1943). Effects of irradiation on oocytes of *Sciara*. *Genetics*, **28**, 71.

BRATTAIN, K.G. (1938). Decomposition and synthesis of HI by α-particles. *J. Phys. Chem.* **42**, 617.

BRODA, E. (1943). Chemical action of X-rays on a non-aqueous solution. *Nature, Lond.*, **151**, 448, 530.

BRUMFIELD, R.T. (1943). Effect of colchicine pretreatment on the frequency of chromosomal aberrations induced by X-irradiation. *Proc. Nat. Acad. Sci., Wash.*, **29**, 190.

BRUYNOGHE, R. & MUND, W. (1935). Effect of radon on bacterial motility. *C.R. Soc. Biol., Paris*, **92**, 211.

BUSSE, W.F. & DANIELS, F. (1928). Chemical effects of cathode rays on oxygen, air, nitric oxide and carbon dioxide. *J. Amer. Chem. Soc.* **50**, 3271.

CANNON, C.V. & RICE, O.K. (1942). A monochromator using a large water prism. *Rev. Sci. Instrum.* **13**, 513.

CANTI, R.G. & DONALDSON, M. (1926). Effect of γ-rays on mitosis. *Proc. Roy. Soc.* B, **100**, 413.

CANTI, R.G. & SPEAR, F.G. (1927). Effect of γ-rays on mitosis *in vitro*. *Proc. Roy. Soc.* B, **102**, 92.

CANTI, R.G. & SPEAR, F.G. (1929). Effect of γ-rays on mitosis *in vitro*. *Proc. Roy. Soc.* B, **105**, 93.

CARLSON, J.G. (1938a). Effects of X-rays on the neuroblast chromosomes of the grasshopper. *Genetics*, **23**, 596.

CARLSON, J.G. (1938b). Mitotic behaviour of induced fragments lacking spindle attachments. *Proc. Nat. Acad. Sci., Wash.*, **24**, 500.

CARLSON, J.G. (1941a). X-ray-induced single breaks in neuroblast chromosomes of the grasshopper. *Proc. Nat. Acad. Sci., Wash.*, **27**, 42.

CARLSON, J.G. (1941b). Effects of X-radiation on grasshopper chromosomes. *Cold Spr. Harb. Symp.* **9**, 104.

CATCHESIDE, D.G. (1938a). Frequency of induced structural changes in the chromosomes of *Drosophila*. *J. Genet.* **36**, 307.

CATCHESIDE, D.G. (1938b). Frequencies of X-ray-induced interchanges in maize. *J. Genet.* **36**, 321.

CATCHESIDE, D.G. & LEA, D.E. (1943). Effect of ionization distribution on chromosome breakage by X-rays. *J. Genet.* **45**, 186.

CATCHESIDE, D.G. & LEA, D.E. (1945a). Induction of dominant lethals in *Drosophila* sperm by X-rays. *J. Genet.* **47**, 1.

CATCHESIDE, D.G. & LEA, D.E. (1945b). Dominant lethals and chromosome breaks in ring-X chromosomes of *Drosophila*. *J. Genet.* **47**, 25.

CAUCHOIS, Y. (1932). Cathode-ray tube. *J. Phys. Radium*, **3**, 512.

CHAMBERS, H. & RUSS, S. (1912). Bactericidal action of radon. *Proc. Roy. Soc. Med.* **5**, 198.

CLARK, G.L. & COE, W.S. (1937). Photochemical reduction with X-rays, and effect of additive agents. *J. Chem. Phys.* **5**, 97.

CLARK, G.L. & PICKETT, L.W. (1930). Chemical effects of X-rays and energy relations involved. *J. Amer. Chem. Soc.* **52**, 465.

CLAUS, W.D. (1933). Bactericidal action of X-rays in the presence of heavy ions. *J. Exp. Med.* **57**, 335.

COBLENTZ, W.W. & FULTON, H.R. (1924). Bactericidal action of ultraviolet light. *Sci. Pap. U.S. Bur. Stand.* **19**, 641.

Cold Spring Harbor Symposia (1941). Vol. **9**, Genes and Chromosomes.

COMPTON, A.H. & ALLISON, S.K. (1935). *X-rays in Theory and Experiment*. New York.

COOK, E.V. (1939). Recovery of *Ascaris* eggs from X-rays. *Radiology*, **32**, 289.

COOLIDGE, W.D. (1926). Cathode-ray tube. *J. Franklin Inst.* **202**, 693.

COOPER, F.S., BUCHWALD, C.E., HASKINS, C.P. & EVANS, R.D. (1939). Cathode-ray tube for biological experiments. *Rev. Sci. Instrum.* **10**, 73.

CROWTHER, J.A. (1924). Action of X-rays on tissue cells. *Proc. Roy. Soc.* B, **96**, 207.

CROWTHER, J.A. (1926). Action of X-rays on *Colpidium colpoda*. *Proc. Roy. Soc.* B, **100**, 390.

CROWTHER, J.A. (1938). Biological action of radiations. *Brit. J. Radiol.* **11**, 132.

CURIE, P. (1935). *Radioactivity*. Paris.

DALE, W.M. (1940). Effect of X-rays on enzymes. *Biochem. J.* **34**, 1367.

DALE, W.M. (1942). Effect of X-rays on the conjugated protein d-aminoacid oxidase. *Biochem. J.* **36**, 80.

DALE, W.M. (1943a). Effects of X-rays on acetylcholine solutions. *J. Physiol.* **102**, 50.

DALE, W.M. (1943b). Effect of X-rays on aqueous solutions of biologically active compounds. *Brit. J. Radiol.* **16**, 171.

DALE, W.M., MEREDITH, W.J. & TWEEDIE, M.C.K. (1943). Mode of action of ionizing radiations on aqueous solutions. *Nature, Lond.*, **151**, 281.

DARLINGTON, C.D. (1937). *Recent Advances in Cytology*. London.

DARLINGTON, C.D. (1942). Chromosome chemistry and gene action. *Nature, Lond.*, **149**, 66.

DARLINGTON, C.D. & LA COUR, L.F. (1942). *The Handling of Chromosomes*. London.

DEE, P.I. (1932). Attempts to detect the interaction of neutrons with electrons. *Proc. Roy. Soc.* A, **136**, 727.

DEMEREC, M. (1934). The gene and its role in ontogeny. *Cold Spr. Harb. Symp.* **2**, 110.

DEMEREC, M. (1937). Relationship between various chromosomal changes in *Drosophila*. *Cytologia*, Fujii jub. vol., p. 1125.

DEMEREC, M. (1938). Hereditary effects of X-rays. *Radiology*, **30**, 212.

DEMEREC, M. & FANO, U. (1941). Deficiencies in *Drosophila*. *Proc. Nat. Acad. Sci., Wash.*, **27**, 24.

DEMEREC, M., HOLLAENDER, A., HOULAHAN, M.B. & BISHOP, M. (1942). Effects of monochromatic ultra-violet light on *Drosophila*. *Genetics*, **27**, 139.

DEMEREC, M., KAUFMANN, B.P. & SUTTON, E. (1942). Genetic effects produced by neutrons in *Drosophila*. *Genetics*, **27**, 140.

DEMEREC, M., KAUFMANN, B.P., SUTTON, E. & FANO, U. (1941). The gene. *Yearb. Carneg. Instn*, **40**, 225.

DEMEREC, M., KAUFMANN, B.P., FANO, U., SUTTON, E. & SANSOME, E.R. (1942). The gene. *Yearb. Carneg. Instn*, **41**, 190.

DEMPSTER, E.R. (1941a). Induction of dominant lethals, recessive lethals and chromosome structural changes in *Drosophila* by neutrons. *Proc. Nat. Acad. Sci., Wash.*, **27**, 249.

DEMPSTER, E.R. (1941b). Dependence upon intensity of the yield of X-ray-induced chromosome structural changes in *Drosophila*. *Amer. Nat.* **75**, 184.

DESSAUER, F. (1923). Point-heat theory. *Z. Phys.* **20**, 288.

DICKINSON, R.G. (1935). Photochemical reactions in gases and in solutions. *Chem. Rev.* **17**, 413.

DICKINSON, R.G. (1938). Photochemical reactions in liquids and gases. *J. Phys. Chem.* **42**, 739.

DOERR, R. & HALLAUER, C. (1938). *Handbook of Virus Research.* Vienna.

DOZOIS, K.P., WARD, G.E. & HACHTEL, F.W. (1936). Effect of γ-rays on *Bact. coli*. *Amer. J. Roentg.* **35**, 392.

DREYER, G. & CAMPBELL-RENTON, M.L. (1936). Bactericidal action of ultra-violet light. *Proc. Roy. Soc.* B, **120**, 447.

DUANE, W. & SCHEUER, O. (1913). Decomposition of water by α-rays. *Radium*, **10**, 33.

DUBININ, N.P. & SIDOROV, B.N. (1935). The position effect of the hairy gene. *Biol. Zh.* **4**, 555.

DUBININ, N.P., SOKOLOV, N.N. & TINIAKOV, G.G. (1935). Cytogenetic study of the position effect. *Biol. Zh.* **4**, 707.

DUGGAR, B.M. (Editor) (1936). *Biological Effects of Radiation.* New York.

DUGGAR, B.M. & HOLLAENDER, A. (1934a). Action of ultra-violet light on bacteria and viruses. *J. Bact.* **27**, 219.

DUGGAR, B.M. & HOLLAENDER, A. (1934b). Action of ultra-violet light on bacteria and viruses. *J. Bact.* **27**, 241.

DURAND, E., HOLLAENDER, A. & HOULAHAN, M.B. (1941). Ultra-violet absorption spectrum of the abdominal wall of *Drosophila*. *J. Hered.* **32**, 51.

DUSHKIN, M.A. & BACHEM, A. (1933). Time factors concerned in the bactericidal action of ultra-violet light. *Proc. Soc. Exp. Biol., N.Y.*, **30**, 700.

EHRISMANN, O. & NOETHLING, W. (1932). Bactericidal action of monochromatic ultra-violet light. *Z. Hyg. InfektKr.* **113**, 597.

ELLINGER, P. & GRUHN, E. (1930). Bactericidal action of secondary radiations from X-rays. *Strahlentherapie*, **38**, 58.

ELLIS, C.D. & ASTON, G.H. (1930). Absolute intensities and internal conversion coefficients of the γ-rays of RaB and RaC. *Proc. Roy. Soc. A*, **129**, 180.

ELLIS, E.L. & DELBRÜCK, M. (1938). Growth of bacteriophage. *J. Gen. Physiol.* **22**, 365.

EMMONS, C.W. & HOLLAENDER, A. (1939a). Production of variants in *Trichophyton* by ultra-violet light. *Genetics*, **24**, 70.

EMMONS, C.W. & HOLLAENDER, A. (1939b). Production of mutations in *Trichophyton* by ultra-violet light. *Amer. J. Bot.* **24**, 467.

EVANS, T.C., SLAUGHTER, J.C., LITTLE, E.P. & FAILLA, G. (1942). Influence of the medium on radiation injury of sperm. *Radiology*, **39**, 663.

EXNER, F.M. & LURIA, S.E. (1941). Sizes of bacteriophages determined by X-ray inactivation. *Science*, **94**, 394.

EYRING, H., HIRSCHFEIDER, J.O. & TAYLOR, H.S. (1936a). Theoretical treatment of ortho-para hydrogen conversion by α-particles. *J. Chem. Phys.* **4**, 479.

EYRING, H., HIRSCHFELDER, J.O. & TAYLOR, H.S. (1936b). Radiochemical synthesis and decomposition of HBr. *J. Chem. Phys.* **4**, 570.

FABERGÉ, A.C. (1940a). Chromosome fragmentation by X-rays. *J. Genet.* **39**, 229.

FABERGÉ, A.C. (1940b). Equivalent effect of X-rays of different wavelength on *Tradescantia* chromosomes. *J. Genet.* **40**, 379.

FANO, U. (1941). Analysis and interpretation of chromosomal changes in *Drosophila*. *Cold Spr. Harb. Symp.* **9**, 113.

FANO, U. (1942). On the interpretation of radiation experiments in genetics. *Quart. Rev. Biol.* **17**, 244.

FANO, U. (1943a). Mechanism of induction of gross chromosomal rearrangements in *Drosophila* sperms. *Proc. Nat. Acad. Sci., Wash.*, **29**, 12.

FANO, U. (1943b). Neutron-induced lethals in *Drosophila*. *Genetics*, **28**, 74.

FANO, U. (1943c). Production of ion-clusters by X-rays. *Nature, Lond.*, **151**, 698 and **152**, 186.

FRANCK, J. & RABINOWITSCH, E. (1934). Free radicals and the photochemistry of solutions. *Trans. Faraday Soc.* **30**, 120.

FRICKE, H. (1934a). Reduction of O_2 to H_2O_2 by the irradiation of its aqueous solution with X-rays. *J. Chem. Phys.* **2**, 556.

FRICKE, H. (1934b). Chemico-physical actions of X-rays. *Cold Spr. Harb. Symp.* **2**, 241.

FRICKE, H. (1935a). Decomposition of H_2O_2 by the irradiation of its aqueous solution with X-rays. *J. Chem. Phys.* **3**, 364.

FRICKE, H. (1935b). Chemical properties of X-ray-activated molecules. *Cold Spr. Harb. Symp.* **3**, 55.

FRICKE, H. (1938). Denaturation of proteins by ionizing radiations. *Cold Spr. Harb. Symp.* **6**, 164.

FRICKE, H. & BROWNSCOMBE, E.R. (1933a). Effect of X-rays on chromate solutions. *J. Amer. Chem. Soc.* **55**, 2358.

FRICKE, H. & BROWNSCOMBE, E.R. (1933b). Inability of X-rays to decompose water. *Phys. Rev.* **44**, 240.

FRICKE, H. & HART, E.J. (1934). Transformation of formic acid by irradiation of its aqueous solution with X-rays. *J. Chem. Phys.* **2**, 824.

FRICKE, H. & HART, E.J. (1935a). Oxidation of ferrous iron by irradiation of its aqueous solution with X-rays. *J. Chem. Phys.* **3**, 60.

FRICKE, H. & HART, E.J. (1935b). Oxidation of the nitrite to the nitrate ion by X-rays. *J. Chem. Phys.* **3**, 365.

FRICKE, H. & HART, E.J. (1935c). Oxidation of the ferrocyanide, arsenite, and selenite ions by X-rays. *J. Chem. Phys.* **3**, 596.

FRICKE, H. & HART, E.J. (1935d). Decomposition of water by X-rays in the presence of the iodide or bromide ion. *J. Chem. Phys.* **3**, 596.

FRICKE, H. & HART, E.J. (1936). Reactions induced by photoactivation of the water molecule. *J. Chem. Phys.* **4**, 418.

FRICKE, H., HART, E.J. & SMITH, H.P. (1938). Chemical action of X-ray activated water. *J. Chem. Phys.* **6**, 229.

FRICKE, H. & MORSE, S. (1927). Action of X-rays on ferrous sulphate. *Amer. J. Roentg.* **18**, 426.

FRICKE, H. & MORSE, S. (1929). Action of X-rays on ferrous sulphate solution. *Phil. Mag.* **7**, 129.

FRICKE, H. & PETERSEN, B.W. (1927). Action of X-rays on haemoglobin. *Amer. J. Roentg.* **17**, 611.

FRIEDEWALD, W.F. & ANDERSON, R.S. (1940). Inactivation of rabbit papilloma virus by X-rays. *Proc. Soc. Exp. Biol., N.Y.*, **45**, 713.

FRIEDEWALD, W.F. & ANDERSON, R.S. (1941). Influence of protein and virus concentration on the inactivation of rabbit papilloma virus by X-rays. *J. Exp. Med.* **74**, 463.

FRIEDEWALD, W.F. & ANDERSON, R.S. (1943). Effects of X-rays on cell-virus associations. *J. Exp. Med.* **78**, 285.

FRY, H.J. (1936). Time schedule of mitotic changes in developing *Arbacia* eggs. *Biol. Bull. Woods Hole*, **70**, 89.

GATES, F.L. (1929a). Bactericidal action of ultra-violet light. *J. Gen. Physiol.* **13**, 231.

GATES, F.L. (1929b). Bactericidal action of ultra-violet light. *J. Gen. Physiol.* **13**, 249.

GATES, F.L. (1930). Bactericidal action of ultra-violet light. *J. Gen. Physiol.* **14**, 31.

GEDYE, G.R. (1931). Decomposition of nitrous oxide by cathode rays. *J. Chem. Soc.* p. 3016.

GEDYE, G.R. & ALLIBONE, T.E. (1930). Decomposition of ammonia by cathode rays. *Proc. Roy. Soc.* A, **130**, 346.

GEDYE, G.R. & ALLIBONE, T.E. (1932). Synthesis of hydrazine and ammonia by cathode rays. *J. Chem. Soc.* p. 1158.

GILES, N.H. (1940). Induction of chromosome aberrations by neutrons in *Tradescantia* microspores. *Proc. Nat. Acad. Sci., Wash.*, **26**, 567.

GILES, N.H. (1943). Comparative studies of the cytogenetic effects of neutrons and X-rays. *Genetics*, **28**, 398.

GLASS, H.B. (1940). Differential susceptibility of the sexes in *Drosophila* to the effects of X-rays in producing chromosome aberrations. *Genetics*, **25**, 117.

GLOCKER, R. (1932). Quantum physics of the biological action of X-rays. *Z. Phys.* **77**, 653.

GLOCKLER, G. & LIND, S.C. (1939). *Electrochemistry of Gases and other Dielectrics.* New York.

GLÜCKSMANN, A. (1941). Quantitative examination of human biopsy material taken from irradiated carcinomata. *Brit. J. Radiol.* **14**, 187.

GLÜCKSMANN, A. & SPEAR, F.G. (1939). Effect of γ-rays on the cells of fasting tadpoles, and of tadpoles at low temperature. *Brit. J. Radiol.* **12**, 486.

GOWEN, J.W. (1939). Effect of X-rays of different wave-lengths on viruses and genes. *Third Int. Cancer Congr.* p. 17.

GOWEN, J.W. (1940). Inactivation of tobacco mosaic virus by X-rays. *Proc. Nat. Acad. Sci., Wash.*, **26**, 8.

GOWEN, J.W. (1941). Mutations in *Drosophila*, bacteria, and viruses. *Cold Spr. Harb. Symp.* **9**, 187.

GOWEN, J.W. & GAY, E.H. (1933). Gene number, kind, and size in *Drosophila*. *Genetics*, **18**, 1.

GOWEN, J.W. & LUCAS, A.M. (1939). Action of X-rays on vaccinia virus. *Science*, **90**, 621.

GRAY, L.H. (1937). Radiation dosimetry. *Brit. J. Radiol.* **10**, 600, 721.

GRAY, L.H. (1944a). Ionization method of measuring neutron energy. *Proc. Camb. Phil. Soc.* **40**, 72.

GRAY, L.H. (1944b). Dosage-rate in radiotherapy. *Brit. J. Radiol.* **17**, 327.

GRAY, L.H., MOTTRAM, J.C., READ, J. & SPEAR, F.G. (1940). Biological effects of fast neutrons. *Brit. J. Radiol.* **13**, 371.

GRAY, L.H. & READ, J. (1939). Neutron dosimetry. *Nature, Lond.*, **144**, 439.

GRAY, L.H. & READ, J. (1942a). Effect of ionizing radiations on the bean root. *Brit. J. Radiol.* **15**, 11.

GRAY, L.H. & READ, J. (1942b). Lethal action of γ-rays on the bean root. *Brit. J. Radiol.* **15**, 39.

GRAY, L.H. & READ, J. (1942c). Lethal action of fast neutrons on the bean root. *Brit. J. Radiol.* **15**, 72.

GRAY, L.H. & READ, J. (1942d). Lethal action of α-rays on the bean root. *Brit. J. Radiol.* **15**, 320.

GRAY, L.H., READ, J. & POYNTER, M. (1943). Lethal action of X-rays on the bean root. *Brit. J. Radiol.* **16**, 125.

GREEN, R.H., ANDERSON, T.F. & SMADEL, J.E. (1942). Morphological structure of the virus of vaccinia. *J. Exp. Med.* **75**, 651.

GROTH, W. (1937). Quantum yields of gas reactions induced by short-wave ultra-violet light. *Z. Phys. Chem.* B, **37**, 307.

GÜNTHER, P. & HOLZAPFEL, L. (1939a). Decomposition of water vapour by X-rays. *Z. Phys. Chem.* B, **42**, 346.

GÜNTHER, P. & HOLZAPFEL, L. (1939b). The X-ray sensitivity of liquid water and ice. *Z. Phys. Chem.* B, **44**, 374.

GÜNTHER, P. & LEICHTER, H. (1936). Decomposition of HI and synthesis of HBr by X-rays. *Z. Phys. Chem.* B, **34**, 443.

HANSON, F.B. (1928). Effects of X-rays on productivity and sex-ratio in *Drosophila*. *Amer. Nat.* **62**, 352.

HANSON, F.B. & HEYS, F. (1929). Analysis of effects of different rays of radium in producing lethal mutations in *Drosophila*. *Amer. Nat.* **63**, 201.

HANSON, F.B. & HEYS, F. (1932). Radium and lethal mutations in *Drosophila*. *Amer Nat.* **66**, 335.

HASKINS, C.P. (1938). Apparatus for studying the biological effects of cathode rays. *J. Appl. Phys.* **9**, 553.

HEIDT, L.J. (1939). Mercury lamp for ultra-violet light of wave-length 2536 A. *Science*, **90**, 473.

HELFER, R.G. (1940). X-ray-induced and naturally occurring chromosomal variations in *Drosophila pseudoobscura*. *Genetics*, **26**, 1.

HENSHAW, P.S. (1932). Effect of X-rays on time of first cleavage in *Arbacia* eggs. *Amer. J. Roentg.* **27**, 890.

HENSHAW, P.S. (1938). Action of X-rays on nucleated and non-nucleated egg fragments. *Amer. J. Cancer*, **33**, 258.

HENSHAW, P.S. (1940). Action of X-rays on the gametes of *Arbacia punctulata*. *Amer. J. Roentg.* **43**, 899.

HENSHAW, P.S. & COHEN, I. (1940). Action of X-rays on the gametes of *Arbacia punctulata*. *Amer. J. Roentg.* **43**, 917.

HENSHAW, P.S. & FRANCIS, D.S. (1936). Effect of X-rays on cleavage in *Arbacia* eggs. *Biol. Bull. Woods Hole*, **70**, 28.

HENSHAW, P.S. & HENSHAW, C.T. (1933). Changes in susceptibility of *Drosophila* eggs to α-particles. *Biol. Bull. Woods Hole*, **64**, 348.

HENSHAW, P.S., HENSHAW, C.T. & FRANCIS, D.S. (1933). Action of X-rays on *Arbacia* eggs. *Radiology*, **21**, 533.

HERCIK, F. (1933). Action of α-rays on bacteria. *Strahlentherapie*, **47**, 374.

HERCIK, F. (1934a). Bactericidal action of α-rays. *Strahlentherapie*, **49**, 438.

HERCIK, F. (1934b). Temperature coefficient of the bactericidal action of α-rays. *Strahlentherapie*, **49**, 703.

HERCIK, F. (1936). Bactericidal action of ultra-violet light. *J. Gen. Physiol.* **20**, 589.

HINSHELWOOD, C.N. (1940). *Kinetics of Chemical Change*. Oxford.

HIRSCHFELDER, J.O. & TAYLOR, H.S. (1938). α-particle reactions in CO, O_2, and CO_2 systems. *J. Chem. Phys.* **6**, 783.

HOFFMANN, J.G. & REINHARD, M.C. (1934). Recovery from the effects of irradiation. *Radiology*, **23**, 738.

HOLIDAY, E.P. (1936). Absorption spectra of tyrosine, tryptophan and their mixtures. *Biochem. J.* **30**, 1795.

HOLLAENDER, A. & CLAUS, W.D. (1936). Bactericidal action of ultra-violet light. *J. Gen. Physiol.* **19**, 753.

HOLLAENDER, A. & DUGGAR, B.M. (1936). Action of ultra-violet light on *Bact. coli* and tobacco mosaic virus. *Proc. Nat. Acad. Sci., Wash.*, **22**, 19.

HOLLAENDER, A. & EMMONS, C.W. (1941). Production of mutations in fungi by ultra-violet light. *Cold Spr. Harb. Symp.* **9**, 179.
HOLWECK, F. & LACASSAGNE, A. (1930). Action on yeasts of soft X-rays. *C.R. Soc. Biol., Paris*, **103**, 60.
HORNBECK, G. & HOWELL, I. (1941). Production of secondary electrons by β-rays. *Proc. Amer. Phil. Soc.* **84**, 33.
HOUSTON, R.A. (1911). Absorption of light by inorganic salts. *Proc. Roy. Soc. Edinb.* **31**, 521.

JAFFÉ, G. (1913). Recombination of ions in α-particle columns. *Ann. Phys., Lpz.*, **42**, 303.
JORDAN, P. (1938a). Physical structure of genes and viruses. *Naturwissenschaften*, **26**, 693.
JORDAN, P. (1938b). Biological action of radiations. *Radiologica*, **2**, 16.
JORDAN, P. (1938c). Methods and results of radio-biology. *Radiologica*, **3**, 157.
JUNGERS, J.C. (1932). Decomposition and synthesis of ammonia by α-rays. *Bull. Soc. chim. Belg.* **41**, 377.
JUUL, J. & KEMP, T. (1933). Effect of X-rays, γ-rays, ultra-violet light and heat on chick tissue in culture. *Strahlentherapie*, **48**, 31.

KANNE, W.R. (1937). Preparation of polonium sources. *Phys. Rev.* **52**, 380.
KARA-MICHAILOVA, E. & LEA, D.E. (1940). Ionization measurements in gases at high pressure. *Proc. Camb. Phil. Soc.* **36**, 101.
KAUFMANN, B.P. (1939). Distribution of induced breaks along the X chromosome of *Drosophila*. *Proc. Nat. Acad. Sci., Wash.*, **25**, 571.
KAUFMANN, B.P. (1941a). Effect of intermittent exposure in the production of chromosome aberrations in *Drosophila*. *Proc. Nat. Acad. Sci., Wash.*, **27**, 18.
KAUFMANN, B.P. (1941b). Induced chromosomal breaks in *Drosophila*. *Cold Spr. Harb. Symp.* **9**, 82.
KAUFMANN, B.P. (1943). Complex induced rearrangement of *Drosophila* chromosomes. *Proc. Nat. Acad. Sci., Wash.*, **29**, 8.
KAUFMANN, B.P. & DEMEREC, M. (1937). Frequency of induced breaks in chromosomes of *Drosophila*. *Proc. Nat. Acad. Sci., Wash.*, **23**, 484.
KAYE, G.W.C. & BINKS, W. (1937). Free air chamber for the measurement of γ-rays. *Proc. Roy. Soc. A*, **161**, 564.
KAYE, G.W.C. & BINKS, W. (1940). Emission and transmission of X- and γ-rays. *Brit. J. Radiol.* **13**, 193.
KERNBAUM, M. (1909). Decomposition of water by β-rays and ultra-violet light. *Radium*, **6**, 225.
KHVOSTOVA, V.V. & GAVRILOVA, A.A. (1938). Relation between number of translocations in *Drosophila* and dosage. *Biol. Zh.* **7**, 381.
KIKUCHI, S. & AOKI, H. (1939). Scattering of fast neutrons. *Proc. Phys.-Math. Soc. Japan*, **21**, 75.
KINSEY, V.E. (1935). Effects of X-rays on glutathione. *J. Biol. Chem.* **110**, 551.
KIRCHNER, F. (1930). General physics of X-rays. *Handb. Exp. Phys.* **24** (i). Leipzig.

KLEIN, O. & NISHINA, Y. (1929). Scattering of radiation by free electrons. *Z. Phys.* **52**, 853.

KLEMPERER, O. (1927). Distribution of ions in α-ray tracks. *Z. Phys.* **45**, 225.

KNAPP, E. & SCHREIBER, H. (1939). Mutations induced in *Sphaerocarpus* by ultra-violet light. *Proc. 7th. Int. Congr. Genet.* p. 175.

KNORR, M. & RUFF, H. (1934). Action of cathode rays on bacteria and bacteriophage. *Arch. Hyg.* **113**, 92.

KOLLER, L.R. (1939). Bactericidal action of ultra-violet light. *J. Appl. Phys.* **10**, 624.

KOLLER, P.C. (1943). Effects of radiation on pollen grain development, differentiation and germination. *Proc. Roy. Soc. Edinb.* B, **61**, 398.

KOLLER, P.C. & AHMED, I.A.R.S. (1942). X-ray-induced structural changes in chromosomes of *Drosophila pseudoobscura*. *J. Genet.* **44**, 53.

KOLLER, P.C. & AUERBACH, C.A. (1941). Chromosome breakage and sterility in the mouse. *Nature, Lond.*, **148**, 501.

KOLUMBAN, A.D. & ESSEX, H. (1940). Effect of electric fields on the decomposition of N_2O by α-rays. *J. Chem. Phys.* **8**, 450.

KOTVAL, J.P. (1944). Production of chromosome structural changes in *Tradescantia* microspores by radiations. Thesis, Cambridge.

KRUEGER, A.P. (1930). Estimation of bacteriophage. *J. Gen. Physiol.* **13**, 557.

KRUGER, P.G. (1940). Biological effects of slow neutrons. *Proc. Nat. Acad. Sci., Wash.*, **26**, 181.

KÜSTNER, H. & TRUBESTEIN, H. (1937). Analysis of X-ray absorption into photoelectric absorption and Compton scattering. *Ann. Phys., Lpz.*, **28**, 385.

LACASSAGNE, A. & HOLWECK, F. (1928). Bactericidal action of X-rays. *C.R. Acad. Sci., Paris*, **186**, 1316, 1318.

LACASSAGNE, A. & HOLWECK, F. (1929a). Bactericidal action of soft X-rays. *C.R. Acad. Sci., Paris*, **188**, 197, 200.

LACASSAGNE, A. & HOLWECK, F. (1929b). Action of soft X-rays on *B. prodigiosus*. *C.R. Soc. Biol., Paris*, **100**, 1101.

LACASSAGNE, A. & HOLWECK, F. (1930). Action of X-rays, α-rays and ultra-violet light on yeast. *C.R. Acad. Sci., Paris*, **190**, 524, 527.

LANNING, F.C. & LIND, S.C. (1938). Chemical action of α-particles on aqueous solutions. *J. Phys. Chem.* **42**, 1229.

LASNITZKI, I. (1940). Effects of X-rays on chick tissue in culture. *Brit. J. Radiol.* **13**, 279.

LASNITZKI, I. (1943a). Effects of X-rays on chick tissue in culture. *Brit. J. Radiol.* **16**, 61.

LASNITZKI, I. (1943b). Effects of X-rays on chick tissue in culture. *Brit. J. Radiol.* **16**, 137.

LAVIN, G.I. (1943). Simplified ultra-violet microscopy. *Rev. Sci. Instrum.* **14**, 375.

LAVIN, G.I., THOMPSON, R.H.S. & DUBOS, R.J. (1938). Ultra-violet absorption spectra of fractions isolated from pneumococci. *J. Biol. Chem.* **125**, 75.

LEA, D.E. (1938a). Time-intensity factor. *Brit. J. Radiol.* **11**, 489.

LEA, D.E. (1938b). Delay in cellular division. *Brit. J. Radiol.* **11**, 554.
LEA, D.E. (1940a). Radiation method for determining number of genes in *Drosophila*. *J. Genet.* **39**, 181.
LEA, D.E. (1940b). Sizes of viruses and genes by radiation methods. *Nature, Lond.*, **146**, 137.
LEA, D.E. & CATCHESIDE, D.G. (1942). Induction by radiation of chromosome aberrations in *Tradescantia*. *J. Genet.* **44**, 216.
LEA, D.E. & CATCHESIDE, D.G. (1945a). Recessive lethals, dominant lethals and chromosome aberrations in *Drosophila*. *J. Genet.* **47**, 10.
LEA, D.E. & CATCHESIDE, D.G. (1945b). Bearing of radiation experiments on the size of the gene. *J. Genet.* **47**, 41.
LEA, D.E. & HAINES, R.B. (1940). Bactericidal action of ultra-violet light. *J. Hyg., Camb.*, **40**, 162.
LEA, D.E., HAINES, R.B. & BRETSCHER, E. (1941). Bactericidal action of X-rays, neutrons and radioactive radiations. *J. Hyg., Camb.*, **41**, 1.
LEA, D.E., HAINES, R.B. & COULSON, C.A. (1936). Bactericidal action of radioactive radiations. *Proc. Roy. Soc.* B, **120**, 47.
LEA, D.E., HAINES, R.B. & COULSON, C.A. (1937). Actions of γ-rays on bacteria. *Proc. Roy. Soc.* B, **123**, 1.
LEA, D.E. & SALAMAN, M.H. (1942). Inactivation of vaccinia virus by radiations. *Brit. J. Exp. Path.* **23**, 27.
LEA, D.E. & SMITH, K.M. (1940). Inactivation of plant viruses by radiations. *Parasitology*, **32**, 405.
LEA, D.E. & SMITH, K.M. (1942). Inactivation of plant viruses by γ-rays, X-rays and α-rays. *Parasitology*, **34**, 227.
LEA, D.E., SMITH, K.M., HOLMES, B. & MARKHAM, R. (1944). Direct and indirect actions of radiation on viruses and enzymes. *Parasitology*, **36**, 110.
LEVIN, B.S. & LOMINSKI, I. (1935). Bactericidal action of soft X-rays. *C.R. Acad. Sci., Paris*, **200**, 863.
LEVIN, B.S. & LOMINSKI, I. (1936). Inactivation of fowl plague virus by X-rays. *C.R. Acad. Sci., Paris*, **203**, 287, 350.
LEWIS, B. (1928). Photochemical decomposition of HI. *J. Phys. Chem.* **32**, 270.
LIND, S.C. (1928). *Chemical Effects of α-particles and Electrons*. New York.
LIND, S.C. & BARDWELL, D.C. (1928). Synthesis of ammonia by α-rays. *J. Amer. Chem. Soc.* **50**, 745.
LIND, S.C. & BARDWELL, D.C. (1929). Ozonization, and interaction of oxygen with nitrogen, under α-rays. *J. Amer. Chem. Soc.* **51**, 2751.
LIND, S.C., BARDWELL, D.C. & PERRY, J.H. (1926). Chemical action of α-rays on gaseous unsaturated carbon compounds. *J. Amer. Chem. Soc.* **48**, 1556.
LIND, S.C. & LIVINGSTON, R. (1932). Photochemical polymerization of acetylene. *J. Amer. Chem. Soc.* **54**, 94.
LIND, S.C. & LIVINGSTON, R. (1936). Radiochemical synthesis and decomposition of HBr. *J. Amer. Chem. Soc.* **58**, 612.
LIND, S.C. & OGG, E.F. (1931). Temperature coefficient of the synthesis of HBr by α-rays. *Z. Phys. Chem.* Bodenstein festband, p. 801.
LIVINGSTON, M.S. & BETHE, H.A. (1937). Nuclear dynamics: experimental. *Rev. Mod. Phys.* **9**, 245.

LOEB, J. (1910). Prevention of the toxic action of various agencies upon the fertilized egg through suppression of oxidation in the cell. *Science*, **32**, 411.

LOEB, J. & WASTENEYS, H. (1911). Respiration rate of marine invertebrate eggs. *Biochem. Z.* **36**, 345.

LORENZ, K.P. & HENSHAW, P.S. (1941). Bactericidal action of X-rays. *Radiology*, **36**, 471.

LUDFORD, R.J. (1932). Cytological changes after irradiation of malignant growths. *Imp. Cancer Res. Fund Rep.* **10**, 125.

LURIA, S.E. (1939). Action of X-rays and α-rays on bacteria. *C.R. Acad. Sci., Paris*, **209**, 604.

LURIA, S.E. & ANDERSON, T.F. (1942). Electron micrography of bacteriophages. *Proc. Nat. Acad. Sci., Wash.*, **28**, 127.

LURIA, S.E. & EXNER, F.M. (1941). Direct and indirect actions of X-rays on bacteriophage. *Proc. Nat. Acad. Sci., Wash.*, **27**, 370.

LUYET, B.J. (1932). Effects of ultra-violet light, X-rays, and cathode rays on fungal spores. *Radiology*, **18**, 1019.

LUYET, B.J. (1934). Effect of X-rays on algae. *C.R. Soc. Biol., Paris*, **116**, 878.

MACKENZIE, K. & MULLER, H.J. (1940). Mutation effects of ultra-violet light in *Drosophila*. *Proc. Roy. Soc. B*, **129**, 491.

MCCLINTOCK, B. (1938). Production of deficiencies by aberrant behaviour of ring-shaped chromosomes. *Genetics*, **23**, 315.

MCCLINTOCK, B. (1939). Behaviour in successive nuclear divisions of a chromosome broken at meiosis. *Proc. Nat. Acad. Sci., Wash.*, **25**, 405.

MCCLINTOCK, B. (1941a). Stability of broken ends of chromosomes in maize. *Genetics*, **26**, 234.

MCCLINTOCK, B. (1941b). Association of mutants with homozygous deficiencies in maize. *Genetics*, **26**, 542.

MCKINLEY, E.B., FISHER, R. & HOLDEN, M. (1926). Inactivation of bacteriophage and of animal viruses by ultra-violet light. *Proc. Soc. Exp. Biol., N.Y.*, **23**, 408.

MANO, G. (1934). Absorption of α-rays. *Ann. Phys., Paris*, **1**, 407.

MARINELLI, L.D., NEBEL, B.R., GILES, N.H. & CHARLES, D.R. (1942). Induction by X-rays of chromosomal aberrations in *Tradescantia* microspores. *Amer. J. Bot.* **29**, 866.

MARKHAM, R., SMITH, K.M. & LEA, D.E. (1942). The sizes of viruses. *Parasitology*, **34**, 315.

MARKHAM, R., SMITH, K.M. & LEA, D.E. (1944). Size of the Shope rabbit papilloma virus. *Parasitology*, **35**, 178.

MARQUARDT, H. (1938). Roentgen-pathology of mitosis. *Z. Bot.* **32**, 401.

MARSHAK, A. (1935). Sensitive volume of chromosomes determined by X-irradiation. *Proc. Nat. Acad. Sci., Wash.*, **21**, 227.

MARSHAK, A. (1937). Effect of X-rays on chromosomes in mitosis. *Proc. Nat. Acad. Sci., Wash.*, **23**, 362.

MARSHAK, A. (1938a). Chromosome aberrations induced by X-rays, and the effect of ammonia. *Proc. Soc. Exp. Biol., N.Y.*, **38**, 705.

MARSHAK, A. (1938b). Stage of mitosis at which chromosomes are rendered less sensitive to X-rays by ammonia. *Proc. Soc. Exp. Biol., N.Y.*, **39**, 194.

MARSHAK, A. (1939a). Effects of fast neutrons on chromosomes in mitosis. *Proc. Soc. Exp. Biol., N.Y.*, **41**, 176.

MARSHAK, A. (1939b). Nature of chromosome division and duration of nuclear cycle. *Proc. Nat. Acad. Sci.*, **25**, 502.

MARSHAK, A. (1942a). Relative effects of X-rays and neutrons on chromosomes in different parts of the resting stage. *Proc. Nat. Acad. Sci., Wash.*, **28**, 29.

MARSHAK, A. (1942b). Effects of X-rays and neutrons on mouse lymphoma chromosomes in different stages of the nuclear cycle. *Radiology*, **39**, 621.

MARSHAK, A. & HUDSON, J.C. (1937). Effect of X-rays on chromosomes. *Radiology*, **29**, 669.

MARSHAK, A. & MALLOCH, W.S. (1942). Effect of fast neutrons on chromosomes in meiosis. *Genetics*, **27**, 576.

MATHER, K. & STONE, L.H.A. (1933). Effects of X-rays on chromosomes. *J. Genet.* **28**, 1.

MAVOR, J.W. & DE FOREST, D.M. (1924). Relative susceptibilities to X-rays of eggs and sperm of *Arbacia*. *Proc. Soc. Exp. Biol., N.Y.*, **22**, 19.

MAYNEORD, W.V. (1934). Physical basis of biological effects of ionizing radiations. *Proc. Roy. Soc. A*, **146**, 867.

MAYNEORD, W.V. (1940). Energy absorption of X- and γ-rays in tissue. *Brit. J. Radiol.* **13**, 235.

MAYNEORD, W.V. & HONEYBURNE, J. (1938). Depth dose with γ-rays. *Brit. J. Radiol.* **11**, 741.

MAYNEORD, W.V. & ROBERTS, J.E. (1937). Measurement of γ-rays in roentgens. *Brit. J. Radiol.* **10**, 365.

MELVILLE, H.W. (1936). Mercury lamp for wave-length 2536A. *Trans. Faraday Soc.* **32**, 1525.

METZ, C.W. & BOZEMAN, M.L. (1940). Induced chromosome changes in *Sciara*. *Proc. Nat. Acad. Sci., Wash.*, **26**, 228.

MICHL, W. (1914). Range of α-particles in liquids. *S.B. Akad. Wiss. Wien*, **123**, 1965.

MICKEY, G.H. (1938). Effect of temperature on frequency of translocations produced by X-rays in *Drosophila*. *Genetics*, **23**, 160.

MILLER, G.L. & STANLEY, W.M. (1941). Acetyl and phenylureido derivatives of tobacco mosaic virus. *J. Biol. Chem.* **141**, 905.

MITCHELL, J.S. (1942). Disturbance of nucleic acid metabolism by X- and γ-rays. *Brit. J. Exp. Path.* **23**, 285.

MOELWYN-HUGHES, E.A. (1933). *Kinetics of Reactions in Solution*. Oxford.

MOHLER, F.L. & TAYLOR, L.S. (1934). Bactericidal effects of X-rays. *Bur. Stand. J. Res.* **13**, 677.

MOORE, H.N. & KERSTEN, H. (1937). Inactivation of encephalitis virus by soft X-rays. *J. Bact.* **33**, 615.

MORNINGSTAR, O., EVANS, R.D. & HASKINS, C.P. (1941). Cathode-ray tube for biological experiments. *Rev. Sci. Instrum.* **12**, 358.

MOTTRAM, J.C. (1926). Effects of β-rays on *Colpidium*. *J. Roy. Micr. Soc.* **46**, 127.

MOTTRAM, J.C. (1927). Early change in nucleus of cells of tumours exposed to β-rays. *Brit. J. Radiol.* **32**, 61.

MOTTRAM, J.C. (1932). Life history of nucleus and nucleolus and effects of β-rays upon them. *J. Roy. Micr. Soc.* **52**, 362.

MOTTRAM, J.C. (1933a). Increase in nuclear size following irradiation. *J. Roy. Micr. Soc.* **53**, 213.

MOTTRAM, J.C. (1933b). Radio-sensitivity of the non-dividing cell. *Brit. J. Radiol.* **6**, 615.

MOTTRAM, J.C. (1935a). Alteration in sensitivity of cells to radiation, produced by cold and anaerobiosis. *Brit. J. Radiol.* **8**, 32.

MOTTRAM, J.C. (1935b). Effect of chemicals on radiosensitivity of bean roots. *Brit. J. Radiol.* **8**, 643.

MOTTRAM, J.C., SCOTT, G.M. & RUSS, S. (1926). Effects of β-rays upon division and growth of cancer cells. *Proc. Roy. Soc. B*, **100**, 326.

MULLER, H.J. (1928). Problem of genic modification. *Z. indukt. Abstamm.- u. VererbLehre*, Suppl. **1**, 234.

MULLER, H.J. (1929). Gene as basis of life. *Proc. Int. Congr. Plant Sci.* **1**, 897.

MULLER, H.J. (1930). Radiation and genetics. *Amer. Nat.* **64**, 220.

MULLER, H.J. (1935). Dimensions of chromosomes and genes in dipteran salivary glands. *Amer. Nat.* **69**, 405.

MULLER, H.J. (1938). Biological effects of radiation with special reference to mutation. *Act. Sci. Ind.* no. 725, p. 477.

MULLER, H.J. (1940). Analysis of process of structural change in chromosomes of *Drosophila*. *J. Genet.* **40**, 1.

MULLER, H.J. (1941). Induced mutations in *Drosophila*. *Cold Spr. Harb. Symp.* **9**, 151.

MULLER, H.J. & MACKENZIE, K. (1939). Discriminatory effects of ultraviolet light on mutation in *Drosophila*. *Nature, Lond.*, **143**, 83.

MULLER, H.J. & PROKOFYEVA, A.A. (1935). The individual gene in relation to the chromomere and chromosome. *Proc. Nat. Acad. Sci., Wash.*, **21**, 16.

MÜLLER, I. (1938). Mass absorption coefficients of water and of aqueous solutions. *Ann. Phys., Lpz.*, **32**, 625.

MUND, W. & JUNGERS, J.C. (1931). Polymerization of acetylene by β-rays. *Bull. Soc. chim. Belg.* **40**, 158.

NAGAI, M.A. & LOCHER, G.L. (1938). Induction of mutations in *Drosophila* by neutrons. *Genetics*, **23**, 179.

NAIDU, R. (1934). Ionization curves of α-rays. *Ann. Phys., Paris*, **1**, 72.

NEBEL, B.R. (1937). Effects of X- and γ-rays on *Tradescantia* chromosomes. *Amer. J. Bot.* **24**, 365.

NEEDHAM, J. (1931). *Chemical Embryology*. Cambridge.

NEWCOMBE, H.B. (1942a). Effects of X-rays on chromosomes. *J. Genet.* **43**, 145.

NEWCOMBE, H.B. (1942b). Effects of X-rays on chromosomes. *J. Genet.* **43**, 237.

NISHINA, Y. & MORIWAKI, D. (1939). Induction of sex-linked lethals in *Drosophila* by neutrons. *Sci. Pap. Inst. Phys. Chem. Tokyo*, **36**, 419.

NISHINA, Y. & MORIWAKI, D. (1941). Induction of sex-linked lethals in *Drosophila* by neutrons. *Sci. Pap. Inst. Phys. Chem. Tokyo*, **38**, 371.

NORD, F.F. & WERKMAN, C.H. (editors) (1941–3). *Advances in Enzymology*. New York.

NORTHROP, J.H. (1939). *Crystalline Enzymes*. New York.

NOYES, W.A. (1937). Photochemical decomposition of N_2O. *J. Chem. Phys.* **5**, 807.

NURNBERGER, C.E. (1934). Effects of α-particles on water and on ferrous sulphate solution. *J. Phys. Chem.* **38**, 47.

NURNBERGER, C.E. (1936a). Decomposition of air-free water by α-rays. *J. Phys. Chem.* **41**, 431.

NURNBERGER, C.E. (1936b). Production of H_2O_2 in water by α-rays. *J. Chem. Phys.* **4**, 697.

NURNBERGER, C.E. (1937). Ionization theory and radiobiological reactions. *Proc. Nat. Acad. Sci., Wash.*, **23**, 189.

OLDENBURG, O. (1924). A light filter for mercury line 2536A. *Z. Phys.* **29**, 328.

OLITSKY, P.K. & GATES, F.L. (1927). Effect of ultra-violet light on *S. aureus* and on the virus of vesicular stomatitis. *Proc. Soc. Exp. Biol., N.Y.*, **24**, 431.

OLIVER, C.P. (1932). The effect of varying X-ray dose on the frequency of mutation in *Drosophila*. *Z. indukt. Abstamm.- u. VererbLehre*, **61**, 447.

ORR, W.J.C. & BUTLER, J.A.V. (1935). Rate of diffusion of deuterium hydroxide in water. *J. Chem. Soc.* p. 1273.

PACKARD, C. (1926). Killing of *Drosophila* eggs by X-rays. *J. Cancer Res.* **10**, 319.

PACKARD, C. (1935). Variation with age of sensitivity of *Drosophila* eggs to X-rays. *Radiology*, **25**, 223.

PARKER, R.F. (1938). The infectivity of vaccinia virus. *J. Exp. Med.* **67**, 725.

PATTERSON, J.T. (1932). Lethal mutations and deficiencies produced by X-rays in the X chromosome of *Drosophila*. *Amer. Nat.* **66**, 193.

PEEL, G.N. (1939). Quartz mercury lamps. *Brit. J. Radiol.* **12**, 99.

PEKAREK, J. (1927). Influence of X-rays on nuclear and cell division in bean root-tips. *Planta*, **4**, 299.

PESKOFF, N. (1919). Gaseous filters for ultra-violet light. *Z. wiss. Photogr.* **18**, 235.

PHILIPP, K. (1923). Stopping of α-particles in liquids and vapours. *Z. Phys.* **17**, 23.

PODDUBNAJA-ARNOLDI, V. (1936). Germination of pollen on artificial medium. *Planta*, **25**, 502.

PONTECORVO, G. (1941). Induction of chromosome losses in *Drosophila* sperm, and their linear dependence on dosages of radiation. *J. Genet.* **41**, 195.

PONTECORVO, G. (1942). Dominant lethals in *Drosophila*. *J. Genet.* **43**, 295.

PONTECORVO, G. & MULLER, H.J. (1941). Lethality of dicentric chromosomes in *Drosophila*. *Genetics*, **26**, 165.

PRICE, W.C. & GOWEN, J.W. (1937). Inactivation of tobacco mosaic virus by ultra-violet light. *Phytopathology*, **27**, 267.

PUGSLEY, A.T., ODDIE, T.H. & EDDY, C.E. (1935). Bactericidal action of X-rays. *Proc. Roy. Soc.* B, **118**, 276.

QUIMBY, E.H. & MACCOMB, W.S. (1937). Rate of recovery of human skin from X- or γ-irradiation. *Radiology*, **29**, 305.

RAJEWSKY, B.N., KREBS, A. & ZICKLER, H. (1936). Mutations by cosmic rays. *Naturwissenschaften*, **24**, 619.

RAJEWSKY, B.N. & TIMOFÉEFF-RESSOVSKY, N.W. (1939). Cosmic rays and mutations. *Z. indukt. Abstamm.- u. VererbLehre*, **77**, 488.

RASETTI, F. (1937). *Elements of Nuclear Physics*. London.

RAY-CHAUDHURI, S.P. (1944). Production of mutations in *Drosophila* by γ-rays at low intensity. *Proc. Roy. Soc. Edinb.* B, **62**, 66.

RENTSCHLER, H.C., NAGY, R. & MOUROMSEFF, G. (1941). Bactericidal effect of ultra-violet light. *J. Bact.* **41**, 745.

REYNOLDS, J.P. (1941). Induction of chromosome abnormalities in *Sciara* by X-rays. *Proc. Nat. Acad. Sci., Wash.*, **27**, 204.

RICK, C.M. (1940). X-ray-induced deletions in *Tradescantia* chromosomes. *Genetics*, **25**, 466.

RILEY, H.P. (1936). Effects of X-rays on chromosomes in meiosis and mitosis. *Cytologia, Tokyo*, **7**, 131.

RISSE, O. (1929). X-ray photolysis of H_2O_2. *Z. Phys. Chem.* A, **140**, 133.

RIVERS, T.M. & GATES, F.L. (1928). Inactivation of vaccinia virus by ultra-violet light. *J. Exp. Med.* **47**, 45.

ROBERTSON, M. (1932). Effect of γ-rays on *Bodo caudatus*. *Quart. J. Micr. Sci.* **75**, 511.

ROBINOW, C.F. (1942). Nuclear apparatus of bacteria. *Proc. Roy. Soc.* B, **130**, 299.

ROBINOW, C.F. (1944). Cytological observations on bacteria. *J. Hyg., Camb.*, **43**, 413.

RUTHERFORD, E., CHADWICK, J. & ELLIS, C.D. (1930). *Radiations from Radioactive Substances*. Cambridge.

SAX, H.J. & SAX, K. (1935). Chromosome structure and behaviour in mitosis and meiosis. *J. Arnold Arbor.* **16**, 423.

SAX, K. (1938). Induction by X-rays of chromosome aberrations in *Tradescantia* microspores. *Genetics*, **23**, 494.

SAX, K. (1939). Time factor in X-ray production of chromosome aberrations. *Proc. Nat. Acad. Sci., Wash.*, **25**, 225.

SAX, K. (1940). X-ray-induced chromosomal aberrations in *Tradescantia*. *Genetics*, **25**, 41.

SAX, K. (1941a). Behaviour of X-ray-induced chromosomal aberrations in onion root-tips. *Genetics*, **26**, 418.

SAX, K. (1941b). Types and frequencies of chromosomal aberrations induced by X-rays. *Cold Spr. Harb. Symp.* **9**, 93.

SAX, K. (1942). Mechanisms of X-ray effects on cells. *J. Gen. Physiol.* **25**, 533.

SAX, K. (1943). Effect of centrifuging upon production of chromosomal aberrations by X-rays. *Proc. Nat. Acad. Sci., Wash.*, **29**, 18.

SAX, K. & ENZMANN, E.V. (1939). Effect of temperature on X-ray-induced chromosome aberrations. *Proc. Nat. Acad. Sci., Wash.*, **25**, 397.

SAX, K. & MATHER, K. (1939). Analysis of progressive chromosome splitting. *J. Genet.* **37**, 483.

SAX, K. & SWANSON, C.P. (1941). Differential sensitivity of cells to X-rays. *Amer. J. Bot.* **28**, 52.

SCOTT, C.M. (1937). Biological actions of X- and γ-rays. *Spec. Rep. Ser. Med. Res. Coun., Lond.*, no. 223.

SHEARIN, P.E. & PARDUE, T.E. (1942). Production of δ-rays by fast electrons. *Proc. Amer. Phil. Soc.* **85**, 243.

SIDKY, A.R. (1940). Translocation between sperm and egg chromosomes in *Drosophila*. *Amer. Nat.* **74**, 475.

SINGLETON, W.R. (1939). Cytological observations on deficiencies produced by treating maize pollen with ultra-violet light. *Genetics*, **24**, 109.

SINGLETON, W.R. & CLARK, F.J. (1940). Cytological effects of treating maize pollen with ultra-violet light. *Genetics*, **25**, 136.

SIZOO, G.J. & WILLEMSEN, H. (1938). Absorption of radium γ-rays. *Physica*, **5**, 100.

SLIZYNSKI, B.M. (1938). Salivary chromosome studies of lethals in *Drosophila*. *Genetics*, **23**, 283.

SLIZYNSKI, B.M. (1942). Deficiencies induced by ultra-violet light in *Drosophila* chromosomes. *Proc. Roy. Soc. Edinb.* B, **61**, 297.

SMITH, C. & ESSEX, H. (1938). Effect of electric fields on the decomposition of ammonia by α-rays. *J. Chem. Phys.* **6**, 188.

SMITH, K.M. (1940). *The Virus, Life's Enemy*. Cambridge.

SNELL, G.D. (1935). Induction by X-rays of hereditary changes in mice. *Genetics*, **20**, 545.

SNELL, G.D. (1939). Induction of hereditary sterility in mice by neutrons. *Proc. Nat. Acad. Sci., Wash.*, **25**, 11.

SNELL, G.D. (1941). Production by X-rays of hereditary changes in mice. *Radiology*, **36**, 189.

SNELL, G.D. & AMES, F.B. (1939). Hereditary changes in the descendants of female mice exposed to X-rays. *Amer. J. Roentg.* **41**, 248.

SONNENBLICK, B.P. (1940). Cytology and development of embryos of X-rayed *Drosophila*. *Proc. Nat. Acad. Sci., Wash.*, **26**, 373.

SPEAR, F.G. (1935). Tissue culture and radiological research. *Brit. J. Radiol.* **8**, 68, 280.

SPEAR, F.G. (1944). Action of neutrons on bacteria. *Brit. J. Radiol.* **17**, 348.

SPEAR, F.G. & GLÜCKSMANN, A. (1938). Effect of γ-rays on cells of the tadpole. *Brit. J. Radiol.* **11**, 533.

SPEAR, F.G. & GLÜCKSMANN, A. (1941). Effect of γ-rays on cells of the tadpole; spaced irradiation. *Brit. J. Radiol.* **14**, 65.

SPEAR, F.G., GRAY, L.H. & READ, J. (1938). Biological effects of fast neutrons. *Nature, Lond.*, **142**, 1074.

SPEAR, F.G. & GRIMMETT, L.G. (1933). Dependence upon intensity of the effects of γ-rays on chick tissue in culture. *Brit. J. Radiol.* **6**, 387.

SPENCER, R.R. (1935). Action of β- and γ-rays on bacteria. *U.S. Publ. Hlth Rep.* **50**, 1642.

STADLER, L.J. (1928). Genetic effects of X-rays in maize. *Proc. Nat. Acad. Sci., Wash.*, **14**, 69.

STADLER, L.J. (1930a). Recovery following genetic deficiency in maize. *Proc. Nat. Acad. Sci., Wash.*, **16**, 714.

STADLER, L.J. (1930b). Some genetic effects of X-rays in plants. *J. Hered.* **21**, 3.

STADLER, L.J. (1931). Experimental modification of heredity in crop plants. *Sci. Agric.* **11**, 557.

STADLER, L.J. (1939). Genetic studies with ultra-violet radiation. *Proc. 7th Int. Congr. Genet.* p. 269.

STADLER, L.J. (1941). Comparison of ultra-violet and X-ray effects on mutation. *Cold Spr. Harb. Symp.* **9**, 168.

STADLER, L.J. & SPRAGUE, G.F. (1936a). Genetic effects of unfiltered ultra-violet light on maize. *Proc. Nat. Acad. Sci., Wash.*, **22**, 572.

STADLER, L.J. & SPRAGUE, G.F. (1936b). Genetic effects of filtered ultra-violet light on maize. *Proc. Nat. Acad. Sci., Wash.*, **22**, 579.

STADLER, L.J. & SPRAGUE, G.F. (1936c). Genetic effects of nearly monochromatic ultra-violet light (2536A.) on maize. *Proc. Nat. Acad. Sci., Wash.*, **22**, 584.

STADLER, L.J. & SPRAGUE, G.F. (1937). Contrasts in the genetical effects of ultra-violet light and X-rays. *Science*, **85**, 57.

STADLER, L.J. & UBER, F. (1938). Genetic effects of ultra-violet light on maize. *Genetics*, **23**, 171.

STADLER, L.J. & UBER, F. (1942). Comparison of genetic effects of different wave-lengths of ultra-violet light on maize. *Genetics*, **27**, 84.

STAHEL, E. & JOHNER, W. (1934). Number of γ-ray quanta emitted by radium. *J. Phys. Radium*, **5**, 97.

STEACIE, E.W.R. & PHILLIPS, N.W.F. (1938). Mercury lamp for 2536A. *Canad. J. Res.* B, **16**, 219.

STENSTRÖM, W. & LOHMANN, A. (1928). Effects of X-rays on solutions of tyrosine and cystine. *J. Biol. Chem.* **79**, 673.

STRANGEWAYS, T.S.P. & HOPWOOD, F.L. (1926). Effects of X-rays on mitosis. *Proc. Roy. Soc.* B, **100**, 283.

STRANGEWAYS, T.S.P. & OAKLEY, H.E.H. (1923). Effects of X-rays on chick tissue in culture. *Proc. Roy. Soc.* B, **95**, 373.

STURTEVANT, A.H. & BEADLE, G.W. (1940). *An Introduction to Genetics.* Philadelphia.

SUTTON, E. (1943). Cytological analysis of Bar eye in *Drosophila*. *Genetics*, **28**, 97.

SVEDBERG, T. & PEDERSEN, K.O. (1940). *The Ultracentrifuge.* Oxford.

SWANSON, C.P. (1940). Induction of chromosome aberrations by ultra-violet light and X-rays in *Tradescantia*. *Proc. Nat. Acad. Sci., Wash.*, **26**, 366.

SWANSON, C.P. (1942). Effects of ultra-violet light and X-rays on pollen tube chromosomes of *Tradescantia*. *Genetics*, **27**, 491.

SWANSON, C.P. (1943). Differential sensitivity of prophase pollen tube chromosomes to X-rays and ultra-violet light. *J. Gen. Physiol.* **26**, 485.

SYVERTON, J.T., BERRY, G.P. & WARREN, S.L. (1941). X-ray inactivation of Shope papilloma virus. *J. Exp. Med.* **74**, 223.

TANG, P.S. (1931). Oxygen tension: oxygen consumption curve of unfertilized *Arbacia* eggs. *Biol. Bull. Woods Hole*, **60**, 242.

TANSLEY, K., SPEAR, F.G. & GLÜCKSMANN, A. (1937). Effect of γ-rays on cell division in the rat retina. *Brit. J. Ophthal.* **21**, 273.

THODAY, J.M. (1942). Effects of neutrons and X-rays on chromosomes of *Tradescantia*. *J. Genet.* **43**, 189.

THOMAS, L.B. (1941). Monochromatic source of mercury resonance radiation. *Rev. Sci. Instrum.* **12**, 309.

TIMOFÉEFF-RESSOVSKY, N.W. (1933*a*). Back mutation and gene mutability in different directions. *Z. indukt. Abstamm.- u. VererbLehre*, **64**, 173.

TIMOFÉEFF-RESSOVSKY, N.W. (1933*b*). Back mutation and gene mutability in different directions. *Z. indukt. Abstamm.- u. VererbLehre*, **65**, 278.

TIMOFÉEFF-RESSOVSKY, N.W. (1933*c*). Back mutation and gene mutability in different directions. *Z. indukt. Abstamm.- u. VererbLehre*, **66**, 165.

TIMOFÉEFF-RESSOVSKY, N.W. (1937). *Mutationsforschung*. Dresden.

TIMOFÉEFF-RESSOVSKY, N.W. (1939). Relation between gene and chromosome mutation. *Chromosoma*, **1**, 310.

TIMOFÉEFF-RESSOVSKY, N.W. & DELBRÜCK, M. (1936). Radiation researches on visible mutations and the mutability of single genes in *Drosophila*. *Z. indukt. Abstamm.- u. VererbLehre*, **71**, 322.

TIMOFÉEFF-RESSOVSKY, N.W. & ZIMMER, K.G. (1938). Induction of mutations by neutrons in *Drosophila*. *Naturwissenschaften*, **26**, 108 and **27**, 362.

TIMOFÉEFF-RESSOVSKY, N.W. & ZIMMER, K.G. (1939). Radiation genetics. *Strahlentherapie*, **66**, 684.

TIMOFÉEFF-RESSOVSKY, N.W., ZIMMER, K.G. & DELBRÜCK, M. (1935). Gene mutation and gene structure. *Nachr. Ges. Wiss. Göttingen*, **1**, 189.

UBER, F.M. (1939). Ultra-violet transmission by maize pollen. *Amer. J. Bot.* **26**, 799.

UBER, F.M. (1940). Mercury lamp source for monochromators. *Rev. Sci. Instrum.* **11**, 300.

UBER, F.M. (1941). Quantum yield of inactivation of tobacco mosaic virus by ultra-violet light. *Nature, Lond.*, **147**, 148.

UBER, F.M. & ELLS, V.R. (1941). Ultra-violet absorption spectrum of ribonuclease. *J. Biol. Chem.* **141**, 229.

UBER, F.M., HAYASHI, T. & ELLS, V.R. (1941). Ultra-violet transmission by vitelline membrane of hen's egg. *Science*, **93**, 22.

UBER, F.M. & JACOBSOHN, S. (1938). Large quartz monochromator for biophysical research. *Rev. Sci. Instrum.* **9**, 150.

UBER, F.M. & MCLAREN, A.D. (1941). Photochemical yield for inactivation of trypsin. *J. Biol. Chem.* **141**, 231.

VAUGHAN, W.E. & NOYES, W.A. (1930). Quantum efficiency of ozone formation by short-wave ultra-violet light. *J. Amer. Chem. Soc.* **52**, 559.

VICTOREEN, J.A. (1943). X-ray mass absorption coefficients. *J. Appl. Phys.* **14**, 95.

WADDINGTON, C.H. (1939). *Introduction to Modern Genetics*. London.

WARD, F.D. (1935). Induction of mutations in Drosophila by α-rays. *Genetics*, **20**, 230.

WEISS, J. (1944). Radiochemistry in aqueous solutions. *Nature, Lond.*, **153**, 748.

WELLS, W.F. (1940). Bactericidal action of ultra-violet light. *J. Franklin Inst.* **229**, 347.

WHITAKER, D.M. (1931). Rate of oxygen consumption by fertilized and unfertilized eggs. *J. Gen. Physiol.* **15**, 167.

WHITAKER, D.M. (1933). Rate of oxygen consumption by fertilized and unfertilized eggs. *J. Gen. Physiol.* **16**, 474.

WHITAKER, M.D., BJORKSTEAD, W. & MITCHELL, A.C.G. (1934). Preparation of polonium sources. *Phys. Rev.* **46**, 629.

WHITE, M.J.D. (1935). Effects of X-rays on mitosis in spermatogonia of *Locusta*. *Proc. Roy. Soc.* B, **119**, 61.

WHITE, M.J.D. (1937). Effect of X-rays on first meiotic division in three species of *Orthoptera*. *Proc. Roy. Soc.* B, **124**, 183.

WIIG, E.O. (1935). Quantum yield for decomposition of ammonia by ultra-violet light. *J. Amer. Chem. Soc.* **57**, 1559.

WILHELMY, E., TIMOFÉEFF-RESSOVSKY, N.W. & ZIMMER, K.G. (1936). Induction of mutations in *Drosophila* by soft X-rays. *Strahlentherapie*, **57**, 521.

WILLIAMS, E.J. (1930). Rate of loss of energy by β-rays in matter. *Proc. Roy. Soc.* A, **130**, 310.

WILSON, C.T.R. (1923). Ionization by X-rays and β-rays. *Proc. Roy. Soc.* A, **104**, 1, 192.

WOLLMAN, E., HOLWECK, F. & LURIA, S.E. (1940). Inactivation of bacteriophage by X-rays and α-rays. *Nature, Lond.*, **145**, 935.

WOLLMAN, E. & LACASSAGNE, A. (1940). Evaluation of the dimensions of bacteriophages by means of X-rays. *Ann. Inst. Pasteur*, **64**, 5.

WOURTZEL, E. (1919). Chemical actions of α-rays. *Radium*, **11**, 289, 332.

WYCKOFF, R.W.G. (1930*a*). Bactericidal action of X-rays. *J. Exp. Med.* **52**, 435.

WYCKOFF, R.W.G. (1930*b*). Action of X-rays of various wave-lengths on *Bact. coli*. *J. Exp. Med.* **52**, 769.

WYCKOFF, R.W.G. (1932). Action of ultra-violet light on *Bact. coli*. *J. Gen. Physiol.* **15**, 351.

WYCKOFF, R.W.G. & LUYET, B.J. (1931). Effects of X-rays, cathode rays, and ultra-violet light on yeast. *Radiology*, **17**, 1171.

WYCKOFF, R.W.G. & RIVERS, T.M. (1930). Bactericidal action of cathode rays. *J. Exp. Med.* **51**, 921.

ZAHL, P.A. & COOPER, F.S. (1941). Physical and biological considerations in the use of slow neutrons for cancer therapy. *Radiology*, **37**, 673.

ZIMMER, K.G. (1934). Dependence of mutation rate on dose of radiation. *Strahlentherapie*, **51**, 179.

ZIMMER, K.G. & TIMOFÉEFF-RESSOVSKY, N.W. (1936). Production of mutations by α-rays. *Strahlentherapie*, **55**, 77.

ZIMMER, K.G. & TIMOFÉEFF-RESSOVSKY, N.W. (1938). Production of mutations in *Drosophila* by neutrons. *Strahlentherapie*, **63**, 528.

ZIRKLE, R.E. (1935). Killing of fern spores by α-rays. *Amer. J. Cancer*, **23**, 558.

ZIRKLE, R.E. (1940). Killing of bacteria, mould spores, and yeast by α-rays. *J. Cell. Comp. Physiol.* **16**, 221.

ADDITIONS TO BIBLIOGRAPHY

BONET-MAURY, P., PERAULT, R. & ERICHSEN, M.L. (1944). L'action bactériostatique des rayons X et ultra-violets. *Ann. de l'Inst. Pasteur*, **70**, 250.

CATCHESIDE, D.G., LEA, D.E. & THODAY, J.M. (1946a). Types of chromosome structural change induced by the irradiation of *Tradescantia* microspores. *J. Genet.* **47**, 113.

CATCHESIDE, D.G., LEA, D.E. & THODAY, J.M. (1946b). The production of chromosome structural changes in *Tradescantia* microspores in relation to dosage, intensity and temperature. *J. Genet.* **47**, 137.

CATSCH, A. & RADU, G. (1943). Über die Abhängigkeit der röntgen-induzierten Translokationsrate vom Reifezustand der bestrahlten Gameten bei *Drosophila melanogaster*—♂♂. *Naturwissenschaften*, **31/32**, 368.

DEMEREC, M. (1936). Frequency of 'cell-lethals' among lethals obtained at random in the X-chromosome of *Drosophila Melanogaster*. *Proc. Nat. Acad. Sci., Wash.*, **22**, 350.

FORSSBERG, A. (1945). Action of X-rays on catalase and its biological significance. *Ark. Kemi Min. Geol.* A, **21**, no. 7, p. 1.

FORSSBERG, A. (1946). Action of röntgen rays on the enzyme catalase. *Acta radiol., Stockh.*, **27**, 281.

HEIDENTHAL, G. (1945). The occurrence of X-ray-induced dominant lethal mutations in *Habrobracon*. *Genetics*, **30**, 197.

KAUFMANN, B.P. (1944). Cytology. (In *Ann. Rep. Dept. Genet.*) *Yearb. Carneg. Instn.* no. 43, p. 115.

KAUFMANN, B.P. (1946a). Organisation of the chromosome. I. Break distribution and chromosome recombination in *Drosophila melanogaster*. *J. Exp. Zool.* **102**, 293.

KAUFMANN, B.P. (1946b). Modification of the frequency of chromosomal rearrangements induced by X-rays in *Drosophila*. III. Effect of supplementary treatment at the time of chromosome recombination. *Genetics*, **31**, 449.

KAUFMANN, B.P. & GAY, H. (1947). The influence of X-rays and near infra-red rays on recessive lethals in *Drosophila melanogaster*. *Proc. Nat. Acad. Sci., Wash.*, **33**, 366.

KAUFMANN, B.P. & HOLLAENDER, A. (1946). Modification of the frequency of chromosomal rearrangements induced by X-rays in *Drosophila*. II. Use of ultraviolet radiation. *Genetics*, **31**, 368.

KAUFMANN, B.P., HOLLAENDER, A. & GAY, H. (1946). Modification of the frequency of chromosomal rearrangements induced by X-rays in *Drosophila*. I. Use of near infra-red radiation. *Genetics*, **31**, 349.

KING, E.D. (1947). The effect of low temperature upon the frequency of X-ray-induced mutations. *Genetics*, **32**, 161.

KOTVAL, J.P. & GRAY, L.H. (1947). Structural changes produced in microspores of *Tradescantia* by α-radiation. *J. Genet.* **48**, 135.

LATARJET, R. (1942). La loi de réciprocité dans l'irradiation d'un bactériophage avec les rayons X. *Ann. Inst. Pasteur*, **68**, 561.

LATARJET, R. (1944). Etude expérimentale de la loi de réciprocité dans l'effet biologique primaire des radiations. *C.R. Acad. Sci., Paris*, **218**, 294.

LATARJET, R. (1946). L'effet biologique primaire des radiations et la structure des microorganismes. *Rev. canad. Biol.* **5**, 9.

LATARJET, R. & WAHL, R. (1945). Précisions sur l'inactivation des bactériophages par les rayons ultraviolets. *Ann. Inst. Pasteur*, **71**, 336.

MAKHIJANI, J.K. (1944). The ineffectiveness of temperature in influencing the production of mutations by X-rays. *J. Univ. Bombay*, B, **13**, (3), 1.

MEDVEDEV, N.N. (1935). The contributory effect of cold with irradiation in the production of mutations. *C.R. Acad. Sci. U.R.S.S.* **4** (IX), 283.

MEDVEDEV, N.N. (1938). The contributory effect of heat with irradiation in the production of mutations. *C.R. Acad. Sci. U.R.S.S.* **19**, 301.

MULLER, H.J. (1941). Induced mutations in *Drosophila*. *Cold Spr. Harb. Symp. Quant. Biol.* **11**, 151.

PROKOFYEVA, A.A. & KHVOSTOVA, V.V. (1939). Distribution of breaks in the X-chromosome of *Drosophila melanogaster*. *C.R. Acad. Sci. U.R.S.S.* **23**, 270.

ROUYER, M. & LATARJET, R. (1946). Augmentation du nombre de bactériophages en présence de Bactéries stérilisées par irradiation. *Ann. Inst. Pasteur*, **72**, 89.

SLIZYNSKI, B.M. (1945). 'Ectopic' pairing and the distribution of heterochromatin in the X-chromosome of salivary gland nuclei of *Drosophila melanogaster*. *Proc. Roy. Soc. Edinb.* **62**, B II, 114.

SWANSON, C.P. (1944). X-ray and ultraviolet studies on pollen tube chromosomes. I. The effect of ultraviolet (2357A.) on X-ray-induced chromosomal aberrations. *Genetics*, **29**, 61.

TIMOFEEFF-RESSOVSKY, N.W. (1937). Über Mutationsraten in reifen und unreifen Spermien von *Drosophila melanogaster*. *Biol. Zbl.* **57**, 309.

VILLARS, D.S. (1926). Transmission of the Oldenberg chlorine filter for λ2537. *J. Amer. Chem. Soc.* **48**, 1874.

WHITING, A.R. (1945). Dominant lethality and correlated chromosome effects in *Habrobracon* eggs X-rayed in diplotene and in late metaphase. I. *Biol. Bull., Woods Hole*, **89**, 61.

WHITING, A.R. (1946). Motherless males from irradiated eggs. *Science*, **103**, 219.

AUTHOR INDEX

Where the reference is to a figure, the page number is in italics (*113*); where the reference is to a table, the page number is in heavy type (**118**).

Aebersold, P.C., 20, 375
Ahmed, I.A., **214**, 218, 384
Alberti, W., 193, 295, *299*, 337, 375
Allen, S.J.M., 347, 349
Allibone, T.E., 37, 380
Allison, S.K., 347, 377
Allsopp, C.B., 33, 46, 375
Alper, T., 351, 375
Altenburg, E., 181, 375
Ames, F.B., 341, 391
Anderson, R.S., **46**, *199*, 111, **118**, 375, 380
Anderson, T.F., 102, **118**, 122, 314, 381, 386, Plate IV D
Aoki, H., 350, 383
Aston, G.H., 14, 379
Atwood, K., 39, 375
Auerbach, C.A., 341, 384
Auger, P., 10

Bachem, A., 316, 378
Bächström, H.L.J., 5, 375
Badian, J., 326, 375
Baker, S.L., 316, 375
Baldinger, E., 350, 375
Bardwell, D.C., 37, 385
Bauer, H., 159, 167, *168*, 170, *171*, 215, 218, 229, 231, 237, 375
Bawden, F.C., 5, 100, 375
Beadle, G.W., 126, 392
Belgovsky, M.L., **234**, 375
Benford, F., 4, 375
Berry, G.P., *113*, **118**, 393
Bethe, H.A., 349, 350, 351, 376
Binks, W., 14, 15, 383
Bishop, D.W., 190, 192, 197, 376
Bishop, M., 182, 206, 242, 378
Bjorksteadt, W., 17, 394
Bonét-Maury, P., 316, 376
Bonner, T.W., 20, 376
Bowen, E.J., 5, 376
Bozeman, M.L., 190, 196, 213, 224, 365, 387
Bragg, W.H., 351
Brattain, K.G., **37**, 376
Bretscher, E., 316, 318, **321**, **324**, 326, 385
Bridges, P.N., 183
Broda, E., 63, 376

Brownscombe, E.R., 41, **46**, 380
Brubaker, W.M., 20, 376
Brumfield, R.T., 220, 376
Bruynoghe, R., 316, 325, 376
Buchwald, C.E., 15, 377
Busse, W.F., **37**, 376
Butler, J.A.V., 50, 389

Campbell-Renton, M.L., 316, 378
Cannon, C.V., 4, 376
Canti, R.G., 293, 295, *296*, *297*, *299*, 376
Carlson, J.G., 192, 193, 195, **198**, 199, 200, 201, 208, **209**, **218**, 225, 229, *230*, 293, 294, 295, *299*, 337, **338**, 376, Plate IV J
Catcheside, D.G., 134, 154, 158, 159, 160, *161*, 162, 165, 166, 167, *168*, 169, 170, *171*, 172, 173, 191, 192, 200, 202, **204**, **207**, **209**, **211**, **212**, **214**, 218, 220, *221*, 226, **227**, 229, *230*, **238**, **239**, 240, **241**, 245, 253, **255**, 263, 266, 269, 376, 377, 385, Plate III
Cauchois, Y., 15, 377
Chadwick, J., 17, 346, 390
Chambers, H., 316, 377
Charles, D.R., 226, **227**, 229, *231*, **241**, 263, 386
Clark, F.J., 185, 187, 391
Clark, G.L., 40, **46**, 377
Claus, W.D., 316, 366, 382
Coblentz, W.W., 316, 137, 377
Coe, W.S., 40, **46**, 377
Cohen, I., 284, *287*, 382
Compton, A.H., 11, 345, 347, 377
Cook, E.V., 309, 312, 377
Coolidge, W.D., 15, 377
Cooper, F.S., 15, 18, 366, 395
Coulson, C.A., 71, 81, 82, 88, 300, *301*, 316, *318*, 320, 321, 322, 324, 385
Crowther, J.A., 71, 81, 125, 377
Curie, P., 10, 377

Dale, W.M., 42, 44, 45, **46**, 54, 55, 56, 57, 61, 63, 65, 377
Daniels, F., **37**, 376
Darlington, C.D., 126, 189, 193, 343, 377
Dee, P. I., 10, 377, Plate I B
Delbrück, M., 100, 104, 136, *143*, **144**, 146, *147*, 154, 172, 379, 393

AUTHOR INDEX

Demerec, M., *143*, **149**, 158, 159, 163, 167, *168*, 170, 176, 180, 182, 183, 206, 215, 218, 223, 229, *231*, 237, 242, 375, 378, 383
Dempster, E.R., **149**, **228**, 238, 378
Dessauer, F., 3, 67, 378
Dickinson, R.G., 39, 378
Doerr, R., 102, 103, 378
Donaldson, M., 293, 376
Dozois, K.P., 316, 378
Dreyer, G., 316, 378
Duane, W., **37**, 40, 378
Dubinin, N.P., 139, 378
Dubos, R.J., **5**, 384
Duggar, B.M., 9, 124, 316, 378, 382
Dushkin, M.A., 316, 378

Eddy, C.E., 316, 317, 390
Ehrismann, O., 316, 379
Elford, W.J., 102, 103, 120
Ellinger, P., 316, 379
Ellis, C.D., 14, 17, 346, 379, 390
Ellis, E.L., 104, 379
Ells, V.R., **5**, 393
Emmons, C.W., 183, 368, 383
Enzmann, E.V., **214**, *221*, 222, 266, 391
Essex, H., 35, **37**, 391
Evans, R.D., 15, 377, 387
Evans, T.C., 285, 312, 379
Exner, F.M., 108, 113, *114*, **118**, 120, 379, 386
Eyring, H., 34, 36, 379

Fabergé, A.C., 222, 240, 253, 379
Failla, G., 285, 312, 379
Fano, U., 92, *143*, **149**, 150, 158, 159, 163, 166, 170, 183, 206, 218, 219, 223, 276, 378, 379
Fisher, R., 124, 386
Forest, D.M. de, 285, 387
Francis, D.S., 284, 285, *286*, 289, 292, 311, 382
Franck, J., 39, 379
Fricke, H., 40, 41, 42, *43*, 45, **46**, 49, 55, *58*, 61, 379, 380
Friedewald, W.F., *109*, 111, **118**, 380
Fry, H.J., 287, 380
Fulton, H.R., 316, 317, 377

Gates, F.L., 4, 5, 124, 316, 319, 327, 380, 389, 390
Gavrilova, A.A., **234**, 383
Gay, E.H., 146, 148, 170, *171*, 180, 381
Gedye, G.R., **37**, 380
Giles, N.H., **149**, 206, 226, *227*, 229, *230*, *231*, 233, 235, **238**, **241**, 263, 272, 381, 386

Glass, H.B., 213, 229, 329, 381
Glocker, R., 71, 81, 381
Glockler, G., 36, 381
Glücksmann, A., 295, 296, *299*, *309*, 310, 338, 381, 391, 392, 393
Gowen, J.W., 110, *113*, **115**, 120, 124, 140, 141, *143*, **144**, 146, 148, 170, *171*, 180, 381, 390
Gray, L.H., 15, 18, 20, **22**, **46**, 67, 71, **204**, **211**, **214**, 232, 233, 239, **241**, **255**, 303, *304*, 305, 306, **312**, 335, 336, 351, 381, 392
Green, R.H., 102, 314, 381, Plate IV D
Grimmett, L.G., 298, 306, 392
Groth, W., **37**, 381
Gruhn, E., 316, 379
Günther, P., **37**, 40, 41, 381, 382

Hachtel, F.W., 316, 378
Haines, R.B., 71, 81, 82, 88, 300, *301*, 316, 317, *318*, 320, **321**, 322, **324**, 326, 385, Plate IV A, B
Haldane, J.B.S., 166
Hallauer, C., 102, 103, 378
Hanson, F.B., 146, 170, *171*, 382
Harrison, B., **46**, 375
Hart, E.J., 41, 42, *43*, 45, **46**, 49, 55, *58*, 380
Haskins, C.P., 15, 377, 382, 387
Hayashi, T., **5**, 393
Heidt, L.J., 4, 5, 382
Helfer, R.G., **214**, 215, 229, 382
Henshaw, C.T., 284, 289, 292, 311, 331, 382
Henshaw, P.S., 284, 285, *286*, *287*, *288*, 289, 290, 292, 293, 294, 311, 316, 331, 382, 386
Hercik, F., 316, *318*, 322, 382
Heys, F., 146, 382
Hinshelwood, C.N., 52, 382
Hirschfelder, J.O., 34, 36, 379, 382
Hoffmann, J.G., 283, 382
Holden, M., 124, 386
Holiday, E.P., **5**, 382
Hollaender, A., **5**, 124, 182, 183, 223, 242, 316, 378, 379, 382, 383
Holmes, B., *38*, **46**, *109*, **110**, 111, **177**, 385
Holtzapfel, L., **37**, 40, 41, 381
Holweck, F., 71, 81, 113, *114*, **115**, 117, **118**, 120, 300, 307, 316, 320, **321**, 322, 383, 384, 394
Honeyburne, J., 15, 387
Hopwood, F.L., 295, 392
Hornbeck, G., 351, 383
Houlahan, M.B., **5**, 182, 242, 378
Houston, R.A., 5, 383

AUTHOR INDEX

Howell, I., 351, 383
Huber, P., 350, 375
Hudson, J.C., 197, 337, 387, Plate IVk

Jacobsohn, S., 4, 393
Jaffé, G., 50, 52, 88, 383
Jensen, C.O., 295
Johner, W., 14, 392
Jordan, P., 2, 3, 67, 82, 88, 125, 383
Jungers, J.C., **37**, 372, 388
Juul, J., 295, 337, 383

Kailan, A., 37
Kanne, W.R., 17, 383
Kara-Michailova, E., 50, 383
Kaufmann, B.P., *143*, **149**, 159, 163, 167, *168*, 170, 183, 215, 218, 219, 223, **228**, 229, *231*, 237, 375, 378, 383
Kaye, G.W.C., 14, 15, 383
Kemp, T., 295, 337, 383
Kernbaum, M., 41, 42, 383
Kersten, H., **118**, 387
Khvostova, V.V., **234**, 383
Kikuchi, S., 350, 383
Kinsey, V.E., **46**, 49, *58*, 383
Kirchner, F., 346, 383
Klein, O., 14, 346, **348**, 349, 384
Klemperer, O., 49, 88, 384
Knapp, E., *188*, 384
Knorr, M., 316, 384
Koller, L.R., 316, 384
Koller, P.C., 193, **214**, 218, 332, 333, *334*, 335, 337, 341, 384
Kolumban, A.D., **37**, 384
Kotval, J.P., **204**, **211**, **214**, 229, *230*, 232, 233, **239**, 240, **241**, **255**, 384
Krebs, A., 181, 390
Krueger, A.P., 105, 384
Kruger, P.G., 18, 384
Küstner, H., 349, 384

Lacassagne, A., 71, 81, 113, *114*, 117, **118**, 300, 307, 316, 320, 321, 322, 383, 384, 394
La Cour, L.F., 189, 377
Lanning, F.C., 40, 41, **46**, 49, 384
Lasnitzki, I., 192, 201, 295, 296, *299*, 307, 310, 337, 338, 344, 384, Plate IV e–i
Lavin, G.I., **5**, 384
Lawrence, J.H., 20, 375
Lea, D.E., *38*, **46**, 50, 71, 81, 82, 83, 85, 86, 88, 102, 106, *107*, 108, *109*, 110, 111, *112*, *113*, *114*, **115**, 117, **118**, 120, **121**, **123**, 124, 125, 154, 158, 159, 160, *161*, 162, 165, 166, *168*, 169, 170, *171*, 172, 173, **177**,
179, 191, 200, 202, **204**, **207**, **209**, 211, 212, 214, 218, *221*, 226, *227*, 229, *230*, **238**, **239**, 240, **241**, 242, 245, 253, **255**, 263, 266, 269, 283, 284, 289, 292, 298, 300, *301*, 313, 315, 316, 317, *318*, 320, **321**, 322, **324**, 325, 326, 327, 376, 377, 383, 384, 385
Leichter, H., **37**, 382
Levin, B.S., **118**, 316, 385
Lewis, B., **37**, 385
Lind, S.C., *33*, 34, 36, **37**, 40, 41, **46**, 49, 381, 384, 385
Little, E.P., 285, 312, 379
Livingston, M.S., 351, 385
Livingston, R., **37**, 385
Locher, G.L., 150, 388
Loeb, J., 291, 386
Lohmann, A., **46**, 49, *58*, 59, 392
Lominski, I., **118**, 316, 385
Lorenz, K.P., 316, 386
Lucas, A.M., *113*, 381
Ludford, R.J., 300, 386
Luria, S.E., 108, 113, *114*, **115**, 117, **118**, 120, 122, 316, 325, 379, 386, 394
Luyet, B.J., 300, 307, 386, 394

McClintock, B., 187, 200, 205, 328, 386
MacComb, W.S., 283, 390
Mackenzie, K., 182, 183, 242, 386, 388
McKinley, E.B., 124, 386
McLaren, A.D., **5**, 125, 394
Makhijani, J.K., *221*, 222, 228
Makki, A.I., **234**
Malloch, W.S., *236*, 237, 387
Mano, G., 351, 386
Marinelli, L.D., 226, *227*, 229, *231*, **241**, 263, 386
Markham, R., *38*, 100, 102, 105, *109*, 110, 111, **177**, 385, 386, Plate II c–e
Marquardt, H., 190, 193, 196, 197, 225, 337, 338, 386
Marshak, A., 190, 192, 197, 223, 225, 235, *236*, 237, **238**, 336, 337, 343, 386, 387, Plate IVk
Mather, K., 198, 200, **209**, 216, 391
Mavor, J.W., 285, 387
Mayneord, W.V., 15, **22**, 81, 387
Melville, H.W., **4**, 387
Mendel, G.J., 126
Meredith, W.J., 45, 61, 377
Metz, C.W., 190, 387
Michl, W., 17, 352, 387
Mickey, G.H., *221*, 222, 387
Miller, G.L., 175, 387
Mitchell, A.C.G., 17, 394

Mitchell, J.S., 5, 283, 387
Moelwyn-Hughes, E.A., 52, 387
Mohler, F.L., 88, 89, 387
Moore, H.N., **118**, 387
Moriwaki, D., 150, 389
Morningstar, O., 15, 387
Morse, S., 45, 380
Mottram, J.C., 22, 295, 300, 303, *304*, 306, **312**, 335, 336, 381, 388
Mott-Smith, L.M., 181
Mouromseff, G., 316, **321**, 322, 390
Muller, H.J., 132, 133, 135, 139, 141, 162, 164, 165, 170, *171*, 172, 173, 176, 180, 181, 182, 183, 197, 209, 210, *221*, 222, 223, **228**, 229, 233, **234**, 239, 242, 329, 386, 388
Müller, I., 349, 388
Mund, W., 36, **37**, 316, 325, 376, 388

Nagai, M.A., 150, 388
Nagy, R., 316, **321**, 322, 390
Naidu, R., 17, 388
Nanavutty, S.H., 316, 375
Nebel, B.R., 201, 226, *227*, 229, *231*, 241, 263, 386, 388
Needham, J., 291, 388
Newcombe, H.B., 198, 206, **209**, **212**, **214**, 229, *230*, 231, **241**, 257, 258, **261**, *308*, 388
Nishina, Y., 14, 150, 346, **348**, 349, 384, 389
Noethling, W., 316, 379
Northrop, J.H., 100, 389
Noyes, W. A., **37**, 389, 394
Nurnberger, C.E., 40, 41, **46**, 59, 389

Oakley, H.E.H., 295, 392
Oddie, T.H., 316, 317, 390
Ogg, E.F., 37, 385
Oldenberg, O., 5, 389
Olitsky, P.K., 124, 389
Oliver, C.P., 146, 156, *157*, 158, 389
Olivier, H.R., 316, 376
Orr, W.J.C., 50, 389

Packard, C., 331, 389
Pardue, T.E., 351, 391
Parker, R.E., 106, 389
Patterson, J.T., 141, 176, 389
Pedersen, K.O., 5, 392
Peel, G.N., **4**, 389
Pekarek, J., 193, 296, 300, 301, 337, 389
Perry, J.H., **37**, 385
Peskoff, N., 389
Petersen, B.W., 45, **46**, 380
Philipp, K., 17, 352, 389

Philipps, N.W.F., 4, 392
Pickett, L.W., **46**, 377
Poddubnaja-Arnoldi, V., 308, 389
Poisson, S.D., 150, 166, 217, **218**, 248, 256
Politzer, G., 193, 295, *299*, 337, 375
Pontecorvo, G., 163, 164, 165, 200, 209, *221*, 222, 223, 389, 390
Poynter, M., **312**, 335, 336, 381
Preston, G.D., 314, Plate II E
Price, W.C., 124, 390
Prokofyeva, A.A., 135, 388
Pugsley, A.T., 316, 317, 390

Quimby, E.H., 283, 390

Rabinowitsch, E., 39, 379
Rajewsky, B.N., 181, 390
Rasetti, F., 316, 300
Ray-Chaudhuri, S.P., 146, *147*, 228, 239, 390
Read, J., 18, 22, 71, 303, *304*, 306, **312**, 335, 336, 381, 392
Reinhard, M.C., 283, 382
Rentschler, H.C., 316, **321**, 322, 390
Reynolds, J.P., 190, 224, 390
Rice, O.K., 4, 376
Rick, C.M., 206, 213, **218**, 221, 229, *231*, 235, **241**, 252, 390
Riley, H.P., 198, 390
Risse, O., 41, 390
Rivers, T.M., 124, 316, 390, 394
Roberts, J.E., 15, 387
Robertson, M., 300, 390
Robinow, C.F., 325, 326, 390, Plate IV c
Rollefson, G.K., 39, 375
Ruff, H., 316, 384
Russ, S., 295, 316, 366, 388
Rutherford, E., 17, 346, 390

Salaman, M.H., 86, 106, 108, *113*, *114*, **115**, 117, **118**, **121**, **123**, **177**, 313, 314, 315, 385, Plate II A
Sax, H.J., 252, 271, 390
Sax, K., 190, 191, 193, 197, **198**, 199, 200, **201**, 205, 208, **209**, **214**, 216, **220**, *221*, 222, 225, 226, *227*, 229, *230*, *231*, 232, 235, 245, 252, 253, 257, **261**, 264, *266*, 267, *268*, 271, 337, 338, 390, 391
Scheuer, O., **37**, 40, 378
Schlesinger, M., 103
Schreiber, H., *188*, 384
Scott, C.M., 69, 391
Scott, G.M., 295, 388
Shearin, P.E., 351, 391

Sidky, A.R., 229, **234**, 391
Sidorov, B.N., 139, 378
Singleton, W.R., 185, 187, 391
Sizoo, G.J., 14, 391
Slaughter, J.C., 285, 312, 379
Slizynski, B.M., 160, 178, 183, 242, 391
Smadel, J.E., 102, 314, 381, Plate IV D
Smith, C., 35, **37**, 391
Smith, H.P., *43*, 45, **46**, 49, 55, *58*, 380
Smith, K.M., *38*, 100, 102, 105, *107*, *109*, **110**, 111, *112*, 113, **115**, 117, **118**, 124, 125, **177**, 385, Plate II C-E
Snell, G.D., 341, 391
Sokolov, N.N., 139, 378
Sonnenblick, B.P., 162, 329, 330, 331, 391
Spear, F.G., 22, 71, 294, 295, *296*, *297*, 298, *299*, 303, *304*, 306, *309*, 310, 312, 316, 338, 365, 381, 391, 392
Spencer, R.R., 300, 316, 392
Sprague, G.F., 183, 220, 392
Stadler, L.J., 154, 183, 185, 186, *188*, 220, *243*, 340, 392
Stahel, E., 14, 392
Stanley, W.M., 175, 387
Steacie, E.W.R., 4, 392
Stenström, W., **46**, 49, *58*, 59, 392
Stone, L.H.A., 198, 387
Strangeways, T.S.P., 295, 392
Sturtevant, A.H., 126, 392
Sutton, E., 140, *143*, **149**, 159, 163, 183, 223, *231*, 237, 378, 392
Svedberg, T., 5, 392
Swanson, C.P., 191, 193, 197, 201, 202, **203**, **209**, 216, **224**, 225, 240, 242, *243*, 391, 392, 393
Syverton, J.T., *113*, **118**, 393

Tang, P.S., 291, 393
Tansley, K., 295, 296, *299*, *309*, 310, 338, 393
Taylor, H.S., 34, 36, 379, 382
Taylor, L.S., 9, 88, 89, 387
Thoday, J.M., **204**, **207**, **209**, 211, 212, 214, 218, *221*, 226, 227, 229, *230*, *231*, 232, 235, *236*, **238**, **241**, 253, **255**, 266, 269, 393
Thomas, L.B., 5, 393
Thompson, R.H.S., 5, 384
Timoféeff-Ressovsky, N.W., 126, 136, 140, 143, **144**, *145*, 146, *147*, *148*, **149**, 152, **154**, 157, **158**, 172, 176, 181, **234**, 393, 394, 395
Tiniakov, G.G., 139, 378
Trubestein, H., 349, 384
Tweedie, M.C.K., 45, 61, 377

Uber, F.M., **4**, **5**, 125, 183, 186, *188*, 243, 392, 393, 394

Vaughan, W.E., **37**, 394
Victoreen, J.A., 20, 348, 394

Waddington, C.H., 126, 394
Ward, F.D., 152, 153, 182, 394
Ward, G.E., 316, 378
Warren, S.L., *113*, **118**, 393
Wasteneys, H., 291, 386
Weigert, F., **46**
Weiss, J., 47, 394
Wells, W.F., 316, 394
Whitaker, D.M., 291, 292, 394
Whitaker, M.D., 17, 394
White, M.J.D., 192, 197, **199**, 337, 394
Wiig, E.O., **37**, 394
Wilhelmy, E., 146, *147*, 148, 394
Willemsen, H., 14, 391
Williams, E.J., 24, 394
Wilson, C.T.R., 10, 27, 49, 88, 394, Plate I C-F
Wollman, E., 113, *114*, **115**, 117, **118**, 120, 394
Wourtzel, E., **37**, 394
Wyckoff, R.W.G., 81, 300, 307, 316, *323*, **324**, 327, 394

Zahl, P.A., 18, 395
Zickler, H., 181, 390
Zimmer, K.G., 136, 146, *147*, *148*, **149**, 152, **154**, 393, 395
Zirkle, R.E., 17, 316, *318*, **324**, 395

SUBJECT INDEX

Where the reference is to a figure, the page number is in italics (*113*); where the reference is to a table, the page number is in heavy type (**118**). An asterisk attached to a page number (127*) indicates that the term indexed is defined on that page.

α-rays or particles, 16*
  absorption of, 15, 17
  associated volume of, 358, **360**, 361
  bactericidal action of, 316, *318*, **321**, **322**, **324**, 326, **327**
  chemical effects of, **37**, 41, 42, **46**. 47, 57
  chromosome changes by, **204**, 210, **211**, 232, 233, 239, 240, **241**, 254, **255**, 256, 259, **261**, 262, 269, **270**, 271, 272, **277**, 278, 279, 280, 281
  column of ions of, 27, 49–52, 88, 89
  δ-rays produced by, **28**, **30**, **32**, 88–89, 272
  delayed division by, 303, *304*, 305, **312**
  dosimetry of, 18, **22**
  efficiency compared to X-rays, 47, 59, 60, 72, **87**, **115**, **121**, **153**, **179**, **241**, **270**, *304*, **312**, **315**, **324**, **327**
  energy dissipation per micron path, **25**, 351, 352
  energy dissipation per r., 8, 349
  energy of, 17
  mutations by, 152, **153**, **179**
  range of, 17, **25**, **32**, 240, 352
  rapidly dividing tissues, effects on, **312**, 335, 336
  sources of, 17, 18
  stopping powers, additive law of, 351, 352
  target size for, 88–92, *97*, 98, 122, **123**
  viruses, effect on, *113*, *114*, **115**, 116, 117, *119*, **121**, **123**, 313, 314, **315**
  Wilson chamber photographs of, 49, Plate I A
Absorption spectra, 35, *188*
Acenaphthene, 191
Acentric chromosome, 137*
Acetone, 55
Acetylene, **37**
Achromatic lesion (in chromatid), 201*
Activated water, 44*, 47–65, 109, 305, **312**
Activation energy for mutation, 136
Active deposit, 17
Affinity of solute for activated water, **56**

Air
  electrons per gram in, **347**
  X-ray absorption in, **347**, 349
Alanine, **56**
Algae, 300
Allelomorphs or alleles, 130*
Allium cepa (onion), 190, 193, 199, **201**, 208, **214**, 220, 223, 337, 338, Plate IV K
Alloxazin-adenine-dinucleotide, 55, **56**
Ammonia, **37**, 223
Amphibia, 295, 296, *299*, 310, 337
Anaphase, 129*, Plates III *l*, IV K
Antirrhinum, 181
Arbacia (sea-urchin), 284–95, 297, 298, 311
Arsenite, **46**
Ascaris (a parasitic worm), 308, 312
Ascorbic acid, **46**
Associated volume, 83–89, 356–63
Asymmetrical exchange between chromosomes, 137*
Attached-X stock of *Drosophila*, 142, *143*, 165
Aucuba virus, **115**, *143*, **144**
Auger effect, 10
Autosome, 132*

β-rays, 15*
  absorption of, 13, 24, 41
  bactericidal action of, 316, *318*, **322**, **324**. **327**
  chemical effects of, **37**, 41, 42
  dosimetry of, 16, **22**
  energy dissipation per mc. radon, 1?
  energy dissipation per r., 8
  energy distribution of, **16**
  mutation by, 147, *148*
  sources of, 15, 18
Bacteria
  chromosomes of, 326, Plate IV c
  clumping of, 78, 317, 318, 319, *320*
  inhibition of division of, 300, 30? Plate IV B
  killed by single ionization, 69, 71, 7? 77, 91, 120, 325
  killing by radiation, 316–27

SUBJECT INDEX 405

Bacteria (*cont.*)
  motility of, 325
  mutation of, 141, **144**
  spores of, *318*, **321**, **322**, 324, 325
  target size and number, 120, 326
  ultra-violet absorption by, 5
Bacteriophage
  density of, 103
  estimation of, 104, Plate II B
  inactivation by radiation, 86, **87**, 108, *114*, **115**, *117*, **118**, *119*, **121**, **177**
  size of, **103**, **118**
  structure of, 122
Barley, 154
Be + D neutrons, 20, **21**
Bean, 71, 190, 225, *236*, **238**, 281, 296, 300, 301, *302*, 303, *304*, 305, **312**, 335–37, 342
Bellevalia romana, 193, 225, 338
Bethe formula, 349–52
Birds, 128 (*see also* Chick)
Boron, 19, 21
Bragg additive law, 351
Breakage of chromosomes (*see* Chromosome structural changes)
Bromide, 41
Bromine, 36, **37**
Broth, 105, 108, 109
Butterflies, 128

χ^2 test of goodness of fit, 146*
Caging effect of solvent, 39
Cancer, 307, 342
Caproic acid, 55
Carbon dioxide, 36, **37**, 42, **46**, 55
Carbon monoxide, 36, **46**
Carboxypeptidase, 44
Cathode rays, 15*, **37**
Centrifugation, 102, 121, **220**, 285
Centromere, 129*
Ceric sulphate, **46**, 48
Chaetopterus, 311
Chain reactions, 36, 63, 175
Chemical bond, 1
Chemical effects of radiations
  biological effect due to, 1, 64
  comparison of different radiations, **37**, 57–60
  competition between solutes, 55–57
  decomposition of water, 40–42
  direct and indirect action, 60–64
  free atoms and radicals, 47–48
  gas reactions, 34–37
  indirect action in aqueous solution, 42–64

ion-density, effect of, 57–60
ionic yields, 33*, **34**, **37**, **46**
kinetics of indirect action, 52–54
liquids and solids, 37–39
protective action, 55–57
recombination of active radicals, 48–52
reduced yield in dilute solution, 57–60
solutions, 39–64
spatial distribution of active radicals, 48–52
ultra-violet light, 2, 36, **37**, 39
Chick, 201, 293, 295, *296*, *297*, *304*, 306, 307, 310, 337, Plate IV E–I
Chlorine, 36
Chortophaga viridifasciata (grasshopper), 192, 193, **198**, 199, 201, 208, **209**, 217, **218**, 225, *230*, 248, 293, 294, *299*, 338, Plate IV J
Chromatid breaks, *195*, 201*, **203**, **218**, *221*, **224**, 226, *227*, 229, *230*, **238**, **239**, 240, **241**, 242, *243*, 258, 259, 334, Plate III *d, e*
Chromatid exchanges, *195*, **198**, 207–14, 217, **218**, **220**, *221*, **224**, 226, *227*, 232, 233, 240, **241**, 279, 280, 334, Plate III *f, g, k*
Chromatid-isochromatid exchanges, 260
Chromatids, 129*
  achromatic lesions in, 201
  attraction between, 201
  half-chromatid breaks, 201
  independent breakage of, 186
  separation of, 203, **241**, **261**, 262
  sister-union and non-union, 164, 187, 200, **204**, 210, 225, 245, 253, 254, **255**, 280
Chromomeres, 134*
Chromosomes, 127*
  absorption of ultra-violet light by, 6, 134
  acentric, 129, 137*, 163, 164, 197, **199**, 205, 206, 208, 216, 236, 328, 333, 336, 337, 343, Plate III *j*
  behaviour, normal, 127–31
  bridges, 137, 163, 165, 192, *194*, *195*, 197, 200, 201, 205, 208, 236, 328, 336, 337, 342, 343, Plates III *l*, IV G, H, I, K
  clumping of, 192–97, **343**, Plate IV E–K
  dicentric, 137*, 163, 164, 171, 200, 208, 236, 328, Plate III *i*
  dimensions of, 67, 134, 135, 215, 252, 271, 274, 278

SUBJECT INDEX

Chromosomes (cont.)
  duplications in, 137, 200, *339*, 340
  loss of, 164, 165, 185, 209
  maps of, 131, 134, 138, 139
  matrix of, 190*, 193, 201, 224
  physiological effect of radiation on, 192*–97, 223, 225, 237, 239, 336–38, 343, 344, Plate IV E–K
  polarization in, 210
  repulsion between, 196
  ring, 137, *171*, 172, 193, *194*, *195*, 205
  salivary, 134*, 135, 138, 139, 155, 159, 160, 161, 163, 183, 189, 190, 191, 200, 206, 209, **212**, 214, 215, 217, 224, 228, 233, 246, Plate III a, b
  spiralization of, 128, 203, 224
  stickiness of (see physiological effect of radiation on)
  time of split, 128, 197, 198, 199, 203–4
Chromosome structural changes
  behaving as dominant lethals, 164–72
  behaving as recessive lethals, 154–60
  behaviour at cell division, 129, 137, 163, 164, 165, 171, *194*, *195*, 199, 200, 205, 206, 208, Plate III l
  behaviour at meiosis, 185, *339*, 340, 341
  breaks or terminal deletions, 187, *194*, *195*, 197, 199–205, **218**, 229, *230*, 236, **238**, **241**, 242, *243*, 258
  cyclical exchanges, 218–20
  exchange between paternal and maternal chromosomes, 229
  exchanges (intrachanges or interchanges), 155–72, 187, *194*, *195*, **198**, 205–14, **218**, **219**, **220**, *221*, **224**, 226, *227*, **228**, 229, *231*, 232, 233, **234**, **241**, 242, 248, 249, 250, 251, 252, 256, 257, *266*, 279, 328, 338–41, Plate III a, b, f, g, i, j, k
  healing, 246
  hereditary partial sterility, 338–41
  incomplete exchanges, 187*, 210, **211**, 242, 253, 254, **255**, 280
  interval between breakage and joining, 191, 197, 223, 228, 229, 246, 263, 279
  inversions and deletions, 159, 163, *194*, *195*, 205, 206, **207**, 212
  joinability of breakage ends, 204, 205, 210, 239, 253, 257, 258, **261**, 262, 278, 280
  minute interstitial deletions, 141, 155, 156, 158, 159, 163, 178, 183, 187, 206, 207, 213, 217, **218**, 221, *231*, 233, **241**, 242, 252, 256, 328
  phenotypical effects, 138, 139, 142
  random or non-random union, 165, 166, 219, 245, 250, 251, 252
  restitution of breaks, 158, 159, 160, 171, 172, 187, 200, 201, 216, 220, 221, 223, 224, 226, 232, 246, 253–69, 279, 280
  separation of breaks which exchange, 249–52, 259, 280
  spontaneous, **239**, **241**, 243
  symmetrical and asymmetrical, 137*, 163, *194*, *195*, **207**, 208, **209**, 210
  transition from chromosome to chromatid types, **198**, 222
  translocations (see exchanges)
  types of, 136–38, *194*, *195*, 197–211, Plate III
  viability, effect on, 138, 163, 166, 185, 209, 225, 328, 329, 333, 340
Chromosome structural changes; induction by radiation
  centrifugation, effect of, **220**, 224
  colchicine, effect of, 220, 221, 224
  comparison of different radiations, **204**, **211**, 226, *227*, 233, 237–40, 254, **255**, **261**, 262, 269–81, 336, 337
  distribution of aberrations among cells, 217–20
  distribution of breaks in chromosomes, 162, 170, 214–16
  dose dependence, 156, *157*, **158**, 167, *168*, **169**, **170**, 173, 229–37
  energy needed to break chromosome, 276
  geometrical factor (g) for isochromatid breaks, 260*, **261**, 262, 278, 279
  intensity dependence, **211**, **214**, 225–29, *231*, 232, 233, 241, 246, 247, 248, 262–69, 279
  mathematical theory of, 166–69
  modifying factors, 220–25, 246, 247
  number of ionizations needed to break chromosome, 245, 246, 270, 276, 280, 281
  probability of ionizing particle breaking chromosome, 260, 261, 262, 270, *271*, 272, 273, 276, 277, 278, 279, 280
  produced by single ionizing particle, 11, 66, 69, 71, 155, 164, 173, 202, 226, 227, 228, 232, 233, 234, 246–49, 250, 260, 280, 282

SUBJECT INDEX

Chromosome structural changes; induction by radiation (*cont.*)
  produced by two ionizing particles, 155, 158, 164, 187, 226, 248, 250, 280, 282
  relative frequency of different types, 196, 197, 211–14
  ring chromosomes, structural change in, *171*, 172
  sensitivity at different stages, 190, 196, **198**, 203, 213, 223–25
  temperature, effects of, **207**, **209**, **211**, *221*, 222, 223, 246, 247, 266
  theory of (summarized), 279–81
  ultra-violet light, 182, 183, 242–44
  yield of breaks per unit dose, 159, 160, 167, 169, 170, **241**, 256, **261**, 262, *270*, *275*, 280, 330, 332
Clam, 311
ClB method in *Drosophila*, 133, 142*, *143*, 144
Cleavage delay in fertilized eggs, 284–95, 311
Cluster theory, 36
Clusters of ions, **27**, 80, *84*, 85, 86, 95, 174, Plate I D
Colchicine, 191, 220, 221, 224
Collisions between
  H and OH radicals, 48–52, 57–60
  radicals and solute molecules, *53*, 54
  solute and solvent molecules, 39
Colour blindness, 127, 130
Competition between solutes, 45, 55–57
Compton effect, 11, 13, 345–49
Concentration, effect on
  bactericidal action, 317
  chemical reactions in solution, 43, 44, 49, 57–64
  sperm inactivation, 312
  virus inactivation, 108, *109*, **110**, 111
Condensation of chromosomes, 128*
Cosmic rays and mutations, 180–81
Crossing-over, 131*, 133, 142
Cumingia (clam), 292, 311
Cumulative dose, 283*, 285, 286, 289, *290*, 292, 295, 297
Cumulative type of action, 76, 77, 79
Cyclotron, 19, 20

δ-rays, 27*, **28–32**, 82, 88–89, 179, 272, 276, 279, 351, 352, 356, 357, 358, 361, Plate I A
d-amino-acid-oxidase, **46**, 55, **56**
D+D neutrons, 8, 20, **21**
Deactivation efficiency, **56**, 62, 63, 111

Deficiency (of gene or genes), 137*, 141, 155, 156, 158, 159, 163, 176, 178, 183, 184, 185, 186, 187
Degenerate cells, *309*, 310
Delayed division (*see* Division)
Deletions (*see under* Chromosome structural changes, *and* Deficiency)
Density (of viruses or proteins), 89, 103
Deuterium hydroxide, 50
Dicentric chromosome, 137*
Dichromate, **46**
Diffusion
  of radicals, 50
  of deuterium hydroxide, 50
Dilute solution, reduced yield in, 57–60
Diploid, 130*
Direct and indirect actions of radiation, 60–64, 107–11, 317
Dissociation of a gas molecule, 35
Dissolved oxygen affecting reactions in solution, 41, 42, **46**, 47
Distribution of radiosensitivity among cells, 77, 78
Division, delay, inhibition, and prolongation of
  bacteria, 300, *301*
  cleavage delay in fertilized eggs, 284–95
  comparison of different radiation, 303–6, **312**
  cumulative dose, 283, 285, 286, 289, *290*, 292, 295, 297
  direct or indirect effect, 285
  dose dependence, 282, *288*, 289, *299*, *304*, 305, 306
  intensity and fractionation, effect of, 292, *297*, 298
  nucleic acid, 283, 304
  nucleolus, 304
  number of ionizations required, 304
  oxygen uptake, relation with, **291**, 292
  rapidly dividing tissues, 71, 194, 196, 205, 223, 225, 237, 295–300
  recovery, 282, 283, 287, 289, 291, 292
  recovery time (τ), 289, 290, **291**, 292
  size increase (of cells or nuclei), 300–2
  stage most sensitive to radiation, 288, 289, 293–95
  stage subject to delay, 286, 287, 293–95
  temperature, effect of, *290*, **291**
Dominant, 126*
Dominant lethal, 161–72, 184, 329–32
Dose-rate (*see* Intensity)

SUBJECT INDEX

Dosimetry
 α-rays, 18
 β-rays, 16
 γ-rays, 14
 interconversion of units, **22**
 neutrons, 19, 20
 protons, 19
 ultra-violet light, 4
 X-rays, 6
Drosophila (fruit fly), 4, **5**, 90, 128, 130, 131, 132, 133, 134, 135, 138, 139, 140, 142, *143*, 144–72, 175, 176, 180, 181, 182, 183, 187, 189, 191, 196, 197, 199, 200, 206, 208, 209, **212**, 213, **214**, 215, 216, 217, 218, **219**, 220, *221*, 222, 223, 227, **228**, *231*, 233, **234**, 237, 242, 243, 245, 246, 247, 315, 325, 328, 329–32, 342, 343
Duplications (of portions of chromosomes), 137*, 200, *339*, 340

$\bar{E}$ (effective mean ionization potential), 351, 352
Egg
 and sperm, relative sensitivity, 285, 329
 delayed cleavage of, 284–93, 311
 failure to hatch, 162, *168*, 169, 308, 309, 329–33
 half-eggs, 285
 irradiation of, 152, 153, 181, 213, 229, 284, 285, *286*, *287*, 288, 289, *290*, **291**, 292, 308, 309, 329–33, 341
 nucleus (in seed plants), 183*
Electrons (*see also* β-rays)
 absorption of, 24
 associated volumes of, 356–59
 attachment to molecules, ions, or radicals, 2, 35, 36, 47, 48, 88
 chromosome breakage by, 210, 211, 249, 272, 273
 δ-rays produced by, **28**, **29**, **32**
 energy dissipation by, **24**, 349, 351, 352
 energy distribution of, **12**
 micrographs, 122, 314, Plates II E, IV D
 number crossing a cell, 249
 number per unit volume, **32**, 346, **347**
 primary ionization by, **25**, 351, 352
 range of, **24**, 27, **32**
 tail, 273, *274*, 275, 276, 280, Plate I c
 volt (eV.), 1
Embryo (in seed plant), 183*
 absence of, 184
 no sister-union in, 187
 restitution in, 187
Encephalitis virus, **118**
Endosperm, 184*
 chromosome deficiencies in, 184–88, *243*
 defective, 184
 mutations affecting, 184
 sister-union of chromatids in, 187, 200, 328
Energy dissipation
 by ionization and excitation, 1, 2, 35
 by ultra-violet light, 3, 6
 in solute and solvent, 39, 44
 per ion-pair, 7, 34
 per micron path, **24**, **25**, **26**, 351, 352
 per roentgen or v-unit, **8**, 22, 34–50
 per unit volume, **8**, 13
Energy-unit, **8**, **22**
Enzymes, 37, *38*, 42, 44, 45, **46**, 54, 65, 174, 176, **177**
Equivalents (chemical), 45, **46**
Euchromatin, 135*, 138, 139, 140, 215, 216, **234**
Excitation, 1*
 biological effects of, 2
 chemical effects of, 35, 47
 energy of, 2, 34
 in solute and solvent, 39, 40, 42, 43
Exponential dose relation, *38*, 45, 61, 72–78, *107*, 110, *112*, *113*, *114*, 124, 164, 235, 305, 306, *318*, 319, *320*, 324, 327, 330
Exponential function, table of, **75**
Extinction coefficient (of ultra-violet light), 4.

F (overlapping function), 85*, **86**, 353–54
F_1 and F_2 (first and second filial generations), 126*
f (proportion of chromosome breaks which are unjoinable), 253–62
Ferric sulphate, **46**
Ferricyanide, **56**
Ferrocyanide, **46**, **56**
Ferrous sulphate, 45, **46**, 47, 48, 60
Fibroblasts, irradiation of, 201, 293, 295, *296*, 297, *304*, 306, 307, 310, 337, Plate IV E–I
Filters for
 γ-rays, **14**
 ultra-violet light, 5
 X-rays, 9
Filtration of viruses, 102, **103**, 119, 121
First haploid mitosis, 191*

SUBJECT INDEX

Flies (see Drosophila, Sciara)
Formaldehyde, **46**, 49, 55
Formate, **56**
Formic acid, *43*, 49, 55
Four-strand pachytene, 131*
Fowl plague virus, **118**
Fractional deficiency (in a chromosome), 186*
Fractionation of dose, effect of, 78, 147, 226, **228**, *266*, 267
Free atoms and radicals
  collisions with solute molecules, *53*, 54
  concentration of, 50–52
  diffusion of, 50–52
  energy to form, 47
  explaining radiochemical actions in solution, 47–48
  formation from H_2O, 47
  in gases, 35
  in water, 47–54
  recombination of, 48–52, 57–60, 305
  spatial distribution of, 47, 48–52, 68
Frog, 296, 310
Fructose, **56**
Fungal spores, *143*, 183, *188*, 300

G (function of time-intensity theory), 264, **265**
g (geometrical factor for isochromatid breaks), 260, **261**, 262, 278, 279
Gametes, 126*
Gametophyte, 185*
Gamma (γ-) rays, 13*
  absorption of, 14, 15
  bactericidal action of, 316, 317, *318*, 322, **324**, 326, **327**
  chemical effects, 37
  chromosome structural changes by, **211**, **228**, **239**, 240
  delayed division by, 295, *296*, *297*, 300, *301*, *304*, 305, 306, *309*, **312**
  dosimetry, 14, **22**
  electrons liberated by, **12**, 13, **14**, 15
  energy dissipation per r., 8
  filtration of, **14**
  intensity at 1 cm. from 1 mg. radium, 15
  mutation by, *147*, **148**, **179**
  rapidly dividing tissues, effects on, *309*, **312**, 335, 336
  scattering of, 14
  sources of, 13
  spectra, 14
  target volume determined by, 91–92
  viruses, effects on, *112*, **115**, 116, **118**, **121**, 122, **123**, 314, **315**

Gas-free water, 41
Gas reactions, 34–37
Gelatin, 108, 109, **110**
Generative nucleus (of pollen), 332*
Genes
  nature of. 101, 130, 133–36
  number of, 90, 134, 135, 179, 180, 315, 326, 342
  reproduction of, 139
  size of, 121, 135, 172–80, 315, 326
  viruses and, 101, 121, 124, 136, 174, 175, 177
Genetical effects of radiation (see also under Mutations, and under Chromosome structural changes)
  cosmic rays, 180–81
  deductions concerning size of gene, 172–80
  dominant lethals in Drosophila, 161–72
  recessive lethals in Drosophila, 144–54
  relation between recessive lethals and chromosomal changes, 154–60
  ultra-violet light, 181–88
  visible mutations, 140–44
Genotype, 126*
Germ (see Bacteria, Egg, Embryo, Sperm)
Globulin, 134
Glucose, **56**
Glutathione, **46**, 49, *58*, **59**
Glycerine, 63
Glycine, **56**
Gram-roentgen, **22**
Grasshopper, 192, 193, **198**, 199, 201, 208, **209**, 217, **218**, 225, *230*, 248, 293, 294, *299*, 337, 338, Plate IV J
Gross structural change (of chromosomes), 138*

Haemoglobin, *38*, 45, **46**
Haemophilia, 127, 130, 131
Half-eggs, 285
Haploid, 130*
Healing (of chromosomes), 246
Hemizygous, 131*
Heredity, 126–31
Heterochromatin, 135*, 138, 139, 215, 216, 234
Heterozygous, 130*
Hexane, 50
Hippurate, **56**
Homologous (genes or chromosomes), 128*
Hydration of viruses, 102, 103

Hydrazine, 37
Hydriodic acid, 37, 39, 46
Hydrobromic acid, 36, 37, 46
Hydrogen, 36, 37, 40, 41, 42, 43, **46**, 48, 49, 55
  atoms (*see under* Free atoms and radicals)
  nuclei (*see under* Protons)
  peroxide, 40, 41, 42, 45, 46
  sulphide, 36
Hydroxyl radical (*see under* Free atoms and radicals)

Ice, 40
Illegitimate union of chromosomes, 137*
Impurities, effects of traces of, 41, 42, 49, 111
Inactivation dose, 61*, 74, 108
  and solid content of solution, 62, *109*, **110**
  and target size, 80–88, 116–23, 362–63
  of viruses, *109*, **110**, 116–23
Incomplete exchange (of chromosomes), 187*, 210, **211**, 242, 253, 254, **255**, 280
Incomplete linkage (of genes), 130*
Independent assortment (of genes), 127*
Indirect actions of radiations
  on chemicals, 39, 40, 42–64
  on sperm, 285
  on viruses, 63, 64, 107–11
Infra-red radiation, 223, 247
Intensity (or dose-rate), 7*, 72, 78–79, 114, **115**, 116, 124, 144, 145, 146, *147*, 153, **214**, 225–29, 241, 246, 247, 248, 262–69, 279, 292, *297*, 298, **321**, 327
Interchange (between chromosomes), 137*
Interphase, 128*
Interstitial deletion (of a chromosome), 187*
Inversion (of a chromosome segment), 137*
Iodide, 41
Ion-clusters, 27, 80, *84*, 85, 86, 95, 174, Plate ID
Ion-columns, 27, 49–52, 88–89, Plate IA, B
Ion-density (or specific ionization), 23*
  affecting apparent target size, *97*, 98, 355, 356
  affecting biological effects, 72, 79–80,
  86, 90, **115**, 116, **153**, **179**, **204**, 205, 210, **211**, 237–40, 254, **255**, 262, 270, *271*, 272, 273, 277, 303, *304*, 305, 306, **312**, 323, **324**, **327**, 336, 337
  affecting chemical effects, 57–60
  affecting probability of effective 'hit', 354
Ionic yield, 33*, **34**, 36, **37**, 38, 40, 41, 45, **46**, 47, 57, *58*, 62, 63, 64, 65, 110, 111
Ionization, 1*
  chamber, for neutron dosimetry, 19
  energy of, 1, 34, 35
  in air, per r., 7
  in solute and solvent, 39, 40, 42, 43
  potential, 351, 352
  single, actions due to, 38, 71–72, 111–16, 153, 325
  spread of effect of, 67–68, 69, 173
Ionizing particles (*see also under* α-rays, δ-rays, Electrons, Protons)
  number per unit volume, **32**
  number traversing cells, 151, 248, 249, 250
  range per unit volume, **32**
Ions
  negative, 2, 34, 35
  positive, 2, 34, 35
  recombination of, 35, 50
  spatial distribution of, 10, 49, 350–52, Plate I
Isochromatid breaks, *195*, 202*–5, 210, 217, **218**, **220**, *221*, **227**, 229, *230*, 237, **238**, **239**, 240, **241**, 242, 260, **261**, 277, 278, 279, 280, 281, 334, Plate IIIc, d, l

Jensen rat sarcoma, 295

Kinetics (chemical), 52–57, 136
Klein-Nishina formula, 14, 345–49

Lactose, 107
Latin square, 105
Lethal effects of radiation
  ascribed to chromosome physiological change, 337–38, 343–44
  ascribed to chromosome structural change, 328–37, 341–43
  ascribed to lethal mutation, 313–16, 324–27, 342
  bacteria, 316–27, 341, 342
  division related to, 307–13, 341, 343
  eggs, 308, 311, 329–32
  hereditary partial sterility, 338–41

SUBJECT INDEX

Lethal effects of radiation (cont.)
 pollen, 308, 332–35
 rapidly dividing tissues, 309, **312**, 341–44
 root-tips, 335–37
 spermatozoa, 312
 viruses, 100–25, 313–16, 341, 342
Lethal mutation (see Recessive lethal and Dominant lethal)
Leucylglycine, **56**
Linear dose relation, 38, 43, 45, 60, 72, 145, 146, 148, 153, 156, 157, **158**, 165, 181, 183, 229, 230, 231, 232, 233, 235, 237, 238, 241, 243, 244, 248, 249, 280
Linkage, 127*
 altered by chromosome structural change, 188
 group, 127*
 incomplete, 130*
Liquids, chemical effects in, 37–39
Lithium, disintegration by neutrons, 19, 21
Li+D neutrons, 8, 20, **21**
Liverwort spores, 188
Local lesions (due to plant viruses), 105, 140, 176, Plate IIc
Locusta migratoria, **199**
Logarithms, 74, **75**
Lycopersicum esculentum (tomato), 190, **238**
Lymphoma, 235, 236, **238**, 343

Maize, **5**, 127, 130, 181, 183–88, 200, 202, 205, 219, 242, 243, 328, 340
Malpighian tube, 175, 176
Mammalia (see Man, Mouse, Rat)
Man, 127, 128, 130, 300
Mass spectrograph, 34
Matrix (of a chromosome), 190*, 193, 201, 224
Mean lethal dose, 61, 74, 321, 335, 336, 337
Meiosis, 129*
Mercury lamp, 4, 5
Metaphase, 128*, Plate III h, i, j, k
Methaemoglobin, 38, **46**
Methyl alcohol, 49, 55, 58, **59**
Micro-moles per 1000 r., **34**, 44
Micronuclei, 199, **333**, 343
Microspores (see Pollen)
Millicurie (mc.), 15, **18**
Minute structural changes (of chromosomes), 138*
Mitosis, 128*–29
 duration of stages of, 294, 295, 305

inhibition of, 295–300, 302, 303–6, 309, 310
 radiosensitivity of different stages of, 190, **198**, 223–25, 287, 288, 289, 293–95
Molecular
 diameter, 52
 weight, 53, 65, 89, 135, 136, **177**, 178
Mosaic, 132*
Moths, 128
Motility of bacteria, 325
Mouse, 225, 235, 236, **238**, 281, 300, 341, 343
Multi-hit target theory, 71, 322
Multi-target theory, 90–92, 122, 314, 326
Mutations
 chemically induced, 131
 endosperm, affecting, 184
 nature of, 131, 133–36, 154–60
 relation to chromosome structural change, 136–40, 154–60
 spontaneous, 131, 180, 181
 temperature affecting spontaneous rate, 136, 181
 viability affected by, 133, 142, 176, 183
Mutations, induction by radiation
 caused by single ionization, 66, 69, 101, 153, 174
 comparison of different radiations, 145, 147, 148, **149**, **153**, 179
 cosmic rays, 180, 181
 dominant lethals, 161–72, 184, 329–32
 dominant visible mutations, 133, 140, 141, 184
 dose dependence, 143, 144, 145, 146, 156, 157, **158**, 167, 168, **169**, 181, 183
 grouping effect, 150–51
 in bacteria, 141, **144**, 324–27
 in barley, 154
 in Drosophila (see Drosophila)
 in viruses, 140, 141, 143, **144**, 176, 313–16
 intensity not affecting, 145, 146, 147
 relative frequency gene mutation and deficiency, 141, 159, 176, 178, 183, 184, 186
 recessive visible mutations, 141–44, 184
 sex-linked recessive lethals, 133, 140, 144–60, 176, 182–83
 target size for, 90, 120, 178, **179**, 180
 temperature not affecting, **154**
 yield per unit dose, 132, **144**, 177, 178, 180

SUBJECT INDEX

Mutations of *Drosophila melanogaster*
  Bar (eye), 140, 142
  brown (eye), 209
  cubitus interruptus (wing vein), 139, **234**
  ebony (body), 209
  forked (bristles), **144**
  miniature (wings), **144**
  Notch (wings), 138, 206
  scute (bristles), **234**, 242
  white (eye) series, 130, 139, *143*, **144**, 154, 175, 176
  yellow (body), 138, 165, 234, 242

v-unit, 20*
  energy dissipation per, **8**
n-unit, 20*
Natural radiation and mutation, 180-81
Nereis, 311
Neuroblasts, 192, 193, **198**, 201, 208, **209**, 217, **218** *230*, 248, 293, 294, *299*, 337, 338, Plate IV J
Neurospora crassa, *143*, 183
Neutrons, fast
  bactericidal action of. 316, *318*, **324**, **327**
  chromosome structural changes by, 204, 210, **211**, **218**, 226, *227*, *230*, *231*, 233, 234, 235, *236*, 237, **238**, 239, **241**, 248, 249, 255, 256, 259, **261**, 262, 269, **270**, 272, **277**, 278, 279, 280, 281, 336
  division delay by, 303, *304*, 305, 306, **312**
  dosimetry, 19, 20, **22**, 237
  efficiency compared to X-rays, 72, 145, *148*, **149**, **241**, **270**, 336
  energy dissipation per v-unit, **8**, 350
  mutations by, 145, *148*, **149**, 150, 151, **153**, **179**
  nitrogen disintegration by, 19, 350
  nuclear projection by, 19, 20, **21**, 350
  rapidly dividing tissues, effects on, *304*, **312**, 335, 336, 337
  scattering of, 20, 350
  sources of, 20
Neutrons, slow, 19, 21
Nitrate, **46**, **56**
Nitrite, **46**, **56**
Nitrogen, 19, 350
Nitrous oxide, **37**
Nucleate, **56**
Nucleic acid, **5**, 128, 139, *188*, 190, 193, 196, 215, 283, 304, 343
Nucleoprotein, **5**, 101, 122, 123, 124, 127, 134, 135, 136, 188

Nucleus (of cell)
  membrane of, 128, 293, 294
  separation from cytoplasm, 285
  size increased by irradiation, 300-2
  size of, 249, 302

Onion, 190, 193, 199, **201**, 208, **214**, 220, 223, 337, 338, Plate IV K
Oocytes, 190, 224
Oregon-R (stock of *Drosophila melanogaster*), 149
Organic acids, oxidation of, **46**
Overlapping function (F), 85*, **86**, 353-54
Oxalate, **56**
Oxalic acid, 49, 55, *58*, **59**
Oxidation by radiation, 41. 42, 45, **46**, 48
Oxygen, 36, **37**, 40, 41, 42, 48, **291**
Oxyhaemoglobin, **46**
Ozone, **37**

p (probability of collision removing active radical), 54-55, **56**, 57, **59**
p (probability of ionization in target being effective), 93, 95-96, **97**, 122, 172, 174, 175, 176, 180, 355, 356
p (probability of ionizing particle breaking a chromosome), 260, 261, 262, 270, *271*, 272, 273, 276, 278, 279, 280
p (probability of sister-union at a chromosome break), 166, 169, 172
P (probability of solute reacting), 54, 60, 61
pH affecting chemical actions of radiation, 47
Pachytene, 131*
Peas, 126, 127, **238**
Permanganate, **46**, 47, 49
Persulphate, 63
Phenotype, 126*
Photochemical reactions (*see under* Ultra-violet light)
Photoelectric absorption, 10, 345-49
Photoelectrons, 10*
  associated volume of, 87
  energy of, 10, **12**, 273, 275
  number of, **12**, **32**, 249, 273, 274, 275
  range of, **32**, 173, 249, 273, 275
Physiological effect of radiation, 162, 192-97, 223, **225**, 237, 239, 336-38, 343, 344, Plate IV E-K
Plague virus (fowl), **118**
Plaques, bacteriophage, 104, Plate II B

SUBJECT INDEX

Point-heat, 3, 67
Poisons produced by radiation, 64, 78, 317, 319, 320
Poisson distribution, 150, 166, 217, **218**, 248, 256
Polar cap (of egg), 152*, 153, 181
Polar fusion nuclei, 184*
Pollen
  defective, 185, 186
  development of, 332, 333
  dimensions of, 249, 259
  germination failure after irradiation, *308*, 332–35, 342
  germination on artificial medium, 191, 242
  irradiation of, 183–88, 189, 193, **198**, 199, 200, 201, 202, **203**, 205, **207**, **209**, **211**, **212**, 213, **214**, 216, 217, **218**, 219, **220**, *221*, 222, **224**, 225, 226, *227*, *230*, *231*, 232, 233, 235, *236*, **238**, **239**, 240, **241**, 242, *243*, 245–81, *308*, 332–35, 337, 340
  mother cell, 190*, 201, 219
  tube, 191, **203**, 204, **209**, **211**, **212**, **214**, 216, **224**, **239**, 240, **241**, *243*, 255
  ultra-violet absorption by, 5, 244
Polonium, 17
Polypeptide, 210
Position effect, 138*–40, 155–58, 175
Potato virus X, *112*, 116
Primary effect of radiation on chromosomes, 193*
Primary ionization, **25**, **26**, 85, 352
Primordial germ cell, 152*
Probability (*see* p, P)
Prophase, 128*
  prolongation of, 286, 287, 294, 296
Protamin, 134
Protection against effect of radiation in solution, 45, 48, 49, 55–57, 60–65, 108–11, 312
Protein, 38, 45, **46**, 55, **56**, 89, 98, *109*, 110, 111, 112, 134, 135, 190, 312
Protons, 19*
  associated volume of, 358, **361**
  biological effects (*see* Neutrons, fast)
  δ-rays projected by, **31**, **32**, 272
  dosimetry of, 19, **22**
  energy dissipation per r., 8, 349
  number traversing a cell, 151, 248, 249, 250
  projected by neutrons, 19, 20, **21**, **32**
  ranges of, **26**, **32**
  sources of, 19

Protozoa, 71, 300
Pyocyaneus, 71, 320

q (probability of chromosome break not undergoing sister-union), 166–72, 330–32
Quantum energy, 6, 9, **12**, 13
Quantum yield, 36, **37**, **39**, 125

Rabbit fibroma virus, **118**
Rabbit papilloma virus, 102, **103**, *109*, 111, *113*, **118**
Radioactive
  matter in organisms, 180, 181
  radiations (*see* α-, β-, γ-rays)
  sources (*see* Polonium, Radium)
Radiochemical reactions, 33* (*see* Chemical effects of radiations)
Radium
  emanation (*see* Radon)
  γ-ray intensity at 1 cm. from 1 mg., 15
  Ra A, B, C, C′, 15, **16**, 17, **18**, **32**, 34, **241**, **270**, **277**
Radon (radium emanation), 13, 17, 18, **32**, 33, **34**, 41, 46, 152, **239**, **241**, **270**, **277**
Range of
  α-particles, 17, **25**, **32**, 240, 352
  electrons, **24**, 27, **32**
  protons, **26**, **32**
Rat, 225, 295, 296, *299*, *309*, 310
Recessive, 126*
Recessive lethals, 133*
Reciprocal translocation, 137*
Recoil electrons, 11*, **12**, **32**, 345–49
Recoil nuclei, **18**, **34**
Recombination of
  ions, 50, 88
  products of chemical decomposition, 36, 39, 60, 175
  radicals, 48–52, 57–60, 305
Recovery from effects of radiation, 79, 196, 282, 283, 287, 289, 291, 292, 308, 309, 312
Reduction (chemical) by radiation, 40, 41, 42, 45, **46**, 48
Resonance radiation, 5
Resting stage, 128*
Restitution (of chromosome break), 158, 159, 160, 171, 172, 187, 200, 201 216, 220, 221, 223, 224, 226, 232, 246, 253–69, 279, 280
Ribonuclease, 5, *38*, **46**, 177
Ring chromosomes, 137, *171*, 172, 193, *194*, *195*, 205

Roentgen (r.), 6*
  energy dissipation per, 8
  number of ionizing particles per, 32
Root-tips, 190, 193, 197, **201**, 208, 225, *236*, **238**, 281, 296, 300, 301, *302*, 303, *304*, 305, **312**, 335–37, 338, 342

σ, σ_a, σ_s (Compton scattering coefficients), 345, 346, 347, **348**, 349
σ (scattering cross-section for neutrons), 350
Salivary gland chromosomes, 134, 135, 138, 139, 155, 159, 160, 161, 163, 183, 189, 190, 191, 200, 206, 209, **212**, 214, 215, 217, 224, 228, **233**, 246, Plate III a, b
Sciara, 190, 196, 213, 224
Sea-urchin, 284–95, 297, 298, 311
Second haploid mitosis, 191*
Secondary effect (of radiation on chromosomes), 193*
Secondary ionization (*see also* δ-rays), 26, 80, 85
Selenite, **46**
Sex-chromosomes, 128*
Sex-linked, 131*
Sex-ratio distortion, 170, *171*, 172
Shope rabbit fibroma virus, **118**
Shope rabbit papilloma virus, 102, **103**, *109*, 111, *113*, **118**
Sigmoid survival curves, 76, 77, 78, 319, **320**, 331
Single-unit action, 77*
Sister-union (*see under* Chromatids)
Solids, chemical effects of radiation on, 37–39
Solutions, chemical effects of radiation on, 39, 42–64
Somatic cell, 130*
Spatial distribution of
  H and OH radicals, 48–52, 68
  ionizations, 10, 49, 79–90, 92, 93, 350–52, Plate I
Specific ionization (*see* Ion-density)
Sperm (of animals), 129*
  chromosomes in, 162, 191, 196, 197
  irradiation of, 132, 141–83, 189, 199, 200, 206, 209, **212**, 214, 217, 218, **219**, *221*, 223, 227, **228**, **234**, 237, 242, 245, 246, 247, 284, 285, *286*, *288*, **291**, 311, 312, 329, 341
  relative radiosensitivity of sperm and egg, 285, 329
  size of, 151
Sperm nuclei (in pollen), 183*, 332
Spiralization (of chromosomes), 128

Spores (*see under* Bacteria, Fungal, Pollen)
Spread of effect of ionization, 67–68, 69, 173
Square-law dose relation, 232, 233, 235, 236, 241, 248, 256, 257, 279
Sterility, hereditary partial, 328, 338–41
Stickiness of chromosomes, 192–97
Survival curves
  of irradiated bacteria, *318*, 319, *320*, *323*
  of irradiated eggs, 330, 331
  of irradiated viruses, *107*, *112*, *113*, *114*, 124
  significance of shape of, 70, 71, 72–78, 110, 124, 305, 306, 324
Symmetrical exchange (between chromosomes), 137*
Synapsis, 129*

τ (photoelectric absorption coefficient), 345, **347**, **348**, 349
τ (time constant for chromosome union), 263–69, 279
τ (time constant for decay of cumulative dose), 289, 290, **291**, 292
Tadpole (of frog), 296, 310
Tail of electron track, 273, *274*, 275, 276, 280, Plate I c
Target
  area, from α-ray data, 91, 313, 314, 326
  calculations, 353–63
  indefinite boundary to, 93, 96–98, 355–56
  multi-hit theory, 71, 322
  multi-target theory, 90–92, 122, 314, 326
  number, 90–92, 122, 179, 180, 315, 326
  of bacteria, 322, 326, 327
  of genes, 172–80
  of viruses, 116–23, **177**
  overlapping factor, 85*, **86**, 353–54
  shape, 91, 93, 94–95
  size, from 37 % dose, 80–90
  size, from relative efficiency of two radiations, 90, 122, 179, 315, 326
  validity of target theory, 98–99
Telophase, 129*
Temperature
  affecting action of radiation, 72, **154**, 181, 246, 247, 266, *290*, **291**, 309, 310, 312, 321, **322**, 336
  affecting recovery from radiation effects, *290*, **291**, 309, 312
  rise caused by radiation, 2
Terminal deficiency (of a chromosome), 187*

Tertiary electrons, 27
Thiocyanate, **56**
Thirty-seven percent dose, 61*, 74
  and associated volume, 362, 363
  and target number, 90
  and target size, 80–88
  of bacteria, 321
  of chemical reaction, 61–64
  of viruses, *109*, **110**, 116–23
Three-halves power dose relation, 156, *157*, **158**, 232, 233, 267, *268*
Time-intensity factor, 78, 114, **115**, 116, 124, 144, 145, 146, *147*, 153, **214**, 225–29, 241, 246, 247, 248, 262–69, 279, 292, *297*, 298, **321**, 327
Tissue
  culture, 201, 293, 295, *296*, 297, *304*, 306, 307, 310, 337, Plate IV E–I
  electrons per gram in, 347
  elementary analysis of, 7, 11
  energy dissipation in, **8**, **22**
  X-ray absorption in, **347**, 349
Tobacco mosaic virus, 100, 102, *109*, **110**, 111, *112*, **115**, 116, 125, 136, 140, *143*, **144**, 175
Tobacco necrosis virus, 100, *107*, *112*, **115**, **118**, **177**, Plate II c, e
Tobacco ringspot virus, **118**, **177**
Tomato, 190, **238**
Tomato bushy stunt virus, 100, 102, **103**, *107*, **115**, **118**, 135, **177**, Plate II D
Tradescantia, 67, 71, 187, 191, 193, **198**, 199, 200, 201, 202, **203**, **204**, 206, **207**, **209**, 210, **211**, **212**, 213, **214**, 216, 217, **218**, **220**, *221*, **224**, 225, 226, *227*, 228, *230*, *231*, 232, 233, 235, *236*, 237, **238**, **239**, 240, **241**, 245–81, *308*, 329, 332–35, 336, 342, 343
Translocation of chromosomes, 137*
Trichophyton mentagrophytes, 183
Triploid
  Drosophila, 209
  endosperm, 184, 328
Triton (newt), *299*
Trypsin, *5*, 125
Tryptophan, **5**
Tulip, **212**, **214**
Tumours, 225, 235, *236*, **238**, 281, 295, 300, 343
Tyrosine, **5**, 46, 47, 49, *58*, 59

Ultra-violet light
  absorption by biological materials, **5**, 182, *188*, 244

  bactericidal action of, 2, 316, 317, *318*, 319, **321**, 322, 327
  chemical effects of, **37**, 39
  chromosome changes by, 182, 183, 185, 216, 223, 242–44, 247
  compared with ionizing radiations, 5, 6, 36, **37**, **39**, 124, 125, 185–88, 202, 203
  genetical effects of, 181–88, 242–44
  physical properties of, 4, 5, 6
  quantum yield, 36, **37**, 39, 125, 327
  viruses, effect on, *112*, **115**, 124–25
  wave-length dependence of biological effect, 3, 182, *188*

Vaccinia virus, 101, 106, *113*, **115**, **118**, **123**, 313–16, 341, Plates II A, IV D
Vegetative nucleus (of pollen), 332*, 335
Vibrational energy of a molecule, **39**
Vicia faba (bean), 71, 190, 225, *236*, **238**, 281, 296, 300, 301, *302*, 303, *304*, 305, **312**, 335–37, 342
Victoreen dosemeter, 20
Viruses
  chemical change tolerated by, 175
  crystalline, 100, 122, Plate II D
  density of, 103, 121
  drying of, 107
  electrons per gram in, 347
  elementary analysis of, 7
  energy dissipation in, per r., 8
  estimation of, 104–6, Plate II A, B, C
  filtration of, 102, **103**, 119, 121
  genes and, 101, 121, 124, 136, 174, 175, 177
  hydration of, 102, 103, 121
  mutation of, 140, 141, *143*, **144**, 176
  nature of, 100–2, 122–24, 313
  serologically related, 136
  shape of, 116
  sizes of, 102, **103**, 104, **118**, 135, 136
  structure of, 314, Plate IV D
  ultra-violet absorption by, **5**
  X-ray absorption by, **347**
Viruses, inactivation by radiation
  caused by single ionization, 69, 98, 101, 111–24, 313
  comparison of different radiations, **115**, 116, **121**, 122, **123**, 314, **315**
  direct and indirect action, 63, 64, 107–11
  interpreted as lethal mutation, 313–16
  ultra-violet light, 2, 3, 124–25
Vitelline membrane (of egg), **5**

SUBJECT INDEX

Water
  content of, affecting ionic yield, 62, 110
  decomposition by radiation, 36, **37**, 40–42
  electrons liberated in, per gram, **12**
  electrons per gram contained in, **347**
  energy dissipation in, per r., 8
  X-ray absorption coefficients in, **347**, 348, 349
Wave-length of ultra-violet light, variation of effect with, 3, 182, *188*
Wave-length of X-rays, variation of effect with, 47, 72, **115**, 116, 144, 145, 147, *148*, **239**, 240, **241**, 269, **270**, 273, *275*, 276, **277**, 280, 322, *323*, **324, 327**
Wild-type, 130*
Wilson chamber, 10, 49, 88, Plate I

ξ (number of primary chromosome breaks per r.), 256–62, 269, **270**
ξ (in target theory), 85, 353
X-chromosome, 128*
X/O males, 164
$X.X/Y$ (attached-X) females, 142, *143*, 165
X-rays
  absorption of, 9, 240, 345–49
  bactericidal action of, 316–27
  chemical effects of, 36–68
  chromosome changes by, 189–281, 328–40
  delayed division caused by, 282–306
  dosimetry of, 6, **22**
  effective wave-length of, 9
  electrons liberated by, 10, 11, **12**, 13
  energy dissipation in tissues by, 8, 9, 345–49
  genetical effects of, 140–88
  lethal effects of, 307–44
  physical properties of, 6–13
  relative efficiency of different wave-lengths of, 47, 72, **115**, 116, 144, 145, 147, *148*, **239**, 240, **241**, 269, **270**, 273, *275*, 276, **277**, 280, 322, *323*, **324, 327**
  scattering of, 11, 345–49
  virus inactivation by, 106–23
  wave-length distribution, 9
Xenon, **37**

Y-chromosome, 128*
Yeasts, 71, 300, 307

Zea mays (maize), **5**, 127, 130, 181, 183–88, 200, 202, 205, 219, 242, *243*, 328, 340
Zygote, 128*